Nucleotides and Regulation of Bone Cell Function

Nucleotides and Regulation of Bone Cell Function

Edited by

GEOFFREY BURNSTOCK
TIMOTHY R. ARNETT

CRC is an imprint of the Taylor & Francis Group,
an informa business

CRC Press
Taylor & Francis Group
6000 Broken Sound Parkway NW, Suite 300
Boca Raton, FL 33487-2742

Printed in the United States of America on acid-free paper
10 9 8 7 6 5 4 3 2 1

International Standard Book Number-10: 0-8493-3368-7 (Hardcover)
International Standard Book Number-13: 978-0-8493-3368-2 (Hardcover)

Visit the Taylor & Francis Web site at
http://www.taylorandfrancis.com

and the CRC Press Web site at
http://www.crcpress.com

Acknowledgment

The substantial, brilliant, and willing editorial assistance given by our colleague, Dr. Gillian E. Knight, was exceptionally helpful for the preparation of this book.

Preface

ATP has long been recognized as an intracellular energy source, although acceptance of its role as an extracellular signaling molecule has taken considerably longer. The potent actions of ATP on the heart and blood vessels were first described in 1929.[1] Some 40 years later, ATP, a purine, was proposed as a neurotransmitter in nonadrenergic, noncholinergic nerves in the gut and bladder and the word "purinergic" was coined.[2] Separate families of receptors for adenosine (P1) and ATP and ADP (P2) were proposed in 1978.[3] However, it was not until receptors for ATP and its ectoenzymatic breakdown product, adenosine, were cloned in the early 1990s that purinergic signaling became more widely accepted.[4] Currently, four subtypes of P1 receptors, seven subtypes of P2X ionotropic receptors, and eight subtypes of P2Y metabotropic receptors are recognized.[5–7]

The field is now expanding rapidly and it is clear that receptors for purines and pyrimidines are widely distributed not only in the nervous system, but also in many nonneuronal cells.[8] Both short-term purinergic signaling in neurotransmission, neuromodulation, exocrine and endocrine secretion, platelet aggregation, vascular endothelial cell-mediated vasodilatation, and nociceptive mechanosensory transduction; and long-term (trophic) purinergic signaling of cell proliferation, differentiation, migration, and death in embryological development, neural regeneration, cell turnover of epithelial cells in skin and visceral organs, wound healing, aging, and cancer have been described.[9] The therapeutic potential of purinergic signalling for a wide variety of diseases is being explored.[10]

The work of Leong et al.[11] demonstrated that extracellular ATP stimulated the breakdown of nasal cartilage in explant cultures, an effect that appeared to be mediated via P2 receptors.[12] A role for extracellular purines in bone biology was first recognized in 1992, when Reimer and Dixon showed that ATP, ADP, and UTP elevated intracellular Ca^{2+} in osteoblasts (bone forming cells), suggesting mediation via two distinct P2 receptor subtypes.[13] Subsequently, several laboratories have described functional effects of extracellular nucleotides on osteoclasts (bone resorbing cells) as well as on osteoblasts, and it has become apparent that these cells each express a range of both P2X and P2Y receptors.[14–22] Useful reviews of the field are available.[23–26] There has been increased interest and important advances made in this field, and the aim of the present volume is to bring together chapters by the leading figures to provide the reader with an account of the current status of purinergic signaling in bone, including pathophysiology and therapeutic potential.

REFERENCES

1. Drury, A.N. and Szent-Györgyi, A., The physiological activity of adenine compounds with special reference to their action upon the mammalian heart, *J. Physiol.*, 68, 213, 1929.
2. Burnstock, G., Purinergic nerves, *Pharmacol. Rev.*, 24, 509, 1972.

3. Burnstock, G., A basis for distinguishing two types of purinergic receptor, in *Cell Membrane Receptors for Drugs and Hormones: A Multidisciplinary Approach*, Straub, R.W. and Bolis, L., Eds., Raven Press, New York, 1978, p. 107.
4. Ralevic, V. and Burnstock, G., Receptors for purines and pyrimidines, *Pharmacol. Rev.*, 50, 413, 1998.
5. North, R.A., Molecular physiology of P2X receptors, *Physiol. Rev.*, 82, 1013, 2002.
6. Abbracchio, M.P. et al., Characterization of the UDP-glucose receptor (re-named here the $P2Y_{14}$ receptor) adds diversity to the P2Y receptor family, *Trends Pharmacol. Sci.*, 24, 52, 2003.
7. Burnstock, G., Introduction: P2 receptors, *Curr. Top. Med. Chem.*, 4, 793, 2004.
8. Burnstock, G. and Knight, G.E., Cellular distribution and functions of P2 receptor subtypes in different systems, *Int. Rev. Cytol.*, 240, 31, 2004.
9. Burnstock, G., Purinergic signalling—overview, In *Novartis Foundation Symposium* 276 Purinergic Signalling in Neuron-Glial Interactions, John Wiley & Sons, Ltd, Chichester, 2006, p. 26.
10. Burnstock, G., Pathophysiology and therapeutic potential of purinergic signalling, *Pharmacol. Rev.* 58, 58, 2006.
11. Leong, W.S., Russell, R.G., and Caswell, A.M., Extracellular ATP stimulates resorption of bovine nasal cartilage, *Biochem. Soc. Trans.*, 18, 951, 1990.
12. Caswell, A.M., Leong, W.S., and Russell, R.G., Evidence for the presence of P2-purinoceptors at the surface of human articular chondrocytes in monolayer culture, *Biochim. Biophys. Acta*, 1074, 151, 1991.
13. Reimer, W.J. and Dixon, S.J., Extracellular nucleotides elevate $[Ca^{2+}]_i$ in rat osteoblastic cells by interaction with two receptor subtypes, *Am. J. Physiol.*, 263, C1040, 1992.
14. Yu, H. and Ferrier, J., ATP induces an intracellular calcium pulse in osteoclasts, *Biochem. Biophys. Res. Commun.*, 191, 357, 1993.
15. Yu, H. and Ferrier, J., Osteoblast-like cells have a variable mixed population of purino/nucleotide receptors, *FEBS Lett.*, 328, 209, 1993.
16. Bowler, W.B. et al., Identification and cloning of human P2U purinoceptor present in osteoclastoma, bone, and osteoblasts, *J. Bone Min. Res.*, 10, 1137, 1995.
17. Bowler, W.B. et al., $P2Y_2$ receptors are expressed by human osteoclasts of giant cell tumor but do not mediate ATP-induced bone resorption, *Bone*, 22, 195, 1998.
18. Dixon, C.J. et al., Effects of extracellular nucleotides on single cells and populations of human osteoblasts: contribution of cell heterogeneity to relative potencies, *Br. J. Pharmacol.*, 120, 777, 1997.
19. Dixon, S.J. and Sims, S.M., P2 purinergic receptors on osteoblasts and osteoclasts: potential targets for drug development, *Drug Dev. Res.*, 49, 187, 2000.
20. Hoebertz, A. et al., Expression of P2 receptors in bone and cultured bone cells, *Bone*, 27, 503, 2000.
21. Hoebertz, A. et al., Extracellular ADP is a powerful osteolytic agent: evidence for signaling through the $P2Y_1$ receptor on bone cells, *FASEB J.*, 15, 1139, 2001.
22. Hoebertz, A. et al., ATP and UTP at low concentrations strongly inhibit bone formation by osteoblasts: a novel role for the $P2Y_2$ receptor in bone remodeling, *J. Cell. Biochem.*, 86, 413, 2002.
23. Bowler, W.B. et al., Signaling in human osteoblasts by extracellular nucleotides. Their weak induction of the c-fos proto-oncogene via Ca^{2+} mobilization is strongly potentiated by a parathyroid hormone/cAMP-dependent protein kinase pathway independently of mitogen-activated protein kinase, *J. Biol. Chem.*, 274, 14315, 1999.
24. Elfervig, M.K. et al., ATP induces Ca^{2+} signaling in human chondrons cultured in three-dimensional agarose films, *Osteoarthritis Cartilage*, 9, 518, 2001.

25. Hoebertz, A., Arnett, T.R., and Burnstock, G., Regulation of bone resorption and formation by purines and pyrimidines, *Trends Pharmacol. Sci.*, 24, 290, 2003.
26. Costessi, A. et al., Extracellular nucleotides activate Runx2 in the osteoblast-like HOBIT cell line: a possible molecular link between mechanical stress and osteoblasts' response, *Bone*, 36, 418, 2005.

Editors

Geoffrey Burnstock completed a B.Sc. at King's College London and a Ph.D. at University College London. He held postdoctoral fellowships with Wilhelm Feldberg (National Institute for Medical Research), Edith Bulbring (University of Oxford), and C. Ladd Prosser (University of Illinois). He was appointed to a senior lectureship in Melbourne University in 1959 and became professor and chairman of zoology in 1964. In 1975, he became head of the Department of Anatomy and Developmental Biology at University College London and convenor of the Centre of Neuroscience. He has been director of the Autonomic Neuroscience Institute at the Royal Free Hospital School of Medicine since 1997. He was elected to the Australian Academy of Sciences in 1971, the Royal Society in 1986, the Academy of Medical Sciences in 1998, and was made an Honorary Fellow of both the Royal College of Surgeons and the Royal College of Physicians in 1999 and 2000. He was awarded the Royal Society Gold Medal in 2000. He is editor-in-chief of the journals *Autonomic Neuroscience* and *Purinergic Signalling* and serves on the editorial boards of many other journals. Geoffrey Burnstock's major research interest has been autonomic neurotransmission and he is best known for his seminal discovery of purinergic transmission and receptors, their signaling pathways, and their functional relevance. He has personally supervised over 100 Ph.D. and M.D. students and published over 1,200 original papers, reviews, and books. He is first in the Institute of Scientific Information list of most cited scientists in pharmacology and toxicology for the period 1994 to 2004.

Tim Arnett holds a B.Sc. in biological sciences from the University of East Anglia and gained a Ph.D. at the Royal Postgraduate Medical School. He held postdoctoral positions at Columbia University and University College London before taking up a lectureship in the Department of Anatomy and Developmental Biology at UCL in 1986. During 1991 and 1992 he undertook sabbatical work at the University of Texas. He was appointed reader in Mineralised Tissue Biology at UCL in 2001. In addition to his work on the actions of extracellular nucleotides on bone cells, Arnett's research has focused on the control of osteoclast and osteoblast function by extracellular pH and oxygen.

Contributors

Hilary P. Benton
Department of Veterinary Medicine: Anatomy, Physiology and Cell Biology
University of California
Davis, California

Katherine A. Buckley
Human Bone Cell Research Group
University of Liverpool
Liverpool, UK

Francesco Di Virgilio
Department of Experimental and Diagnostic Medicine
Interdisciplinary Center for the Study of Inflammation
University of Ferrara
Ferrara, Italy

S. Jeffrey Dixon
Department of Physiology and Pharmacology and Division of Oral Biology
Schulich School of Medicine & Dentistry
The University of Western Ontario
London, Ontario, Canada

George R. Dubyak
Department of Physiology and Biophysics
Case Western Reserve University School of Medicine
Cleveland, Ohio

James A. Gallagher
Human Bone Cell Research Group
University of Liverpool
Liverpool, UK

Alison Gartland
Department of Cell Biology
University of Massachusetts
Medical School
Worcester, Massachusetts

Niklas R. Jørgensen
Osteoporosis and Metabolic Bone Unit
Departments of Clinical Biochemistry and Endocrinology
Copenhagen University Hospital
Hvidovre, Denmark

Jasminka Korcok
Department of Physiology and Pharmacology and Division of Oral Biology
Schulich School of Medicine & Dentistry
The University of Western Ontario
London, Ontario, Canada

Irma Lemaire
Department of Cellular and Molecular Medicine
Faculty of Medicine
University of Ottawa
Ottawa, Ontario, Canada

Thomas H. Steinberg
Department of Internal Medicine
Washington University
School of Medicine
St. Louis, Missouri

Stephen M. Sims
Department of Physiology and Pharmacology and Division of Oral Biology
Schulich School of Medicine & Dentistry
The University of Western Ontario
London, Ontario, Canada

Jessica M. Stokes
Department of Veterinary Medicine: Anatomy, Physiology and Cell Biology
University of California
Davis, California

Table of Contents

1 Purinergic Signaling in Osteoblasts

Niklas R. Jørgensen and Thomas H. Steinberg

CONTENTS

1.1 NUCLEOTIDES AS SIGNALING MOLECULES IN OSTEOBLASTS

The regulation of bone turnover is a complex and finely tuned process. It is the result of bone-forming and bone-resorbing activities carried out by osteoblasts and osteoclasts, respectively. These two processes are closely coupled and regulated to maintain a constant bone mass and a well-functioning skeleton able to resist fractures. Many factors influence remodeling of the skeleton. Systemic hormonal factors act on skeletal cells directly and indirectly, modulating receptor expression, activation, synthesis, or receptor binding of local factors, resulting in an increased or decreased bone formation or resorption. These local factors such as cytokines, growth factors, and prostaglandins can regulate the sensitivity of bone cells to hormonal factors and can also regulate the activity of bone cells. Abundant evidence exists as to the physiological importance of many of these agents.

Over the last decade, it has become clear that adenosine 5′–triphosphate (ATP) and other nucleotides play important roles as extracellular signaling molecules in addition to their intracellular roles. Extracellular nucleotides function primarily, though probably not exclusively, by binding to and activating two families of receptors termed "P2 receptors." This chapter will focus on purinergic signaling in osteoblasts, which are the bone-forming cells and the main regulators of bone turnover in the bone microenvironment.

Osteoblasts are mononuclear cells of mesenchymal origin. They are derived from stromal cells in the bone marrow and the periosteum from the same stem cells as adipocytes, chondrocytes, and myocytes. Actively secreting osteoblasts are large cuboidal cells with a large protein-synthesizing complex. However, the osteoblasts covering the bone surfaces are termed "lining cells" and have a flat morphology. Osteoblasts synthesize unmineralized bone matrix, also called osteoid, which they subsequently mineralize. During bone formation, approximately 10% of the osteoblasts are embedded in the bone matrix and are transformed to osteocytes, which are highly specialized cells capable of sensing mechanical stimuli and forces and transducing these into biological signals that may result in modulation of bone turnover.[1]

Nucleotides play fundamental roles in energy metabolism, nucleic acid synthesis, and enzyme regulation. Biological responses to ATP have been demonstrated in virtually all tissues and organ systems studied.[2] Nucleotides, predominantly ATP, are present intracellularly in concentrations of 2 to 5 mM. In contrast, extracellular ATP concentrations are extremely low due largely to the presence of ubiquitous ecto-ATPases and other ecto-phosphatases that rapidly hydrolyze extracellular nucleotides to their respective nucleoside 5′–di- and monophosphates, nucleosides, and free phosphates or pyrophosphate.[2]

Nucleotides are released into the extracellular fluid from a variety of sources. A number of cells release ATP upon mechanical stimulation, including tumor cells, platelets, and endothelial and epithelial cells. There are three major mechanisms by which intracellular ATP can be released: (a) cytosolic ATP released from sites of tissue and cell damage, including sites of bone fracture; (b) exocytic release of ATP specifically concentrated within secretory granules or vesicles, with the concentration of ATP in a granule of up to 100 mM; and (c) release of cytosolic ATP via intrinsic plasma membrane channels or pores in the absence of cytolysis,[3] which includes gap junction hemichannel controlled release.[4] Various agonists to P2 receptors can also initiate ATP release, and the ability of these agonists to elicit ATP secretion can be correlated with their abilities to trigger, via different mechanisms, rapid increases in cytosolic calcium concentrations. Like many other cell types, osteoblasts are capable of releasing ATP both in response to different mechanical stimuli, as shown for the human osteoblastic cell line HOBIT,[5] and for the murine osteoblastic cell line MC3T3-E1 in response to fluid flow and shear stress.[6] However, different mechanisms seem to be involved in the release of ATP: gap junction hemichannels are involved in the release process in the HOBIT cells,[7,8] whereas later work from the same group provided evidence for vesicular release of ATP from HOBIT cells;[9] the shear-induced release from MC3T3-E1 cells also seems to be mediated through vesicular fusion and requires Ca^{2+} entry into the cytoplasm through L-type voltage-operated calcium channels.[6] Finally, agonist stimulation has been shown to initiate ATP release from osteoblastic cells, which might be the mechanism for regeneration of the widespread signals involved in intercellular calcium signaling among osteoblasts.[10,11] Furthermore, osteoblastic cells express both ATP-consuming and ATP-generating potential at the cell surface. Studies on the human osteoblastic cell line SaOS-2 have shown the presence of an ecto-nucleoside diphosphokinase on the surface, enabling the cells to interconvert ADP to ATP. Thus, osteoblastic cells are not only capable of releasing ATP but also converting extracellular ADP to ATP.

1.2 PURINERGIC RECEPTORS IN OSTEOBLASTS

Receptors for extracellular nucleotides are termed "purinergic receptors," or P2 receptors, to distinguish them from P1 receptors, which recognize primarily adenosine. Several different families of P2 receptors, called P2X, P2Y, P2Z, P2U, P2T, were inferred primarily from pharmacologic data, but subsequent to the cloning of these receptors in the mid 1990s it is now understood that P2 receptors can be classified into two very different families, now called P2X and P2Y.

P2Y receptors are members of the large family of seven transmembrane domain G protein–coupled receptors, and currently eight different P2Y receptors are recognized. The best characterized P2Y receptors are $P2Y_1$, which recognizes ADP and ATP, and $P2Y_2$,[12] which is activated equally well by ATP and UTP. Both of these receptors have been identified on many different cell types. $P2Y_4$ receptors recognize UTP but not UDP or adenine nucleotides and have a more limited tissue distribution, primarily placenta[13,14] and pancreas.[15] $P2Y_6$ receptors, which like $P2Y_1$ and $P2Y_2$ receptors have a broad tissue distribution, recognize UDP.[16] All four of these P2Y receptors signal by activation of phospholipase C β (PLC β) and subsequent generation of inositol (1,4,5)-trisphosphate (IP_3)-mediated calcium rises and protein kinase C (PKC) activation. Other more recently described P2Y receptors, including $P2Y_{11}$ and $P2Y_{12}$, signal via cyclic adenosine monophosphate (cAMP) pathways and are limited in tissue distribution. The $P2Y_{12}$ receptor, for instance, is the ADP receptor on platelets, which inhibits adenylate cyclase through Gi and in conjunction with $P2Y_1$ is critical for platelet activation and hemostasis.[17]

In 1994, two groups reported the expression cloning of P2X family receptors from rat vas deferens[18] and PC12 cells.[19] These proteins proved to define a novel class of ligand-gated ion channels, with two membrane-spanning domains and a large extracellular loop. Employing a polymerase chain reaction (PCR) strategy with degenerate primers, four other P2X receptors were subsequently cloned. $P2X_{1-6}$ receptors have been studied mostly in neuronal and smooth muscle cells, although at least the $P2X_4$ receptor has a broad tissue distribution.[20,21] Upon activation by ATP, these channels form cation-specific channels that transport sodium and potassium equally well and have some differences in calcium permeability. Ion channel formation probably requires trimers of P2X subunits, and some but not all P2X receptors can mix to form heteromers.[22]

Cockcroft and Gomperts identified an unusual ATP-mediated activity primarily in mast cells[23] and a similar activity was reported in mononuclear phagocytes.[24] In these cells, addition of ATP caused not only ion channel formation and plasma membrane depolarization, but additionally permeabilized the plasma membrane to molecules < 900 Da, including fluorescent dyes such as Lucifer Yellow and YoPro. This activity was termed P2Z and was felt to be distinct from the P2X ion channels until 1996, when the identification of $P2X_7$ receptors was reported.[25] This receptor, found most abundantly in hematopoeitic cells, mediates pore formation as well as cation channel activity in response to ATP and appears to fulfil the requirements necessary to be termed the "P2Z activity." The structure of the $P2X_7$ receptor is similar to that of other P2X receptors except that it has a significantly longer cytoplasmic carboxyl terminus. Removal of this domain yields a receptor that functions as an ion channel but no longer opens highly permeable pores.[25] $P2X_7$ receptors also differ from other P2X receptors in that it appears to form oligomers only with itself, not other P2X receptors.[26] The $P2X_7$ receptor is expressed mostly in bone marrow-derived cells: in macrophages,[25] granulocytes,[27] B lymphocytes,[27] dendritic cells,[28] and osteoclasts.[29,30]

Several *in vitro* studies have shown that ATP and other nucleotides act through P2 purinergic receptors to induce formation of IP_3 and transient elevation of $[Ca^{2+}]_i$ in osteoblastic cells. However, a large heterogeneity in receptor profiles is evident when reviewing the studies, which is undoubtedly due to differences in model systems and, more importantly, in differentiation status of the osteoblastic cells. In rat osteoblastic cells (the rat osteosarcoma cell line UMR-106), at least two different receptor subtypes seem to be involved,[31–34] with proportions varying between individual cells.[35] The pharmacological profiles are identical to the two metabotropic purinergic receptors $P2Y_1$ and $P2Y_2$. Also human bone explant–derived[36,37] and bone marrow stromal cell–derived osteoblasts[11] increase $[Ca^{2+}]_i$ in response to agonist. Studies of single cells and populations of human osteoblasts derived from bone explants revealed a large heterogeneity within the cell culture: two different populations were identified, one responding to ATP and ADP by increasing $[Ca^{2+}]_i$, the other sensitive to ATP alone. This finding may reflect differences in receptor expression during osteoblast differentiation. Furthermore, exposure to ATP desensitizes subsequent receptor stimulation, an effect that is reversible in a time-dependent manner upon removing ATP from the medium. This is true both for UMR-106 cells[35,38] and for human osteoblastic cells.[11]

In osteoblastic cells, nucleotides induce large transient elevations of $[Ca^{2+}]_i$, which are associated with increases in IP_3. Removing Ca^{2+} from the extracellular medium has only minor effects on $[Ca^{2+}]_i$ in response to nucleotide stimulation, thus indicating calcium release from intracellular stores. This is consistent with P2Y receptors being primarily responsible for the effect of nucleotides on $[Ca^{2+}]_i$ in osteoblasts.

The first molecular evidence for the expression of several receptor subtypes in osteoblastic cells came in 1995, where both *in situ* hybridization and RT-PCR showed that human osteoblasts from bone explants express $P2Y_2$ receptors,[39] which was later confirmed by other investigators.[11] P2X receptors also seem to be expressed by primary human osteoblasts. The presence of mRNA coding for $P2X_2$, $P2X_4$, and $P2X_7$ receptors has been demonstrated.[40] The presence of $P2X_7$ receptors is controversial, as conflicting results have been presented. However, it seems that most osteoblasts express the $P2X_7$ mRNA, while only a subset of the osteoblasts express the receptor protein.[40] In these cells, ATP stimulation can induce pore formation, as occurs in macrophages[24] and in human multinucleated osteoclasts.[30] Attempts to demonstrate the expression in rat osteoblasts have failed, but in murine osteoblasts 2′– and 3′–*O*-(4-benzoyl-benzoyl)-ATP (BzATP) induce pore formation in a subpopulation of osteoblasts.[41]

Other studies have investigated the expression of P2 receptors in osteoblastic cell lines and found evidence for the expression of $P2Y_1$, $P2Y_2$, $P2Y_4$, and $P2Y_6$ receptors in the human osteosarcoma cell lines MG-63 and OHS-4,[42] while the rat osteosarcoma cell line ROS 17/2.8 does not express $P2Y_1$ and $P2Y_2$ receptors as does UMR 106-01.[10] In rat osteoblasts the expression of $P2Y_1$, $P2Y_2$, $P2X_2$, and $P2X_5$ receptors has also been demonstrated.[43]

Thus, a number of different P2 receptors are expressed in osteoblastic cells, and Table 1.1 gives an overview of the different studies. However, the expression pattern varies depending on the cell system tested, where both species differences and whether the osteoblast model is based on primary cells or cell lines seem to be important. Furthermore, as for most other proteins, there is a discrepancy between mRNA detected in the cytosol and the expression of the protein. Finally, the expression of P2 receptors is also highly dependent on the differentiation status of the osteoblast,[37,44] suggesting that P2 receptor signaling may play varied roles during osteoblast differentiation.

1.2.1 ATP in Calcium Signaling

Transient and oscillatory elevations of the cytosolic free calcium concentration initiate or modulate many cellular activities, including cell growth, motility, and secretion. Many studies have investigated intracellular calcium homeostasis and the mechanisms by which extracellular signals are translated into intracellular calcium transients, but less attention has been paid to the mechanisms by which groups of cells propagate calcium transients among themselves and coordinate responses to these signals. The propagation of calcium signals among groups of cells seems to be a common mechanism in many tissues and cell types, including bone,[10,11,30,45] and two major mechanisms of intercellular calcium signaling have been identified: the release of a soluble messenger into the extracellular space acting on nearby cells and gap junctional communication. Intercellular calcium signaling by released soluble mediators often involves activation of receptors of the P2 type, with the mediator being extracellular nucleotides such as ATP,[46–48] but the extracellular calcium-sensing receptor has been shown to be involved in some instances.[49] Gap junction–mediated calcium signals are propagated either by allowing IP_3 or potentially other small messenger molecules between cells or by allowing electrical coupling of cells and subsequent activation of voltage-sensitive calcium channels.[50–53]

Many cell types possess the ability to propagate intercellular calcium signals, including respiratory tract epithelial cells,[50–52,54–57] articular chondrocytes,[58–62] glial cells and cell lines,[63–67] glomerular mesangial cells,[68] smooth muscle cells,[69,70] hepatocytes,[47,71] fibroblasts,[72] lens epithelial cells,[73] mast cells,[46] and cardiac muscle cells.[74] Only in a few cell systems have the propagation of intercellular calcium transients been coupled to a specific physiologic function of the tissue. The most extensively characterized model of calcium wave propagation is the ciliated epithelium from the respiratory tract. Mechanical stimulation of one ciliated cell *in vitro* induces a calcium transient that propagates to the nearby cells. This wave is followed by a coordinated movement of the cilia on the cell membrane on adjacent cells. *In vivo* this would lead to clearance of inhaled particles from the airways, thus protecting the organism against potentially hazardous agents.[75] Insulin-secreting cells from the pancreatic islets of Langerhans also coordinate activities by synchronous calcium oscillations, and both primary cells and cell lines are capable of coordinating insulin secretion in this fashion.[76–78] Thus, intercellular calcium signaling is a widely used mechanism for cells to coordinate activities throughout populations of cells in response to different stimuli. Several good reviews are published on this topic.[79–83]

TABLE 1.1
Presence and proposed function in osteoblastic cells of the different P2 receptors

P2 Receptor Subtype	Species	Osteoblast Model	Evidenced by	Proposed Function	Ref.
$P2X_2$	Rat	Primary cells	Immunostaining, *in situ* hybridization		43, 102
$P2X_4$	Rat	Primary cells	*In situ* hybridization		43, 102
	Human	Cell lines: MG-63 and SaM-1	RT-PCR		99, 105
$P2X_5$	Rat	Primary cells	Immunostaining, *in situ* hybridization	Proliferation, differentiation	43, 103
	Human	Cell lines: MG-63 and SaM-1	RT-PCR	Increases proliferation and DNA synthesis (possibly also via other P2XR)	99, 105
$P2X_6$	Human	Cell lines: MG-63 and SaM-1	RT-PCR		99, 105
$P2X_7$	Mouse, rat	Primary cells	RT-PCR, pore formation, *in situ* hybridization, pharmacology	Cell death	41, 43, 102
$P2X_7$	Human	Cell line: MG-63	RT-PCR		99
	Human	Primary cells	Immunostaining, RT-PCR	Cell death	40
P2Y	Rat		Pharmacology	Calcium release	33, 34, 113
	Human	SaM-1	RT-PCR	IL-6 release	105
$P2Y_1$	Rat	Cell line: UMR-106	Pharmacology	Stimulates RANKL expression and bone resorption	106
	Mouse, rat	Primary cells	*In situ* hybridization, immunostaining	Enhances PTH-induced calcium increases	43, 101, 102
	Human	Cell lines: MG-63, OHS-4, and HOBIT	RT-PCR		42, 104
	Human	Primary cells	RT-PCR	Stimulates osteoclast formation and resorption	37, 42, 92

(continues)

TABLE 1.1 (continued)
Presence and proposed function in osteoblastic cells of the different P2 receptors

P2 Receptor Subtype	Species	Osteoblast Model	Evidenced by	Proposed Function	Ref.
$P2Y_2$	Rat	Cell line: UMR-106	*In situ* hybridization, RT-PCR, pharmacology	Inhibition of bone formation, calcium signalling among osteoblasts	42, 43, 106, 114
	Rat	Primary cells	*In situ* hybridization	Inhibits bone formation	43, 102, 103
	Human	Cell lines: MG-63, OHS- 4, MC3T3-E1, HOBIT, and SaM-1	RT-PCR, calcium imaging	Increases Ca^{2+}, acts as mitogens, activates PLD and release of AA and synthesis of PGE_2, increases Egr-1 protein levels	42, 86, 95, 98, 104, 105
	Human	Primary cells	RT-PCR, pharmacology	Calcium signaling among osteoblasts	11, 37, 39, 92
$P2Y_4$	Human	Cell lines: MG-63 and OHS-4	RT-PCR		42
	Human	Primary cells	RT-PCR		42
$P2Y_6$	Human	Cell lines: MG-63, OHS-4, and SaM-1	RT-PCR		42, 105
	Human	Primary cells	RT-PCR		42

Note: RT-PCR = reverse transcriptase polymerase chain reaction; IL-6 = interleukin 6; RANKL = receptor activator of nuclear factor κ ligand; PLD = phospholipase D; AA = arachidonic acid; Egr-1 = early growth response gene 1

1.2.2 Intercellular Calcium Signaling in Bone

In 1992, Xia et al. demonstrated that intercellular calcium waves could be generated in response to mechanical stimulation of a single cell in monolayers of primary cultures of rat calvarial cells and in rat osteosarcoma cells (ROS), and that these calcium waves were dependent on the passage of an unknown signaling molecule through gap junctions and regeneration of the calcium transient in neighboring cells by calcium-induced calcium release.[84] Later studies used other osteoblastic cells, the UMR-106 cell line, to demonstrate the ability of the bone cells to propagate calcium transients among cells. However, striking differences in calcium wave appearance were demonstrated when compared to the ROS cells. UMR-106 cells responded to mechanical stimulation with a fast calcium wave, with kinetics very different from the one seen in ROS cells. While the wave in ROS cells is slow, with a wave extending to 5 to 15 cells, the wave in UMR cells is fast, spreading to more than 50 cells in less than 30 s.[10] The differences in kinetics were found to reflect two very different underlying mechanisms for wave propagation. In the ROS cells the wave was propagated via passage of an unknown messenger between cells through the gap junction channel,[10] depolarizing the plasma membrane of the neighboring cell and inducing opening of voltage-operated calcium channels with subsequent influx of calcium from the extracellular space into the cytoplasm (Figure 1.1).[45] In the UMR cells, the mechanism for wave propagation involved the autocrine action of an extracellular nucleotide, possibly ATP, on $P2Y_2$ receptors on adjacent cells, inducing intracellular IP_3 production and release of intracellular calcium stores (Figure 1.1). Thus, the mechanisms described in these cell types for wave propagation follow the two fundamental mechanisms mentioned in the section above.

The strikingly different mechanisms of wave propagation observed in these two distinct but phenotypically similar cell lines raise the question as to the physiologic meaning of these short-range intercellular calcium signals. Subsequent studies have investigated whether calcium transients in response to mechanical stimulation were inducible in primary cultures of human osteoblastic cells. Interestingly, both mechanisms were activated in human osteoblasts when mechanically induced calcium waves were generated.[11] Upon mechanical stimulation, the human cells propagate a fast wave, with kinetics in between those of UMR and ROS cells (Figure 1.2). These waves are mediated by P2 receptor activation. After ATP desensitization of the receptors, the fast wave disappears and a slow gap junction–mediated wave is uncovered. This wave is similar to the intercellular calcium waves seen in ROS cells.[11] To further prove that gap junctions and $P2Y_2$ receptors are involved in intercellular calcium signaling in osteoblastic cells, UMR cells were transfected with the gap junction protein Cx43 to increase cell–cell coupling. After ATP desensitization, a slow ROS-like wave was seen, proving that the cells were able to propagate a gap junction–mediated wave. Conversely, in ROS cells transfected to express $P2Y_2$ receptors, mechanical stimulation of the cells resulted in a fast UMR-like wave.[10] The relative importance of each of these two mechanisms was investigated during the differentiation of a primary culture of human osteoblasts. In the relatively immature osteoblasts, the purinoceptor-mediated mechanism prevailed, while in the ageing (three to four months in culture) osteoblasts, the gap junction–mediated mechanism

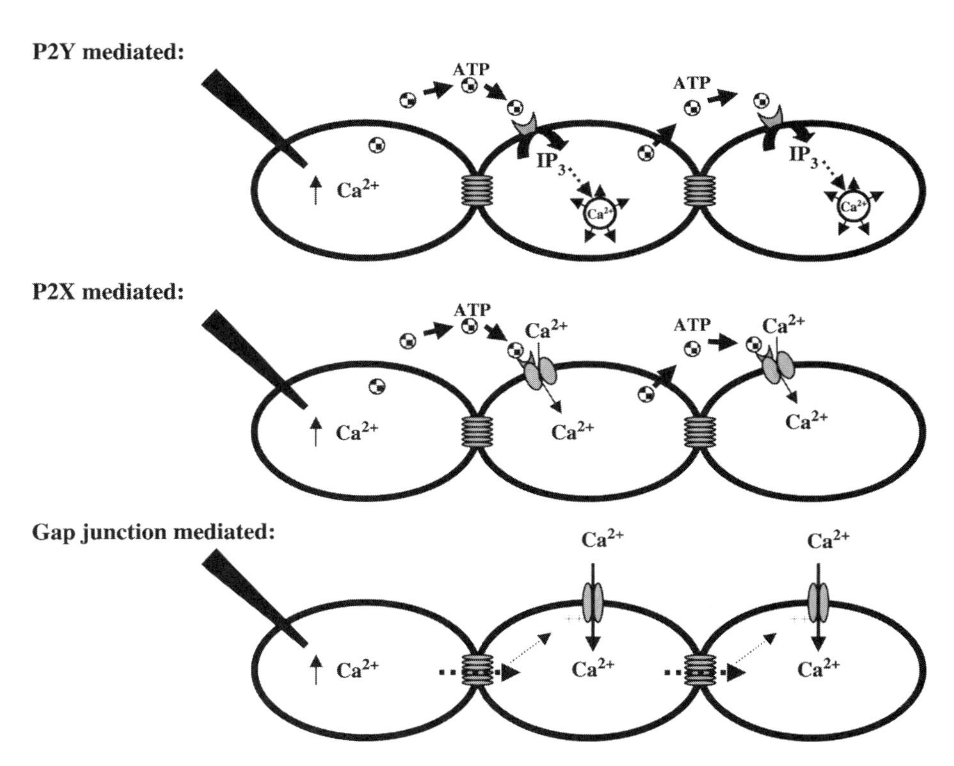

FIGURE 1.1 Mechanisms for the propagation of intercellular calcium signals. Ligand-mediated calcium waves: mechanical stimulation increases intracellular free calcium concentration in the stimulated cell. As a consequence, ATP or related nucleotide is released to the extracellular space and binds to surface receptors on neighboring cells. If ATP binds to P2Y receptors (upper panel), inositol triphosphate (IP_3) is generated, inducing release of calcium from IP_3-sensitive intracellular calcium stores. If ATP binds to P2X receptors (middle panel), conformational changes of the receptor/channel are induced, resulting in the opening of the channel, with subsequent influx of extracellular calcium. In both cases, a calcium wave is generated by successive activation of P2 receptors in neighboring cells. Gap junction–mediated calcium waves (lower panel): the increase in intracellular free calcium concentration caused by mechanical stimulation produces a signaling molecule (IP_3?) that passes through the gap junction channel into adjacent cells, where it induces depolarization of the plasma membrane and subsequent opening of voltage-operated calcium channels with influx of calcium from the extracellular space. The intracellular calcium increase is then propagated to the next cell through the same mechanism, thus producing a calcium wave.

predominated,[85] though the cells still responded to ATP stimulation with an increase in [Ca^{2+}]. This finding might imply that the two different mechanisms are important in different cellular processes during cell differentiation.

Studies using the human HOBIT cell line demonstrated that osteoblasts release ATP in response to a number of mechanical stimuli[5] and that this release is at least partially mediated via gap junction hemichannels.[7,8] Autocrine action of ATP on P2 receptors seems to be involved, because intact intracellular calcium stores are required for the wave propagation, and suramin and apyrase were both able to inhibit the waves.[8]

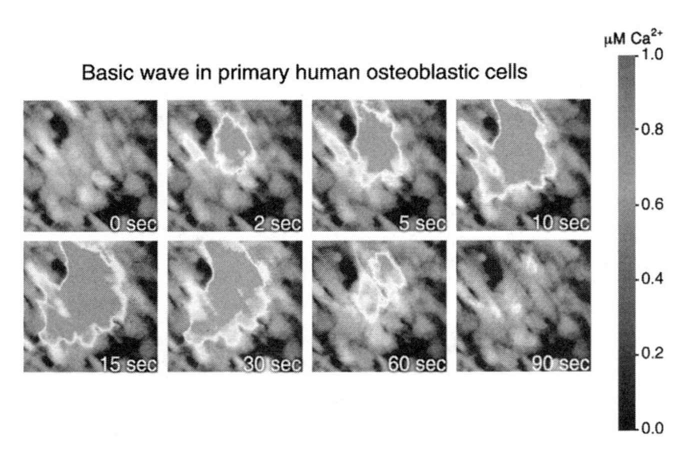

FIGURE 1.2 (See color insert following page 80) Intercellular calcium waves in human primary osteoblastic cells. A single cell is stimulated mechanically, causing an increase in the intracellular calcium concentration that subsequently propagates to adjacent cells. Numbers indicate seconds after mechanical stimulation. Scale bar indicates relation between pseudocolors and intracellular calcium concentrations.

ATP released from osteoblastic cells may act on P2 receptors on osteoclasts and be a mechanism for cross talk between these cells in the control of bone remodeling in addition to other paracrine factors and direct cell-to-cell contact. P2 receptors are present in both osteoblasts and osteoclasts, and studies have shown that calcium transients can also be propagated between these two cell types. In cocultures of osteoblasts and osteoclasts, mechanical perturbation of a single osteoblast induces a calcium transient that propagates not only to other osteoblasts but also to nearby osteoclasts (Figure 1.3). The signal can go both ways, as stimulation of an osteoclast can induce a signal that propagates both to other osteoclasts and to osteoblasts. However, the signaling to osteoblasts involves activation of receptors of the $P2Y_2$ subtype, whereas signaling to osteoclasts and among osteoclasts requires activation of receptors of the $P2X_7$ subtype.[30] This raises the possibility that the activity of the two cell types can be regulated independently by modulation of either the $P2Y_2$ or the $P2X_7$ receptor, thus regulating either formation or resorption without affecting the other.

All the studies mentioned above demonstrate the propagation of intercellular calcium signals among osteoblasts in response to direct mechanical perturbation of a single cell *in vitro*. It has not yet been demonstrated that intercellular calcium signaling occurs between bone cells *in vivo*, but some evidence suggests that two major mechanisms exist by which mechanical forces applied to the skeleton are transformed into biological signals. One mechanism is fluid flow along the osteocyte membrane. Osteocytes embedded in mineralized bone are connected to each other with long processes, and these processes are situated in narrow canals, or canaliculi. Extracellular fluid surrounds the osteocyte processes, and mechanical forces applied to bone are assumed to initiate fluid flow along the cell membrane of the osteocytes, thus converting mechanical stimuli to biological signals. The other mechanism is shear stress of the osteocytes. Again, physical activity leads to microdeformation of

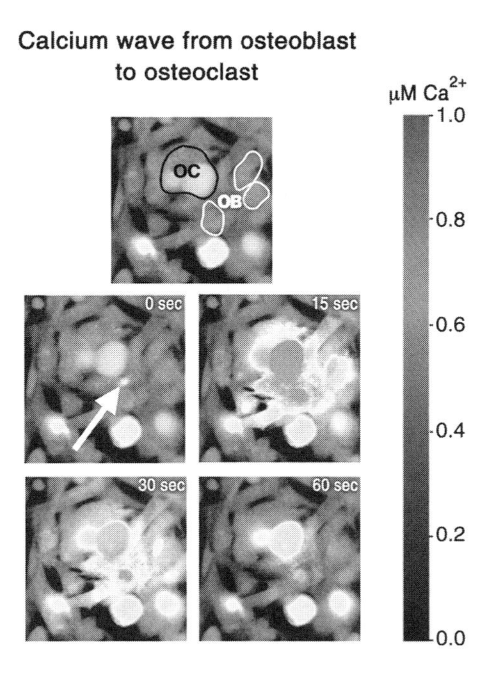

FIGURE 1.3 (See color insert following page 80) Intercellular calcium signaling occurs between osteoblasts and osteoclasts in coculture monolayers. Cocultures of human osteoblasts and osteoclasts were loaded with fura-2, single osteoblasts were mechanically stimulated with a glass pipette, and ratio imaging was performed. Numbers on the pictures indicate time in seconds after mechanical stimulation. Osteoblasts (OB) and osteoclasts (OC) are shown by arrows. Scale bar shows intracellular calcium concentrations in micromoles.

bone, thus stretching or compressing the osteocytes embedded in the bone matrix, initiating signaling to the bone surface. Recent studies have shown that oscillatory fluid flow is able to increase calcium in both the mouse osteoblastic cell line MC3T3-E1 and in ROS cells transfected to express $P2Y_2$ receptors (ROS/$P2Y_2$ cell line), which the parent cells do not express. Apyrase prevented the fluid flow–induced increases in $[Ca^{2+}]_i$, and ATP or UTP, but not ADP, increased $[Ca^{2+}]_i$. This study demonstrated that $P2Y_2$ receptors can mediate the propagation of calcium signals in response to shear stress in osteoblastic cells.[86] It has also been shown that MC3T3-E1 cells release ATP in response to shear stress, and that this release is mediated not through gap junction hemichannels, but through exocytosis granules containing ATP. It is dependent on calcium entry through L-type voltage-operated calcium channels and the fact that the release of ATP is linked to simultaneous release of prostaglandin E_2 (PGE_2).[6]

All the above studies have been performed on osteoblastic cells, while there is general agreement that the osteocyte is the mechanosensor of bone. Studies addressing calcium signaling in osteocytes and the expression and function of P2 receptors in osteocytes are needed, although they are not as easy to perform as those on osteoblasts because osteocytes most often require three-dimensional culture conditions in order to retain the osteocyte phenotype.

1.2.3 ATP as a Cosignaling Molecule—Localizing Systemic Signals

From the studies described above, it is evident that nucleotides act on osteoblastic cells by increasing $[Ca^{2+}]_i$, but the exact role of nucleotides in the regulation of bone turnover has not been elucidated in full. Several studies have addressed the role of nucleotides in the normal physiological regulation of bone remodeling, and it has become increasingly apparent that nucleotides serve important roles as costimulators in bone. Not only do they have individual roles in regulating bone formation and resorption directly, but they also seem to potentiate or sensitize bone cells to systemic signals like parathyroid hormone (PTH). PTH is the major regulator of bone and mineral metabolism and has complex effects on bone formation and resorption. *In vivo*, it is capable of both increasing bone formation and resorption, and it has distinct effects on osteoblast activity and differentiation, as well as on osteoclastic bone resorption and apoptosis. However, activation of basic multicellular units is highly local, and release of extracellular nucleotides in response to, for example, mechanical forces applied to certain parts of the skeleton might be a way systemic signals like PTH are localized.[87] Recent studies suggest that nucleotide release through autocrine signaling may represent a universal key determinant in establishing the set point for activation of certain signal transduction pathways.[88] Thus the effect of PTH and other systemic regulators of bone metabolism may be locally regulated by autocrine nucleotide activity.

Conflicting results exist as to whether PTH increases $[Ca^{2+}]_i$ in osteoblastic cells. However, in UMR cells, PTH activates adenylate cyclase and PLC, thereby increasing cAMP and $[Ca^{2+}]_i$.[89] Further, ATP, UTP, and UDP potentiate PTH-induced increases in $[Ca^{2+}]_i$, but not the effect of PTH on cAMP production,[89,90] which was supported by another study in which PTH(1-31), which activates only G_s, mimicked the actions of PTH(1-34), whereas PTH(3-34), which only activates G_q, failed to potentiate nucleotide-induced increases in $[Ca^{2+}]_i$.[91] Further, UDP was the most effective in potentiating the PTH-induced increase in $[Ca^{2+}]_i$, which was independent of influx of extracellular calcium.[89–91] Thus, the combined action of PTH and nucleotides in increasing $[Ca^{2+}]_i$ is most probably mediated via action of the nucleotides on $P2Y_2$ receptors, the subtype responsible for propagation of mechanically induced signals in osteoblastic cells,[10,11] and PTH by itself might not be able to activate PLC in osteoblasts before P2 receptors are activated.[90] These studies demonstrate a clear costimulatory effect of nucleotides on PTH-induced increases in $[Ca^{2+}]_i$ in osteoblastic cell systems, but other intracellular pathways may be potentiated or synergistically affected when PTH and nucleotides act in common. Activation of P2 receptors in primary human osteoblasts seem to potentiate PTH-induced *c-fos* gene expression.[92] The *c-fos* gene is involved in signal transduction, cell proliferation, and differentiation and is expressed during embryonic bone development. In *c-fos* knockout mice, animals are growth retarded, osteopetrotic, and have deficiencies in bone remodeling and tooth eruption. They also lack osteoblasts along the periosteal and endosteal surfaces.[93] Thus, autocrine action of nucleotides can potentiate strong responses to ubiquitous growth and differentiation factors.[94]

Other pathways also seem to be involved in nucleotide signaling in osteoblasts. In the MC3T3 cells, extracellular ATP has been shown to stimulate phospholipase D (PLD) in a Ca^{2+}/calmodulin-dependent manner, and neither PKC nor GTP binding protein is involved.[95] Further, the ATP-mediated stimulation of PLD induces release of arachidonic acid, which is partly mediated by phosphatidyl choline hydrolysis.[96] However, PKC by itself selectively inhibits the $P2Y_1$ receptor signaling pathway in UMR cells, indicating that endocrine and paracrine factors acting via PKC may regulate the responsiveness of osteoblasts to extracellular nucleotides.[97] The known coordinated actions of PTH and nucleotides on osteoblasts are summarized in Figure 1.4.

Various downstream effects of nucleotide activation of P2 receptors have been reported in osteoblasts, involving effects on differentiation, proliferation, and osteoblast activity. The exact determination of the distinct effects of each P2 receptor subtype has been hampered by the lack of selective P2 antagonists and by the fact that only the $P2X_7$ receptor knockout mouse has been investigated with respect to a bone phenotype.[41]

Most of the described effects are mediated via P2Y receptors, but activation of P2X subtypes seems to be important also. Overall, activation of P2 receptors in osteoblasts seems to induce cell proliferation. ATP stimulation increases DNA synthesis as measured by [^{3}H] thymidine incorporation and cell proliferation. Further, it synergistically increases platelet-derived growth factor (PDGF)- and insulin-like growth factor (IGF)-induced incorporation of [^{3}H] thymidine via activation of the

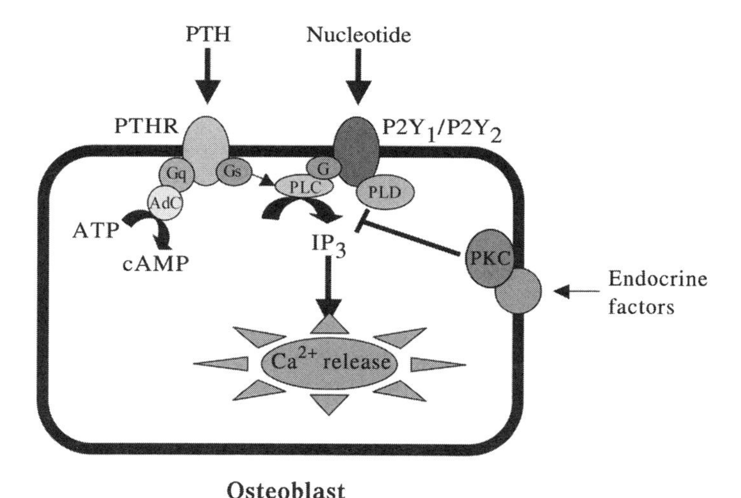

FIGURE 1.4 (See color insert following page 80) Shows the interacting pathways of parathyroid hormone (PTH) and purinergic $P2Y_1$ and $P2Y_2$ receptors. PTH acts via the PTH receptor (PTHR), which is a G protein–coupled receptor. Upon ligand binding, G_q and adenylate cyclase (AdC) are activated, inducing conversion of adenosine triphosphate (ATP) to cyclic adenosine monophosphate (cAMP). PTH binding also activates G_s and the phospholipase C (PLC) pathway, which is sensitized by nucleotide binding to P2Y receptors, enhancing PTH-induced inositol triphosphate (IP_3) generation and subsequent release of calcium from intracellular stores. Endocrine and paracrine factors like estradiol activates protein kinase C (PKC), which inhibits signal transduction from P2Y receptors and downstream. See text for further details.

mitogen-activated protein kinase (MAPK) pathway. This has been shown for both the MC3T3 and the MG-63 cell lines.[98,99] In contrast, the P2 receptor inhibitor, suramin, decreased the number of osteoblasts in cultures of primary human osteoblasts.[100] However, the net effect in primary rat osteoblast cultures is inhibition of bone formation when ATP and UTP are used as agonists, whereas ADP and adenosine have no effects on osteoblast activity, but only on stimulation of bone resorption.[101,102] In a rat calvarial assay of bone formation, ATPγS and ATP decreased bone formation, whereas adenosine did not. In contrast to the *in vitro* study mentioned above, UTP had no effect, or at least only a slight stimulatory effect, on bone formation.[103] These conflicting results may be due to the different model systems used, differences in cell differentiation and culture conditions. Other studies have demonstrated a stimulatory effect on COL1A2 gene expression in the HOBIT cell line in response to ATP stimulation via an increased expression of the early growth response gene (Egr1),[104] as well as an increased gene and protein expression of interleukin-6 in the human osteoblast-like SaM-1 cell line, which can be abolished by suramin or an IP_3 blocker.[105]

Conflicting results have been published on the expression of the $P2X_7$ receptor in osteoblasts. The $P2X_7$ receptor was thought originally to be predominantly expressed in cells of the hematopoietic cell lineage, including macrophages and osteoclasts. However, recent studies have shown the expression of $P2X_7$ receptors in a subpopulation of osteoblasts, both from human bone explants, in the human osteosarcoma cell line SaOS-2, and in primary osteoblasts obtained from mouse calvarial cultures.[40] Further, prolonged ATP stimulation induced pore formation in subpopulations of these osteoblast cultures as measured by ethidium bromide uptake.[40,41] In the SaOS-2 cells, stimulation with the selective $P2X_7$ receptor agonist BzATP not only induced pore formation, but also increased LDH release and initiated apoptotic processes in the cells,[40] indicating that $P2X_7$ receptors might be involved in the regulation of cell survival in osteoblasts. Finally, P2 receptors have also been shown to indirectly regulate the osteoclastic bone resorption via osteoblasts by increasing receptor activator of nuclear factor-κB ligand (RANKL) expression on osteoblasts (UMR-106 cells).[106]

1.3 PURINERGIC SIGNALING IN BONE—FUNCTIONS, THEORIES, AND FUTURE PERSPECTIVES

Though an increasing number of studies document the expression of different purinergic receptor subtypes in osteoblastic cells, as seen in Table 1.1, the relevance of P2 receptors for bone biology remains to be demonstrated. Most studies agree that P2 receptors can mediate osteoblast proliferation and differentiation, in addition to their importance in the regulation of osteoblast activity and indirect regulation of osteoclast activity via osteoblasts. However, studies present conflicting results as to which P2 receptors are expressed in osteoblastic cells. These differences may be accounted for by differences in model systems, cell types, and species that have been used for the investigation of P2 receptor expression. Different osteoblast populations may have distinct receptor profiles, and it is likely that P2 receptor expression changes during osteoblast differentiation. Culture conditions can affect cell differentiation and thereby potentially change receptor expression. Finally, different approaches have been taken to prove the existence of the various P2 receptor subtypes. Some studies have used RT-PCR to examine the expression of mRNA for the different receptors, while other studies have determined expression of receptor protein either *in vitro* by Western blot or by immunocytochemistry or *ex vivo* by *in situ* hybridization. Finally, pharmacologic approaches have also been applied to determine the expression of receptors by using the knowledge of differences in agonist selectivity for the various P2 subtypes.

Many effects have been ascribed to nucleotide action on P2 receptors, but these also have, in some instances, been conflicting, presumably due in large part to the considerations mentioned above. A drawback of many of these studies is that they are based on immortalized cell lines that may have receptor profiles not found in primary osteoblasts and may have abnormal intracellular signaling pathways compared to primary cells. Therefore, to fully understand the function of each of the single P2 receptor subtypes in osteoblasts, more studies are warranted examining the effects *in vivo* and in primary cultures of osteoblasts. Previously, this task has been hampered by the lack of selective agonists and antagonist, but evaluation of genetically altered animals and the use of *in vitro* molecular techniques such as siRNA gene silencing can overcome this hurdle. Currently, only two studies have investigated the bone phenotype in P2 receptor null mice, and both of these studies have focused on the $P2X_7$ receptor.[41] Other P2 receptor–deficient mice have been generated, but the bone phenotype in these animals has not been described. Of course, issues of functional compensation among different P2 family members may make the task of ascribing specific functions to specific proteins a considerable challenge.

However, despite the problems determining the exact roles of the various P2 receptor subtypes in osteoblasts, the mounting data on the effect of nucleotides on osteoblasts suggests a significant role for P2 receptors in the regulation of osteoblast function. P2 receptors are apparently also involved in indirect regulation of osteoclast activity and probably osteoclast numbers via osteoblastic expression of RANKL,[106] in addition to the direct effects on osteoclasts, which are covered elsewhere in this book.

It is important to establish the different effects of nucleotides on osteoblasts *in vitro*, but even more relevant to try to understand the overall importance and function of the receptors in the total skeleton and in bone and mineral metabolism. Only a few studies have shed light on this, but the implications of the sparse data have the potential to be quite interesting. First, nucleotide release in bone can occur from different sources. The most obvious is release in connection with cell or tissue damage, or even fractures, where huge amounts of nucleotides are released into the proximity of the fracture or tissue damage. In this case, the nucleotides may serve functions not only on bone cells, but also on cells of the hematopoietic and immune system, as well as activation of many other cell types in order to normalize the tissue damage. It is also important to consider how P2 receptors might be involved in the normal physiology of bone. In addition to nucleotide release in relation to cell lysis and tissue damage, nucleotides can also be released from intact cells as exocytic release from granules or vesicles, or via intrinsic channels or pores. This has been shown to occur in response to agonist stimulation and mechanical stimulation of osteoblasts and osteoclasts, and P2 receptors are, together with gap junctions, responsible for the propagation of calcium waves in response to mechanical perturbations like direct manipulation of the plasma membrane,[10,11] or presumably more physiologic stimuli like fluid flow and shear stress.[6,86,107] Thus, nucleotide release, P2 receptor activation, and propagation of calcium transients among populations of cells could be the means by which bone cells translate mechanical forces into biological signals and responses.

It is a well-established fact that weight-bearing physical exercise increases bone mineral density in the human skeleton as well as in animal models. Thus, under normal conditions, bone adapts to the different needs through the lifetime, with increasing bone mass and bone strength in response to mechanical stimuli. In contrast, under conditions with decreased mechanical stimuli like weightlessness or inactivity, bone is lost,[108] which is due to the lack of mechanical stimulation of the skeleton. There is general agreement that osteocytes are the mechanosensors of bone, transmitting signals to bone-lining cells covering the bone surface, and thus activating remodeling processes in areas of the skeleton that need maintenance. So far, however, no studies have examined the presence of P2 receptors in osteocytes or the capacity of these cells to propagate calcium signals.

Purinergic signaling in osteoblasts thus seems to be a means by which local bone remodeling processes could be regulated. PTH is the major regulator of bone and mineral metabolism, and intriguing results indicate that nucleotides can sensitize cells to PTH stimulation,[89–91] or that some intracellular pathways cannot be stimulated by PTH without P2 receptor activation at the same time. Thus, nucleotides act as costimulators on bone cells, localizing systemic signals in some areas of the skeleton where activation is needed, while PTH will have no effect in areas not exposed to mechanical load or other signals that activate P2 receptors. Further, PTH induction of *c-fos* gene expression is strongly potentiated by P2 receptor activation, thus augmenting growth and differentiation signals in osteoblasts, as well as proliferation. A speculative model of the mechanotransductory coupling to the PTH-purinergic activation cascade is shown in Figure 1.5.

P2 receptor responsiveness to agonists also seems to be regulated, especially via PKC. Thus endocrine signals acting via the PKC pathway are theoretically able to regulate the responsiveness and "tonus" of the receptors. Only a few studies have addressed this issue, and none have examined the effect of endocrine factors on P2 receptor responsiveness in bone. However, 17β–estradiol seems to activate the PKC pathway in osteoblastic cells and could thus potentially affect P2 receptor function in bone cells.[109] Breast cancer MCF-7 cells respond to ATP by increasing $[Ca^{2+}]_i$ from intracellular stores, a response that can be inhibited by pretreatment with thapsigargin, indicating that the calcium originates from intracellular stores. ATP-induced increases in $[Ca^{2+}]_i$ could further be inhibited by 17β–estradiol.[110] Thus, 17β–estradiol has the potential to regulate P2Y receptor responsiveness to nucleotides. P2X receptors also seem to be regulated by estrogens. Cario-Toumaniantz et al. used patch-clamp techniques to examine the effect of 17β–estradiol on BzATP-induced $P2X_7$ receptor channel function in CV-1 monkey kidney cells transformed by SV40 (COS cells). Results showed that channel function was abolished after estradiol treatment of the cells, which implies that estrogens can systemically regulate the function of at least $P2X_7$

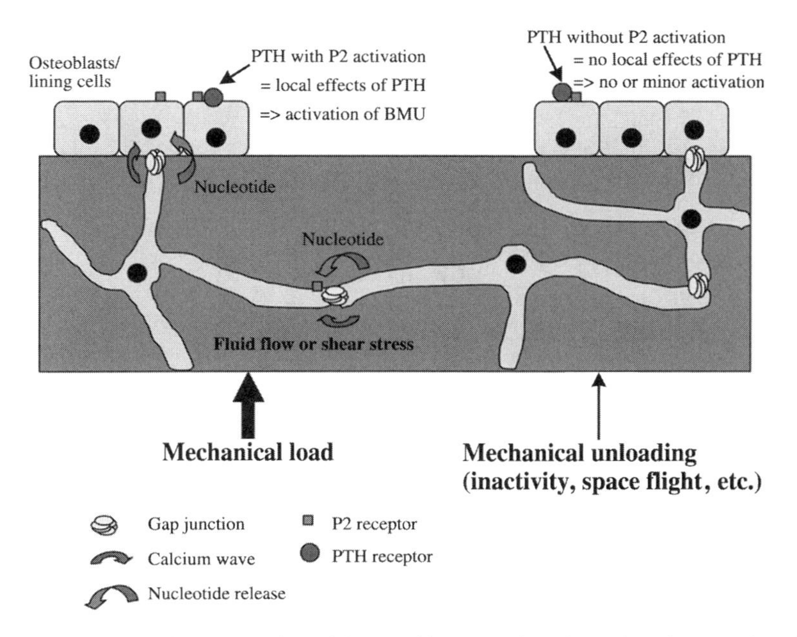

FIGURE 1.5 Schematic presentation of the possible role of calcium signaling, gap junctions, and purinergic receptors in the transduction of mechanical stimuli on bone and the localization of endocrine signals via interactions between parathyroid hormone (PTH) receptors and purinergic P2Y receptors. Mechanical forces are sensed by osteocytes and propagated through the osteocyte network via gap junctions and purinergic receptors to the bone surface. Gap junction– and ATP-mediated calcium signals are then transmitted to lining cells/osteoblasts on the endosteal surface, subsequently involving a larger population of osteoblastic cells. In response to released ATP or other nucleotides, osteoblasts are induced to differentiate, proliferate, and modify activity. Further, osteoblasts are capable of modifying osteoclast number and activity by paracrine action of secreted ATP on osteoclast P2Y and P2X receptors. The sensitivity of the osteocyte network and the calcium signaling system can be superiorly regulated by calciotropic hormones such as estrogen, and the purinergic receptor activity can modulate bone cells' sensitivity to PTH, thereby localizing endocrine signals so that activation frequency can be regulated independently in different regions of the skeleton, depending on the mechanical load on bone locally.

receptors.[111] Other studies have shown that estrogen inhibits $P2X_7$ receptor-mediated apoptosis in human ectocervical epithelial cells.[112] This finding implies that the regulation of purinergic receptors may be different between males and females, which can account for some of the striking differences between genders we see in the $P2X_7$ receptor null mouse model. In our model, the percentage of formative tetracycline double-labeled surfaces compared to total bone surfaces as measured by histomorphometry is significantly reduced in the male knockouts, while no differences can be detected in bone specimens from the female animals (Figure 1.6).

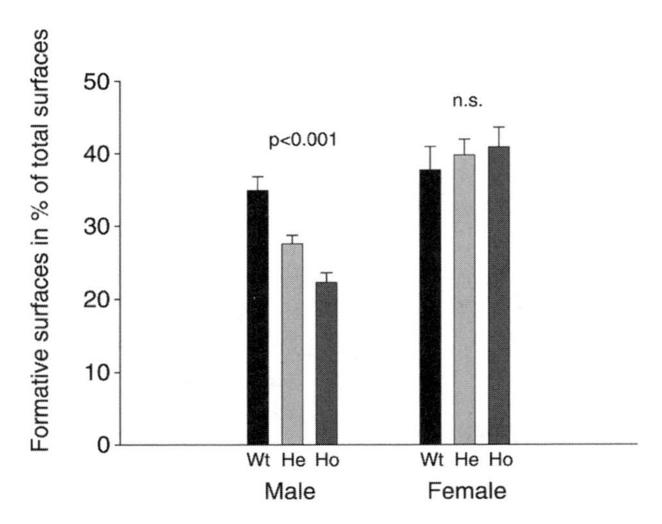

FIGURE 1.6 Four-month-old C57/BL $P2X_7$ null mice were sacrificed and bones examined for dynamic histomorphometric indices of bone formation. Six groups of animals were examined. Wild-type (Wt) animals were compared to heterozygous (He) and homozygous (Ho) animals. Further, groups of males and females were examined. Each of the 6 groups consisted of approximately 10 animals. Tetracycline labeling was performed 2 and 8 days before sacrifice. After embedding and sectioning, the number of formative surfaces in percentage of total bone surfaces was assessed. A clear difference between genotypes was detected in the males, whereas no difference between genotypes was detected in the females.

Further studies are warranted to clarify how P2 receptor activity is regulated by endocrine factors.

Another area that has not garnered much attention is the regulation of nucleotide concentration in the extracellular space. Many cells have ecto-ATPases and ecto-phosphatases on the outer cell membrane, which rapidly hydrolyze extracellular nucleotides to their respective nucleoside 5′–di- and monophosphates, nucleosides, and free phosphates or pyrophosphate. These ecto-enzymes are present on many cell types, but very few studies have addressed this issue in bone. Osteoblasts presumably express these enzymes, and a high activity of nucleotide degradation on the endosteal and periosteal surfaces could be anticipated, particularly the endosteal surface, which is heavily populated. Other areas of bone are less cellular, especially in the mineralized matrix, where osteocytes are situated surrounded by bone, only leaving a small rim of extracellular fluid and no surrounding cells. Thus, low ecto-ATPase activity could be anticipated, potentially allowing nucleotides to remain in the extracellular fluid longer and to act for a longer time and at greater distance. This is, of course, purely speculative, as no studies have addressed the presence or function of P2 receptors in osteocytes. Thus, future research should focus on these highly specialized bone cells that are situated ideally in the bone microenvironment to sense, translate, and propagate mechanical forces applied to bone and to transmit signals via action of nucleotides on P2 receptors on osteoblasts.

In conclusion, mounting evidence documents the expression of P2 receptors in osteoblastic cells, and it is clear that nucleotides acting on the various receptor subtypes are important in regulating osteoblast proliferation, differentiation, and activity.

REFERENCES

1. Mikuni-Takagaki, Y. et al., Distinct responses of different populations of bone cells to mechanical stress, *Endocrinology*, 137, 2028, 1996.
2. Burnstock, G. and Knight, G.E., Cellular distribution and functions of P2 receptor subtypes in different systems, *Int. Rev. Cytol.*, 240, 31, 2004.
3. Bodin, P. and Burnstock, G., Evidence that release of ATP from endothelial cells during increased shear stress is vesicular, *J. Cardiovasc. Pharmacol.*, 38, 900, 2001.
4. Novak, I., ATP as a signaling molecule: the exocrine focus, *News Physiol. Sci.*, 18, 12, 2003.
5. Romanello, M. et al., Mechanically induced ATP release from human osteoblastic cells, *Biochem. Biophys. Res. Commun.*, 289, 1275, 2001.
6. Genetos, D.C. et al., Fluid shear-induced ATP secretion mediates prostaglandin release in MC3T3-E1 osteoblasts, *J. Bone Miner. Res.*, 20, 41, 2005.
7. Romanello, M. and D'Andrea, P., Dual mechanism of intercellular communication in HOBIT osteoblastic cells: a role for gap-junctional hemichannels, *J. Bone Miner. Res.*, 16, 1465, 2001.
8. Romanello, M., Veronesi, V., and D'Andrea, P., Mechanosensitivity and intercellular communication in HOBIT osteoblastic cells: a possible role for gap junction hemichannels, *Biorheology*, 40, 119, 2003.
9. Romanello, M. et al., Autocrine/paracrine stimulation of purinergic receptors in osteoblasts: contribution of vesicular ATP release, *Biochem. Biophys. Res. Commun.*, 331, 1429, 2005.
10. Jorgensen, N.R. et al., ATP- and gap junction-dependent intercellular calcium signaling in osteoblastic cells, *J. Cell Biol.*, 139, 497, 1997.
11. Jorgensen, N.R. et al., Human osteoblastic cells propagate intercellular calcium signals by two different mechanisms, *J Bone Miner. Res.*, 15, 1024, 2000.
12. Lustig, K.D. et al., Expression cloning of an ATP receptor from mouse neuroblastoma cells, *Proc. Natl. Acad. Sci U.S.A.*, 90, 5113, 1993.
13. Pirotton, S. et al., Involvement of inositol 1,4,5-trisphosphate and calcium in the action of adenine nucleotides on aortic endothelial cells, *J. Biol. Chem.*, 262, 17461, 1987.
14. Communi, D. et al., Cloning and functional expression of a human uridine nucleotide receptor, *J. Biol. Chem.*, 270, 30849, 1995.
15. Stam, N.J. et al., Molecular cloning and characterization of a novel orphan receptor (P2P) expressed in human pancreas that shows high structural homology to the P2U purinoceptor, *FEBS Lett.*, 384, 260, 1996.
16. Chang, K. et al., Molecular cloning and functional analysis of a novel P2 nucleotide receptor, *J. Biol. Chem.*, 270, 26152, 1995.
17. Hollopeter, G. et al., Identification of the platelet ADP receptor targeted by antithrombotic drugs, *Nature*, 409, 202, 2001.
18. Valera, S. et al., A new class of ligand-gated ion channel defined by P2x receptor for extracellular ATP, *Nature*, 371, 516, 1994.

19. Brake, A.J., Wagenbach, M.J., and Julius, D., New structural motif for ligand-gated ion channels defined by an ionotropic ATP receptor, *Nature*, 371, 519, 1994.
20. Soto, F., Garcia-Guzman, M., and Stuhmer, W., Cloned ligand-gated channels activated by extracellular ATP (P2X receptors), *J. Membr. Biol.*, 160, 91, 1997.
21. Khakh, B.S. et al., International union of pharmacology. XXIV. Current status of the nomenclature and properties of P2X receptors and their subunits, *Pharmacol. Rev.*, 53, 107, 2001.
22. Torres, G.E., Egan, T.M., and Voigt, M.M., Identification of a domain involved in ATP-gated ionotropic receptor subunit assembly, *J. Biol. Chem.*, 274, 22359, 1999.
23. Cockcroft, S. and Gomperts, B.D., ATP induces nucleotide permeability in rat mast cells, *Nature*, 279, 541, 1979.
24. Steinberg, T.H. and Silverstein, S.C., Extracellular ATP^{4-} promotes cation fluxes in the J774 mouse macrophage cell line, *J. Biol. Chem.*, 262, 3118, 1987.
25. Surprenant, A. et al., The Cytolytic P_{2Z} receptor for extracellular ATP identified as a P_{2X} receptor ($P2X_7$), *Science*, 272, 735, 1996.
26. Subramanian, S. et al., Underlying mechanisms of symmetric calcium wave propagation in rat ventricular myocytes, *Biophys. J.*, 80, 1, 2001.
27. Collo, G. et al., Tissue distribution of the $P2X_7$ receptor, *Neuropharmacology*, 36, 1277, 1997.
28. Mutini, C. et al., Mouse dendritic cells express the $P2X_7$ purinergic receptor: characterization and possible participation in antigen presentation, *J. Immunol.*, 163, 1958, 1999.
29. Naemsch, L.N., Dixon, S.J., and Sims, S.M., Activity-dependent development of $P2X_7$ current and Ca^{2+} entry in rabbit osteoclasts, *J. Biol. Chem.*, 276, 39107, 2001.
30. Jorgensen, N.R. et al., Intercellular calcium signaling occurs between human osteoblasts and osteoclasts and requires activation of osteoclast $P2X_7$ receptors, *J. Biol. Chem.*, 277, 7574, 2002.
31. Kumagai, H. et al., Neurotransmitter regulation of cytosolic calcium in osteoblast-like bone cells, *Calcif. Tissue Int.*, 45, 251, 1989.
32. Kumagai, H., Sacktor, B., and Filburn, C.R., Purinergic regulation of cytosolic calcium and phosphoinositide metabolism in rat osteoblast-like osteosarcoma cells, *J. Bone Miner. Res.*, 6, 697, 1991.
33. Reimer, W.J. and Dixon, S.J., Extracellular nucleotides elevate $[Ca^{2+}]_i$ in rat osteoblastic cells by interaction with two receptor subtypes, *Am. J. Physiol.*, 263, C1040, 1992.
34. Sistare, F.D. et al., Separate P2T and P2U purinergic receptors with similar second messenger signaling pathways in UMR-106 osteoblasts, *J. Pharmacol. Exp. Ther.*, 269, 1049, 1994.
35. Yu, H. and Ferrier, J., Osteoblast-like cells have a variable mixed population of purino/nucleotide receptors, *FEBS Lett.*, 328, 209, 1993.
36. Schofl, C. et al., Evidence for P2-purinoceptors on human osteoblast-like cells, *J. Bone Miner. Res.*, 7, 485, 1992.
37. Dixon, C.J. et al., Effects of extracellular nucleotides on single cells and populations of human osteoblasts: contribution of cell heterogeneity to relative potencies, *Br. J. Pharmacol.*, 120, 777, 1997.
38. Luo, L.C., Yu, H., and Ferrier, J., Differential purinergic receptor signalling in osteoclasts and osteoblastic cells, *Cell Signal*, 9, 603, 1997.
39. Bowler, W.B. et al., Identification and cloning of human P2U purinoceptor present in osteoclastoma, bone, and osteoblasts, *J. Bone Miner. Res.*, 10, 1137, 1995.

40. Gartland, A. et al., Expression of a $P2X_7$ receptor by a subpopulation of human osteoblasts, *J. Bone Miner. Res.*, 16, 846, 2001.
41. Ke, H.Z. et al., Deletion of the $P2X_7$ nucleotide receptor reveals its regulatory roles in bone formation and resorption, *Mol. Endocrinol.*, 17, 1356, 2003.
42. Maier, R. et al., Cloning of $P2Y_6$ cDNAs and identification of a pseudogene: comparison of P2Y receptor subtype expression in bone and brain tissues, *Biochem. Biophys. Res. Commun.*, 237, 297, 1997.
43. Hoebertz, A. et al., Expression of P2 receptors in bone and cultured bone cells, *Bone*, 27, 503, 2000.
44. Orriss, I.R. et al., Osteoblast responses to nucleotides increase during differentiation. *Bone*, 39, 300, 2006.
45. Jorgensen, N.R. et al., Activation of L-type calcium channels is required for gap junction-mediated intercellular calcium signaling in osteoblastic cells, *J. Biol. Chem.*, 278, 4082, 2003.
46. Osipchuk, Y. and Cahalan, M., Cell-to-cell spread of calcium signals mediated by ATP receptors in mast cells, *Nature*, 359, 241, 1992.
47. Schlosser, S.F., Burgstahler, A.D., and Nathanson, M.H., Isolated rat hepatocytes can signal to other hepatocytes and bile duct cells by release of nucleotides, *Proc. Natl. Acad. Sci. U.S.A.*, 93, 9948, 1996.
48. Brake, A.J. and Julius, D., Signaling by extracellular nucleotides, *Ann. Rev. Cell Dev. Biol.*, 12, 519, 1996.
49. Hofer, A.M. et al., Intercellular communication mediated by the extracellular calcium-sensing receptor [see comments], *Nat. Cell Biol.*, 2, 392, 2000.
50. Boitano, S., Dirksen, E.R., and Sanderson, M.J., Intercellular propagation of calcium waves mediated by inositol trisphosphate, *Science*, 258, 292, 1992.
51. Boitano, S., Sanderson, M.J., and Dirksen, E.R., A role for Ca^{2+}-conducting ion channels in mechanically-induced signal transduction of airway epithelial cells, *J. Cell Sci.*, 107, 3037, 1994.
52. Sanderson, M.J., Charles, A.C., and Dirksen, E.R., Mechanical stimulation and intercellular communication increases intracellular Ca^{2+} in epithelial cells, *Cell Regul.*, 1, 585, 1990.
53. Deitmer, J.W., Verkhratsky, A.J., and Lohr, C., Calcium signalling in glial cells, *Cell Calcium*, 24, 405, 1998.
54. Sanderson, M.J., Chow, I., and Dirksen, E.R., Intercellular communication between ciliated cells in culture, *Am. J. Physiol.*, 254, C63, 1988.
55. Hansen, M. et al., Intercellular calcium signaling induced by extracellular adenosine 5′–triphosphate and mechanical stimulation in airway epithelial cells, *J. Cell Sci.*, 106, 995, 1993.
56. Evans, J.H. and Sanderson, M.J., Intracellular calcium oscillations induced by ATP in airway epithelial cells, *Am. J. Physiol.*, 277, L30, 1999.
57. Homolya, L., Steinberg, T.H., and Boucher, R.C., Cell to cell communication in response to mechanical stress via bilateral release of ATP and UTP in polarized epithelia, *J. Cell Biol.*, 150, 1349, 2000.
58. D'Andrea, P. et al., Intercellular Ca^{2+} waves in mechanically stimulated articular chondrocytes, *Biorheology*, 37, 75, 2000.
59. D'Andrea, P. and Vittur, F., Gap junctions mediate intercellular calcium signalling in cultured articular chondrocytes, *Cell Calcium*, 20, 389, 1996.
60. D'Andrea, P. and Vittur, F., Spatial and temporal Ca^{2+} signalling in articular chondrocytes, *Biochem. Biophys. Res. Commun.*, 215, 129, 1995.

61. Guilak, F. et al., Mechanically induced calcium waves in articular chondrocytes are inhibited by gadolinium and amiloride, *J. Orthop. Res.*, 17, 421, 1999.
62. D'Andrea, P. and Vittur, F., Propagation of intercellular Ca^{2+} waves in mechanically stimulated articular chondrocytes, *FEBS Lett.*, 400, 58, 1997.
63. Charles, A.C. et al., Intercellular calcium signaling via gap junctions in glioma cells, *J. Cell Biol.*, 118, 195, 1992.
64. Charles, A.C. et al., Intercellular signaling in glial cells: calcium waves and oscillations in response to mechanical stimulation and glutamate, *Neuron*, 6, 983, 1991.
65. Naus, C.C. et al., Altered gap junctional communication, intercellular signaling, and growth in cultured astrocytes deficient in connexin43, *J. Neurosci. Res.*, 49, 528, 1997.
66. Venance, L. et al., Inhibition by anandamide of gap junctions and intercellular calcium signalling in striatal astrocytes, *Nature*, 376, 590, 1995.
67. Venance, L. et al., Mechanism involved in initiation and propagation of receptor-induced intercellular calcium signaling in cultured rat astrocytes, *J. Neurosci.*, 17, 1981, 1997.
68. Yao, J. et al., Coordination of mesangial cell contraction by gap junction-mediated intercellular Ca^{2+} wave, *J. Am. Soc. Nephrol.*, 13, 2018, 2002.
69. Young, S.H., Ennes, H.S., and Mayer, E.A., Propagation of calcium waves between colonic smooth muscle cells in culture, *Cell Calcium*, 20, 257, 1996.
70. Christ, G.J. et al., Gap junction-mediated intercellular diffusion of Ca^{2+} in cultured human corporal smooth muscle cells, *Am. J. Physiol.*, 263, C373, 1992.
71. Saez, J.C. et al., Hepatocyte gap junctions are permeable to the second messenger, inositol 1,4,5-triphosphate, and to calcium ions, *Proc. Natl. Acad. Sci. U.S.A.*, 86, 2708, 1989.
72. Grierson, J.P. and Meldolesi, J., Shear stress-induced $[Ca^{2+}]_i$ transients and oscillations in mouse fibroblasts are mediated by endogenously released ATP, *J. Biol. Chem.*, 270, 4451, 1995.
73. Churchill, G.C., Atkinson, M.M., and Louis, C.F., Mechanical stimulation initiates cell-to-cell calcium signaling in ovine lens epithelial cells, *J. Cell. Sci.*, 109, 355, 1996.
74. Suadicani, S.O., Vink, M.J., and Spray, D.C., Slow intercellular Ca^{2+} signaling in wild-type and Cx43-null neonatal mouse cardiac myocytes, *Am. J. Physiol.*, 279, H3076, 2000.
75. Sanderson, M.J. and Dirksen, E.R., Mechanosensitivity of cultured ciliated cells from the mammalian respiratory tract: implications for the regulation of mucociliary transport, *Proc. Natl. Acad. Sci. U.S.A.*, 83, 7302, 1986.
76. Sakuma, N. et al., Glucose induces calcium-dependent and calcium-independent insulin secretion from the pancreatic beta cell line MIN6, *Eur. J. Endocrinol.*, 133, 227, 1995.
77. Cao, D. et al., Mechanisms for the coordination of intercellular calcium signaling in insulin-secreting cells, *J. Cell Sci.*, 110, 497, 1997.
78. Bertuzzi, F. et al., Mechanisms of coordination of Ca^{2+} signals in pancreatic islet cells, *Diabetes*, 48, 1971, 1999.
79. Schwiebert, E.M., Extracellular ATP-mediated propagation of Ca^{2+} waves. Focus on "mechanical strain-induced Ca^{2+} waves are propagated via ATP release and purinergic receptor activation," *Am. J. Physiol.*, 279, C281, 2000.
80. Berridge, M.J., The biology and medicine of calcium signalling, *Mol. Cell. Endocrinol.*, 98, 119, 1994.
81. Sanderson, M.J. et al., Mechanisms and function of intercellular calcium signaling, *Mol. Cell. Endocrinol.*, 98, 173, 1994.
82. Rottingen, J. and Iversen, J.G., Ruled by waves? Intracellular and intercellular calcium signalling, *Acta Physiol. Scand.*, 169, 203, 2000.

83. Berridge, M.J., Lipp, P., and Bootman, M.D., The versatility and universality of calcium signalling, *Nature Rev.*, 1, 11, 2000.
84. Xia, S.L. and Ferrier, J., Propagation of a calcium pulse between osteoblastic cells, *Biochem. Biophys. Res. Commun.*, 186, 1212, 1992.
85. Jorgensen, N.R. et al., *In vitro* differentiation of human osteoblast-like cells alters the predominant mechanism of intercellular calcium wave propagation, *J. Bone Miner. Res.*, 14, S476, 1999.
86. You, J. et al., P2Y purinoceptors are responsible for oscillatory fluid flow-induced intracellular calcium mobilization in osteoblastic cells, *J. Biol. Chem.*, 277, 48724, 2002.
87. Bowler, W.B. et al., Extracellular nucleotide signaling: a mechanism for integrating local and systemic responses in the activation of bone remodeling, *Bone*, 28, 507, 2001.
88. Ostrom, R.S., Gregorian, C., and Insel, P.A., Cellular release of and response to ATP as key determinants of the set-point of signal transduction pathways, *J. Biol. Chem.*, 275, 11735, 2000.
89. Kaplan, A.D. et al., Extracellular nucleotides potentiate the cytosolic Ca^{2+}, but not cyclic adenosine 3′, 5′–monophosphate response to parathyroid hormone in rat osteoblastic cells, *Endocrinology*, 136, 1674, 1995.
90. Sistare, F.D., Rosenzweig, B.A., and Contrera, J.G., P2 purinergic receptors potentiate parathyroid hormone receptor-mediated increases in intracellular calcium and inositol trisphosphate in UMR-106 rat osteoblasts, *Endocrinology*, 136, 4489, 1995.
91. Buckley, K.A. et al., Parathyroid hormone potentiates nucleotide-induced $[Ca^{2+}]_i$ release in rat osteoblasts independently of G_q activation or cyclic monophosphate accumulation. A mechanism for localizing systemic responses in bone, *J. Biol. Chem.*, 276, 9565, 2001.
92. Bowler, W.B. et al., Signaling in human osteoblasts by extracellular nucleotides. Their weak induction of the c-fos proto-oncogene via Ca^{2+} mobilization is strongly potentiated by a parathyroid hormone/cAMP-dependent protein kinase pathway independently of mitogen-activated protein kinase, *J. Biol. Chem.*, 274, 14315, 1999.
93. Wang, Z.Q. et al., Bone and haematopoietic defects in mice lacking c-fos, *Nature*, 360, 741, 1992.
94. Jorgensen, N.R., Steinberg, T.H., and Civitelli, R., Osteoblast-osteoclast communication, *Curr. Opin. Orthoped.*, 10, 367, 1999.
95. Suzuki, A. et al., Mechanism of phospholipase D activation induced by extracellular ATP in osteoblast-like cells, *J. Endocrinol.*, 145, 81, 1995.
96. Watanabe, T. et al., Arachidonic acid release induced by extracellular ATP in osteoblasts: role of phospholipase D, *Prostaglandins, Leukotrienes and Essential Fatty Acids*, 57, 335, 1997.
97. Gallinaro, B.J., Reimer, W.J., and Dixon, S.J., Activation of protein kinase C inhibits ATP-induced $[Ca^{2+}]_i$ elevation in rat osteoblastic cells: selective effects on P2Y and P2U signaling pathways, *J. Cell Physiol.*, 162, 305, 1995.
98. Shimegi, S., ATP and adenosine act as a mitogen for osteoblast-like cells (MC3T3-E1), *Calcif. Tissue Int.*, 58, 109, 1996.
99. Nakamura, E. et al., ATP activates DNA synthesis by acting on P2X receptors in human osteoblast-like MG-63 cells, *Am. J. Physiol.*, 279, C510, 2000.
100. Walther, M.M. et al., Suramin inhibits bone resorption and reduces osteoblast number in a neonatal mouse calvarial bone resorption assay, *Endocrinology*, 131, 2263, 1992.
101. Hoebertz, A. et al., Extracellular ADP is a powerful osteolytic agent: evidence for signaling through the $P2Y_1$ receptor on bone cells, *FASEB J.*, 15, 1139, 2001.

102. Hoebertz, A. et al., ATP and UTP at low concentrations strongly inhibit bone formation by osteoblasts: a novel role for the $P2Y_2$ receptor in bone remodeling, *J. Cell Biochem.*, 86, 413, 2002.
103. Jones, S.J. et al., Purinergic transmitters inhibit bone formation by cultured osteoblasts, *Bone*, 21, 393, 1997.
104. Pines, A. et al., Extracellular ATP stimulates the early growth response protein 1 (Egr-1) via a protein kinase C-dependent pathway in the human osteoblastic HOBIT cell line, *Biochem. J.*, 373, 815, 2003.
105. Ihara, H. et al., ATP-stimulated interleukin-6 synthesis through P2Y receptors on human osteoblasts, *Biochem. Biophys. Res. Commun.*, 326, 329, 2005.
106. Buckley, K.A. et al., Adenosine triphosphate stimulates human osteoclast activity via upregulation of osteoblast-expressed receptor activator of nuclear factor-kB ligand, *Bone*, 31, 582, 2002.
107. Saunders, M.M. et al., Fluid flow-induced prostaglandin E2 response of osteoblastic ROS 17/2.8 cells is gap junction-mediated and independent of cytosolic calcium, *Bone*, 32, 350, 2003.
108. Vico, L. et al., Effects of long-term microgravity exposure on cancellous and cortical weight-bearing bones of cosmonauts [see comments], *Lancet*, 355, 1607, 2000.
109. Boyan, B.D. and Schwartz, Z., Rapid vitamin D-dependent PKC signaling shares features with estrogen-dependent PKC signaling in cartilage and bone, *Steroids*, 69, 591, 2004.
110. Rossi, A.M. et al., Evidence on the operation of ATP-induced capacitative calcium entry in breast cancer cells and its blockade by 17beta-estradiol, *J. Cell Biochem.*, 87, 324, 2002.
111. Cario, T. et al., Non-genomic inhibition of human $P2X_7$ purinoceptor by 17β–oestradiol, *J. Physiol.*, 508, 659, 1998.
112. Gorodeski, G.I., Estrogen attenuates $P2X_7$-mediated apoptosis of uterine cervical cells by blocking calcium influx, *Nucleos. Nucleot. Nucleic Acids*, 23, 1287, 2004.
113. Yu, H. and Ferrier, J., ATP induces an intracellular calcium pulse in osteoclasts, *Biochem. Biophys. Res. Commun.*, 191, 357, 1993.
114. Jorgensen, N.R. et al., ATP- and gap junction–dependent intercellular calcium signaling in osteoblastic cells, *J. Cell Biol.*, 139, 497, 1997.

2 P2 Nucleotide Receptor Signaling in Osteoclasts

Jasminka Korcok, Stephen M. Sims, and S. Jeffrey Dixon

CONTENTS

2.1 SCOPE

Extracellular nucleotides bind to P2 receptors on many cell types, including osteoclasts. These large, multinucleated cells are responsible for the resorption of bone and other mineralized tissues during skeletal development, tooth eruption, bone remodeling, and wound healing. In addition, osteoclasts mediate the pathological destruction of bone in metabolic, inflammatory, and neoplastic disorders, including osteoporosis, rheumatoid arthritis, periodontitis, and osteolytic metastases.[1] Osteoclasts express multiple subtypes of both ionotropic (P2X) and metabotropic (P2Y) receptors for extracellular nucleotides. The recently described skeletal phenotype of mice with targeted disruption of the $P2X_7$ receptor points to interesting roles for this receptor in the regulation of bone formation and resorption.[2] Nucleotides released from other cell types in response to mechanical stimulation or injury may serve as paracrine regulators of osteoclast function. Thus, nucleotides and the complex network of P2 receptors in bone cells may underlie the well-recognized responses of skeletal tissues to mechanical stimuli.[3,4] A number of excellent reviews have previously described P2 receptors in osteoclasts.[5–10] This chapter reviews recent work on the identity of P2 receptor subtypes on osteoclasts, their associated signal transduction mechanisms, and their possible roles in regulating bone resorption.

2.2 OSTEOCLAST BIOLOGY

Osteoclasts are resorptive cells, which together with osteoblasts play an essential role in skeletal development and remodeling. Under physiological conditions in adult bone, resorption by osteoclasts is in balance with formation by osteoblasts. Disruption of this balance can lead to various diseases such as osteoporosis. In both normal and pathological conditions, bone resorption can be regulated by changes in the recruitment and differentiation of osteoclast precursors and in the intrinsic resorptive activity and life span of mature osteoclasts. Osteoclasts are large, multinucleated cells of the monocyte/macrophage lineage with the unique ability to resorb bone and other calcified tissues.[11] Some of the markers used to identify mature osteoclasts are tartrate-resistant acid phosphatase (TRAP), $\alpha_v\beta_3$ integrin (vitronectin receptor), calcitonin receptor, carbonic anhydrase II, as well as the lysosomal protease, cathepsin K. An isolated osteoclast visualized by phase-contrast microscopy has multiple nuclei and broad pseudopods (Figure 2.1A). A commonly used, highly selective osteoclast marker is the enzyme TRAP, which can be identified by cytochemistry (Figure 2.1B). However, the definitive characteristic of osteoclasts is the ability to resorb mineralized substrates. In Figure 2.1B, isolated rabbit osteoclasts plated on a hydroxyapatite-coated disc are seen adjacent to clear areas where the substrate has been resorbed. When cultured on slices of dentin, these cells form typical resorption lacunae or “pits” (Figure 2.1C), confirming that these preparations contain authentic osteoclasts. Such preparations of adherent bone marrow cells are often used for *in situ* labeling and functional characterization of P2 receptors on mammalian osteoclasts.

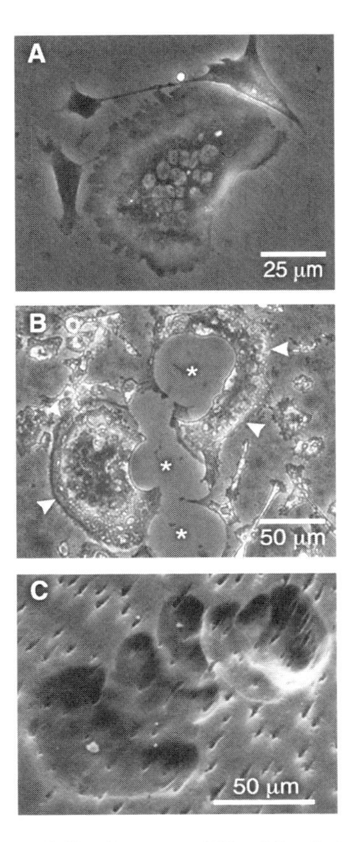

FIGURE 2.1 (See color insert following page 80) Morphology and resorptive activity of isolated osteoclasts. Osteoclasts were isolated from bone marrow of neonatal rats and rabbits. Shown are (A) A micrograph of a representative multinucleated rat osteoclast visualized by phase-contrast microscopy. (B) Isolated rabbit osteoclasts were plated on a hydroxyapatite-coated discs. Two days following isolation, osteoclasts were washed, fixed, and stained for the osteoclast marker tartrate-resistant acid phosphatase (TRAP). TRAP-stained (maroon) rabbit osteoclasts (arrowheads) are associated with resorbed areas (asterisks). (C) A scanning electron micrograph of dentin slice reveals a resorption trail left behind by an osteoclast while resorbing matrix. Isolated rabbit osteoclasts were cultured for 48 h on dentin slices made from human teeth. Cells were removed, and slices were then examined using scanning electron microscopy.

2.2.1 Osteoclast Formation

Osteoclasts form by the fusion of mononucleated precursor cells of the monocyte/macrophage lineage.[12] Osteoclast formation, activation, and survival are dependent on the presence of key signaling molecules, such as the receptor activator of NF-κB (RANK) ligand and macrophage colony stimulating factor (M-CSF). RANK ligand, a tumor necrosis factor (TNF)-related trimeric protein, is synthesized by stromal cells, osteoblasts, and activated lymphocytes in membrane-bound or secreted forms.[12–15] RANK ligand binds to its receptor RANK, a member of the TNF receptor superfamily, to stimulate osteoclastogenesis, activate bone resorption, and enhance osteoclast survival. Alternatively, RANK ligand can be bound by the soluble decoy receptor osteoprotegerin, which prevents its signaling. Upon binding of ligand, RANK activates multiple signaling pathways including mitogen-activated protein kinases, phosphatidylinositol 3-kinase, and Ca^{2+}/calcineurin, as well as the transcription factors AP-1, NF-κB, and NFAT.[15–17]

Like RANK ligand, M-CSF is expressed by marrow stromal cells and osteoblasts in membrane-bound and secreted forms. The important role for M-CSF in osteoclast development was demonstrated in *op/op* mice, which, due to a defect in M-CSF production, lack osteoclasts and exhibit an osteopetrotic phenotype.[18] M-CSF acts on its receptor c-fms on monocytes, followed by RANK ligand acting on RANK, to induce osteoclast development. M-CSF acts synergistically with RANK ligand to induce osteoclastogenesis and activation of mature osteoclasts *in vitro*.[19]

Several transcription factors, including PU.1, c-fos, NFATc1, and NF-κB play essential roles in osteoclast development and function. PU.1 is expressed specifically in the monocytic and B lymphoid lineages[20] and regulates transcription of *c-fms* in early osteoclast precursors.[21] As such, PU.1 is the earliest known marker of osteoclast differentiation. The transcription factor c-fos, a member of the AP family of transcription factors, is downstream of PU.1. Recent evidence suggests that the main role of c-fos during osteoclast formation is transcriptional induction of NFATc1, which in turn activates a number of genes involved in osteoclast development and function.[22] In fact, NFATc1 is thought to be a master switch for regulating the terminal differentiation of osteoclasts.[23,24] RANK ligand activates c-fos and Ca^{2+} signaling, which are essential for NFATc1 expression and activation in osteoclasts.[25,26]

The most extensively studied transcription factor that is essential for osteoclastogenesis is NF-κB. NF-κB is a dimer composed of different combinations of p50, p52, p65, RelB, and c-Rel, the most common being the p50/p65 heterodimer. In the canonical NF-κB pathway, inactive NF-κB is bound to the inhibitory protein IκB, which retains NF-κB in the cytoplasm by masking its nuclear localization sequence.[27] Phosphorylation and subsequent ubiquitination of IκB leads to its proteasomal degradation and release of NF-κB. NF-κB can then translocate into the nucleus to regulate the expression of a variety of genes involved in inflammation and immunity, cell proliferation, responses to stress, and apoptosis.[28] The essential role of NF-κB in osteoclast formation was demonstrated in mice deficient for both the p50 and p52 subunits of NF-κB, which showed a pronounced osteopetrotic phenotype as a result of complete absence of osteoclasts.[29,30]

2.2.2 Mechanism of Resorption

At sites of resorption, osteoclasts attach to the bone creating an isolated extracellular microenvironment (Figure 2.2). The sealing zone (also called "clear zone"),[31] in which the osteoclast membrane is in close proximity to the bone surface, is thought to be established by adhesion proteins possibly integrins. Osteoclasts express at least three integrins: $\alpha_v\beta_3$ (vitronectin receptor), $\alpha_2\beta_1$ (collagen receptor), and $\alpha_v\beta_1$.[32] $\alpha_v\beta_3$

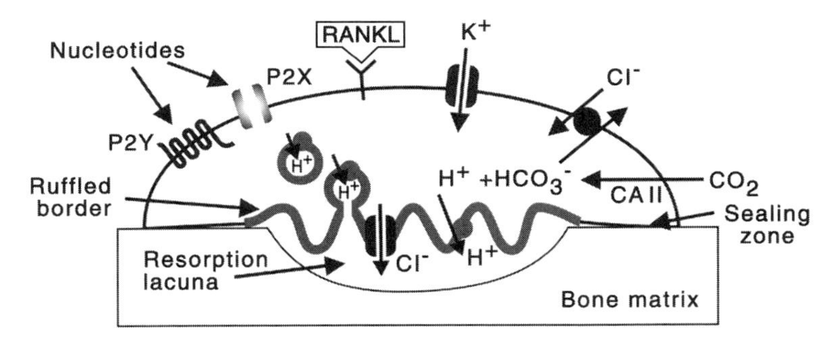

FIGURE 2.2 Mechanism and regulation of osteoclastic bone resorption. The resorbing osteoclast forms a tight seal with bone matrix and creates an isolated extracellular microenvironment, called the "resorption lacuna." Attachment of osteoclasts to the bone matrix prompts the acidifying vesicles, containing proton pumps (H^+-ATPase) in their membrane and degradative enzymes in their lumen, to fuse with the plasma membrane, thus forming the ruffled border. Within the osteoclast, H^+ are produced by type II carbonic anhydrase (CAII) and are extruded across the ruffled border, acidifying the resorption lacuna and causing dissolution of bone mineral. Cl^- efflux and K^+ influx balance electrogenic charge transfer caused by the proton pump. Cytosolic pH is maintained by the Cl^-/HCO_3^- exchanger located on the basolateral surface. The organic component of the bone matrix is broken down by cathepsin K and other degradative enzymes released by exocytosis into the resorption lacuna. Osteoclast formation, resorptive activity, and survival are regulated by receptor activator of NF-κB ligand (RANKL) acting via its receptor RANK. Nucleotides act via P2X and P2Y receptors on osteoclasts.

has the ability to bind a large number of matrix proteins, including fibronectin, bone sialoprotein, and osteopontin.[33]

The extracellular matrix of bone is composed of organic and inorganic phases. The organic phase is mostly type-I collagen, with a small fraction of noncollagenous proteins and proteoglycans. To resorb bone, attached osteoclasts must first demineralize the matrix, exposing the organic components.[34] This process requires acidification of the resorption lacuna, which is achieved by the vacuolar-type proton pump (H^+-ATPase) located in the ruffled border. The ruffled border corresponds to the apical surface of a secretory cell (Figure 2.2). Cl^- efflux through a channel, located on the same membrane, balances electrogenic charge transfer by the proton pump. Within osteoclasts, protons are derived from carbonic acid made by hydration of CO_2, a process catalyzed by carbonic anhydrase. Proton extrusion leaves its conjugate base within the cell, and the Cl^-/HCO_3^- exchanger located on the basolateral surface of the osteoclast is responsible for maintaining cytosolic pH (Figure 2.2).[35] The acidic environment of the resorption lacuna leads to dissolution of mineral, leaving organic components of the matrix exposed for subsequent enzymatic hydrolysis. Degradative enzymes include the lysosomal protease, cathepsin K,[36] metalloproteinases,[37] and TRAP.[38] Degradation products of osteoclastic resorption are removed from the resorption lacunae by transcytosis and released into the extracellular environment.[39,40] After osteoclasts have eroded bone to a particular distance, they can either move to new locations, leaving a characteristic resorption trail (e.g., Figure 2.1C), or they die by apoptosis.[41]

Osteoclast migration can be initiated by soluble agonists and chemotactic agents such as M-CSF and transforming growth factor β (TGF-β).[42–44] Motile osteoclasts are recognized *in vitro* by the presence of broad pseudopodia that actively retract and spread (Figure 2.1A). These pseudopodia are flat structures, devoid of organelles, formed by localized increase in actin polymerization.[45] To migrate, osteoclasts must interact with the matrix, a process mediated by integrins. Thus, in addition to mediating adhesion, integrins are also involved in regulating the shape and motility of osteoclasts.

2.2.3 Regulation of Osteoclast Survival

The predominant form of osteoclast death *in vivo* and *in vitro* is apoptosis.[46] Some distinct morphological features of apoptotic osteoclasts are stronger cytoplasmic TRAP staining than viable cells, and the condensation and fragmentation of all nuclei.[47] *In vitro* (in the absence of supporting cells such as osteoblasts or bone marrow stromal cells, or survival factors such as RANK ligand and M-CSF), osteoclasts undergo apoptosis more rapidly than *in vivo*.[48]

It has been suggested that change in the rate of osteoclast apoptosis may be the most important mechanism regulating bone resorption *in vivo*.[49,50] Some systemic hormones, including sex hormones estrogen and progesterone, stimulate osteoclast apoptosis. Other hormones act indirectly to regulate osteoclast apoptosis. Both parathyroid hormone and 1,25$(OH)_2$ vitamin D3 prevent osteoclast apoptosis, but this most likely occurs through induction of RANK ligand expression by cells of the osteoblast lineage, coupled with downregulation of osteoprotegerin expression.[47] Some bisphosphonates, which are widely used for treatment of osteoporosis, induce osteoclast apoptosis, thereby suppressing osteoclast numbers and, hence, bone resorption.[51]

2.3 OVERVIEW OF P2 RECEPTOR SUBTYPES IN OSTEOCLASTS

Using a variety of techniques, the expression of a number of P2 receptors has been established in osteoclasts. Early studies showed that ATP (an agonist at multiple P2 receptors) induces a transient rise in $[Ca^{2+}]_i$ in isolated rabbit osteoclasts.[52] This rise in $[Ca^{2+}]_i$ was found to be due to both influx of extracellular Ca^{2+} (consistent with the involvement of P2X channels) and release from intracellular stores (consistent with the involvement of G protein–coupled P2Y receptors).[52,53] Permeabilization studies on murine osteoclasts showed that relatively high concentrations of ATP (2 mM), in the absence of divalent cations, induced ethidium bromide uptake[54]—a process now known to be mediated by the $P2X_7$ receptor. In addition, Bowler and coworkers provided evidence for the expression of $P2Y_2$ receptor transcripts in osteoclast-like cells.[55] Electrophysiological studies on isolated rat osteoclasts showed activation of two currents by ATP—an initial P2X receptor–mediated inward current, followed by an outward K^+ current activated in response to P2Y-mediated release of Ca^{2+} from intracellular stores.[56] These studies provided evidence for coexpression of P2X and P2Y receptors on osteoclasts. Soon after, P2 receptors were shown to have potent effects on osteoclast function. Using isolated rat and mouse osteoclasts, it was demonstrated that ATP stimulates osteoclast formation and resorptive activity.[57]

Thus far, the expression of $P2X_2$, $P2X_4$, $P2X_7$, $P2Y_1$, $P2Y_2$, and $P2Y_6$ receptors has been shown in authentic osteoclasts (Table 2.1). Studies of the expression of individual receptors, their signaling pathways, and proposed functions in osteoclasts are reviewed below.

2.4 P2X RECEPTOR SUBTYPES IN OSTEOCLASTS: EXPRESSION, SIGNALING, AND FUNCTION

P2X receptors are ligand-gated cation channels, which upon activation cause membrane depolarization and are, in some cases, permeable to Ca^{2+}.[58] There are seven mammalian P2X receptors, $P2X_{1-7}$. Multiple approaches including *in situ* hybridization and immunocytochemistry have been used to show that isolated osteoclasts express $P2X_2$, $P2X_4$, and $P2X_7$ receptors.[59–61] A wide range of P2X receptors ($P2X_{1, 4, 5, 6, 7}$) was demonstrated using RT-PCR on osteoclast-like cells derived *in vitro* from human peripheral blood monocytes.[62] However, additional studies are necessary to determine if all of these receptors are expressed in authentic osteoclasts and, if so, whether they are functional.

P2X receptors are activated by ATP and more stable analogues of ATP such as adenosine 5′–*O*-(3-thiotriphosphate) (ATPγS). Using ATP and several other agonists at P2X receptors, electrophysiological studies identified the presence of functional $P2X_4$ and $P2X_7$ receptors on isolated rabbit osteoclasts.[60,61] Several previously published reviews have summarized evidence for, and the properties of $P2X_2$, $P2X_4$ and $P2X_7$ receptors on osteoclasts.[7,9] Due to the plethora of recent findings regarding the function of $P2X_7$ receptors in osteoclasts, this review will focus mostly on the $P2X_7$ receptor.

2.4.1 $P2X_2$ Receptors

Studies on the distribution of $P2X_2$ receptors revealed their expression primarily in neural tissue, but also in other tissues such as vascular smooth muscle.[58] The affinity of the $P2X_2$ receptor for ATP was shown to be dramatically enhanced by extracellular acidification (to pH 6.5).[63,64] Using immunocytochemistry and *in situ* hybridization, the expression of $P2X_2$ receptors was demonstrated in isolated rat osteoclasts.[59] Immunocytochemical staining of histological sections of rat long bone also revealed expression of $P2X_2$ on chondrocytes and osteoblasts, indicating its wide expression in skeletal tissues.[59]

An early study showed a stimulatory effect of ATP on bone resorption by rat osteoclasts and on the formation of murine osteoclasts.[57] Interestingly, these effects were greatly amplified in acidified medium (pH 6.9 to 7.0). Notably, the pH dependence of resorption pit formation by rat osteoclasts was similar to the pH dependence for sensitization of the recombinant $P2X_2$ receptor expressed in *Xenopus* oocytes.[65] Thus, it was initially suggested that the stimulatory effect of ATP on bone resorption could be mediated by the $P2X_2$ receptor. However, this effect was later attributed to the $P2Y_1$ receptor, since ADP, a degradation product of ATP, and a $P2Y_1$ agonist

TABLE 2.1
P2 Receptors in Mammalian Osteoclasts

Family	Subtype	Evidence for Expression	Signaling	Function	Ref.
P2X	$P2X_2$	Immunolabeling *In situ* hybridization	Not determined	Not determined	59
	$P2X_4$	Electrophysiology Immunolabeling *In situ* hybridization RT-PCR	Nonselective cation current Depolarization	Not determined	56, 59, 60
	$P2X_7$	Permeabilization Immunolabeling Electrophysiology Ca^{2+} fluorescence	Pore formation Nonselective cation current Ca^{2+} influx Depolarization NF-κB activation	Cytolysis Precursor fusion (controversial Enhanced apoptosis	54, 59, 61, 72, 76, 77
P2Y	P2Y*	Electrophysiology Ca^{2+} fluorescence	Ca^{2+} release from stores Ca^{2+}-activated K^+ current Hyperpolarization	Not determined	52, 56, 103, 124
	$P2Y_1$	Immunolabeling *In situ* hybridization RT-PCR Ca^{2+} fluorescence	Ca^{2+} release from stores	Increased osteoclastogenesis Stimulated resorption	66, 98
	$P2Y_2$	RT-PCR *In situ* hybridization Ca^{2+} fluorescence	Ca^{2+} release from stores	Not determined	55, 59, 98, 102
	$P2Y_6$	RT-PCR Ca^{2+} fluorescence	Ca^{2+} release from stores NF-κB activation	Increased survival	98

* Subtype not defined.

(2-methylthioADP, 2-MeSADP) were more potent stimulators of bone resorption.[66] Moreover, the effects of ADP were blocked by the $P2Y_1$ receptor antagonist MRS2179.

P2X-mediated currents in osteoclasts have been examined using patch-clamp techniques. P2X currents were activated by local application of ATP and the identity of currents was determined based on kinetics, rundown, and pharmacological profile.[60] All P2X receptors activate rapidly, but different subtypes exhibit different inactivation kinetics.[67] $P2X_1$ and $P2X_3$ receptors exhibit rapid current decay, $P2X_4$ intermediate, whereas $P2X_2$, $P2X_5$, and $P2X_7$ desensitize slowly. Electrophysiological studies showed that ATP-induced P2X currents in rabbit osteoclasts had intermediate inactivation kinetics. Moreover, these currents were not inhibited by the P2 receptor antagonists suramin (8-(3-benzamido-4-methylbenzamido)-naphthalene-1,3,5-trisulfonic acid) or Cibacron blue (an isomer of reactive blue 2).[60] However, $P2X_2$ receptors are sensitive to suramin and Cibacron blue. Taken together, these findings were inconsistent with the expression of functional $P2X_2$ receptors on rabbit osteoclasts. However, it is possible that $P2X_2$ receptors are expressed on the plasma membrane, where they activate alternate signaling pathways or are nonfunctional, or are present intracellularly.

2.4.2 $P2X_4$ Receptors

A combination of molecular, immunological, and functional studies supports the expression of $P2X_4$ receptors in osteoclasts. $P2X_4$ mRNA transcripts were identified using RT-PCR in rabbit osteoclasts purified by micromanipulation.[60] The $P2X_4$ receptor was also detected in isolated rat osteoclasts by immunocytochemistry and *in situ* hybridization.[59] In contrast, no evidence for the $P2X_4$ receptor was found on osteoblasts or chondrocytes, using *in situ* hybridization,[59] suggesting a more restricted localization than other P2 receptors in skeletal tissues.

Patch-clamp recording was used to determine whether $P2X_4$ receptors expressed on osteoclasts are functional. Osteoclasts were identified as cells containing three or more nuclei (Figure 2.3A upper panel), and selected samples were stained for TRAP following recording (Figure 2.3A lower panel). ATP activated a biphasic pattern of currents, with inward current that declined, followed by a transient outward current (Figure 2.3B). The initial inward current induced by ATP reversed near 0 mV (Figure 2.3C left panel), consistent with P2X current. The later outward current reversed at more negative potentials (Figure 2.3C right panel) and was identified as K^+ current, activated by P2Y receptor–mediated release of Ca^{2+} from intracellular stores.[56,60] The initial inward current activated by ATP showed several characteristics consistent with those of $P2X_4$-mediated currents. First, reduction of Na^+ concentration in the extracellular fluid shifted the reversal potential to more negative potentials, consistent with a nonselective cation conductance. Second, by examining the response to various agonists, it was determined that, in most osteoclasts, inward current was elicited by ATP, ATPγS, or ADP, but not by adenosine 5′–*O*-(2-thiodiphosphate) (ADPβS), UTP, or α,β–methylene ATP (α,β–meATP)—an agonist profile most consistent with activation of $P2X_4$ receptors. Other observations that helped identify the inward current as $P2X_4$ mediated were that it inactivated with intermediate kinetics during agonist

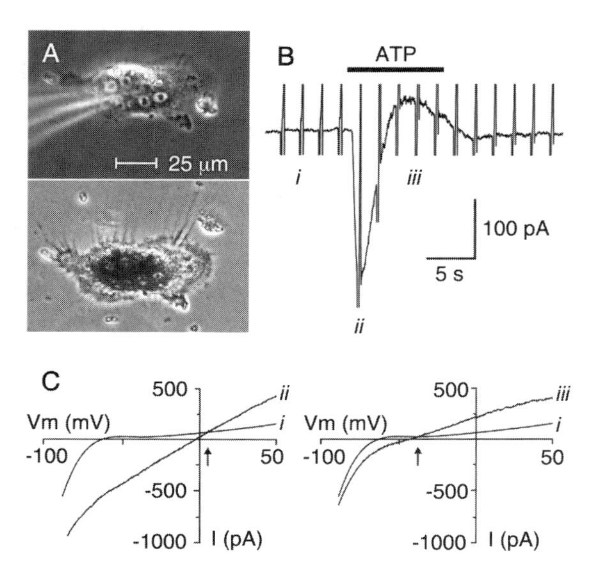

FIGURE 2.3 (See color insert following page 80) Extracellular ATP activates P2X and Ca^{2+}-activated K^+ currents in rabbit osteoclasts. (A) Phase-contrast photomicrograph of a multinucleated rabbit osteoclast (upper panel) used for electrophysiological studies in B and C. After the experiment, the cell was stained for TRAP, confirming its identity as an osteoclast (lower panel). (B) Whole-cell currents were recorded at –30 mV. ATP (100 μM) applied at the time indicated by the bar above the trace evoked an initial inward current (ii) that declined and was followed by a transient outward current (iii). To identify ATP-activated currents, voltage-ramp commands were used to obtain current–voltage (I-V) relationships. (C) Under control conditions, the I-V relationship showed an inwardly rectifying K^+ conductance (i). The initial ATP-activated current (ii) reversed near 0 mV (left panel). Rapid activation of this current is consistent with a ligand-gated P2X channel. An I-V relationship from the same cell shows the later part of the ATP-activated current (iii) reversing at more negative membrane potential (right panel), consistent with a K^+-selective current. Delayed activation of K^+ current suggests the involvement of a signaling cascade, consistent with activation of Ca^{2+}-dependent K^+ channels by P2Y receptor–mediated release of Ca^{2+} from intracellular stores. (Reprinted from Naemsch, L.N. et al., $P2X_4$ purinoceptors mediate an ATP-activated, non-selective cation current in rabbit osteoclasts, *J. Cell Sci.*, 112, 4425, 1999. With permission.)

application, was potentiated by Zn^{2+}, and was not inhibited by suramin or Cibacron blue. Suramin and Cibacron blue are antagonists at $P2X_2$ and $P2X_7$, but not $P2X_4$, receptors in heterologous expression systems.[58,67]

Ca^{2+} permeability of the $P2X_4$ receptor on isolated rat osteoclasts has been examined using a fluorescent indicator dye. Osteoclasts responded to ATP with rise of $[Ca^{2+}]_i$, as well as inward current.[68] The phospholipase C (PLC) inhibitor U-73122 abolished the ATP-induced rise in $[Ca^{2+}]_i$ without markedly affecting $P2X_4$ current. Thus, $[Ca^{2+}]_i$ elevation arises from P2Y receptor–mediated release of Ca^{2+} from intracellular stores and Ca^{2+} entry through $P2X_4$ channels on osteoclasts is negligible.

The role of $P2X_4$ receptors in osteoclast physiology remains unidentified. Activation of this receptor leads to cation influx and depolarization, which may be important signals influencing a number of processes, such as osteoclast-matrix interactions, formation of the ruffled border, and expression of genes regulating bone resorption.

2.4.3 $P2X_7$ RECEPTORS

2.4.3.1 Evidence for Expression of $P2X_7$ Receptors on Osteoclasts

The $P2X_7$ receptor requires relatively high concentrations of ATP for activation in comparison to other members of the P2X receptor family. Recent reports have described an alternative pathway for its activation. The $P2X_7$ receptor was shown to be activated following its ADP ribosylation by a cell surface ADP-ribosyltransferase (ART-2) utilizing NAD as a substrate.[69] In the presence of divalent cations in the extracellular medium, the $P2X_7$ receptor acts as nonselective cation channel. However, when divalent cations are present at low concentration, activation of the $P2X_7$ receptor is accompanied by the formation of pores permeable to hydrophilic molecules as large as 900 Da.[70,71]

The first evidence for the presence of $P2X_7$ receptors in osteoclasts came from a study in which stimulation with a relatively high concentration of ATP (2 mM) in the absence of divalent cations induced ethidium bromide (394 Da) uptake and increased membrane conductance in murine osteoclasts.[54] The effect of ATP on membrane conductance was reversed by addition of Mg^{2+}, consistent with the bifunctional properties of $P2X_7$ receptors. Later, the expression of $P2X_7$ receptors was demonstrated by immunocytochemical staining of authentic rat, rabbit, and human osteoclasts and *in vitro*–derived human osteoclasts.[59,61,72,73]

2.4.3.2 Characterization of $P2X_7$ Receptors by Electrophysiology and Ca^{2+} Fluorescence

$P2X_7$ receptors were studied using patch-clamp techniques to record whole cell currents in isolated osteoclasts[61] (Figure 2.4). A CsCl electrode solution was used to block the K^+ currents previously identified in ATP-stimulated osteoclasts. Initial application of ATP (100 μM, sufficient to activate $P2X_4$ and some P2Y receptors, but not $P2X_7$) evoked an inward $P2X_4$ current that activated rapidly and then declined with successive applications (Figure 2.4A). The current-voltage relationship revealed that the initial ATP-activated current was inwardly rectifying and reversed direction close to 0 mV (Figure 2.4B). In contrast, 2′– and 3′–O-(4-benzoyl-benzoyl)-ATP (BzATP; an agonist at the $P2X_7$ receptor more potent than ATP) evoked the initial inward current followed by an inward current that increased in amplitude with successive stimulations (Figure 2.4C). The later component of BzATP-activated current was inwardly rectifying and reversed close to 0 mV (Figure 2.4D). Furthermore, the later component was activated only by BzATP or higher concentrations of ATP.[61] Thus, in contrast to $P2X_4$ current, which desensitizes with repeated agonist application, the later BzATP-induced current is activity-dependent, consistent with currents observed in cells expressing $P2X_7$ receptors.[70]

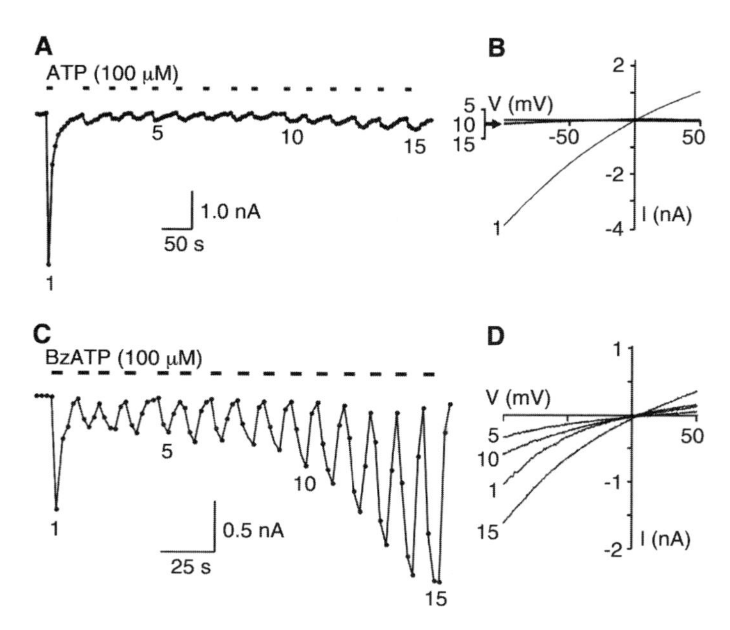

FIGURE 2.4 Nucleotides activate two distinct P2X currents: an initial $P2X_4$ current that desensitizes, followed by activity-dependent $P2X_7$ current. Whole-cell currents were recorded from rabbit osteoclasts at a holding potential of –30 mV with Cs^+ in the electrode solution to block K^+ currents. Voltage ramp commands were used to obtain I-V relationships at times indicated by the numbers. Subtraction of the control current revealed the nucleotide-induced currents. Bars above the current trace represent the length of agonist applications. (A) Initial stimulation of an osteoclast with 100 μM ATP elicited inward $P2X_4$ current, while successive applications of ATP resulted in little response. (B) I-V relationships for the same cell as in panel A. The initial stimulation with ATP (1) induced an inwardly rectifying current that reversed near 0 mV. Little current was elicited upon the 5th, 10th, or 15th stimulation. (C) Another osteoclast was stimulated with the $P2X_7$ receptor agonist BzATP. After the initial response to BzATP (100 μM), successive stimulation led to the progressive development of an inward current of increasing amplitude. D) I-V relationship for the cell in panel C revealed that both early and developing BzATP-induced currents reversed close to 0 mV. Data are consistent with initial activation of $P2X_4$ channels, followed by their desensitization and progressive activation of $P2X_7$ channels. (Reprinted from Naemsch, L.N. et al., Activity-dependent development of $P2X_7$ current and Ca^{2+} entry in rabbit osteoclasts. *J. Biol. Chem.*, 276,39107, 2001. With permission.)

BzATP is an agonist at $P2X_7$ receptors, but it is also known to activate a number of other P2 receptors.[67] Thus, investigators cannot rely on responses to BzATP alone to establish the involvement of $P2X_7$ receptors. In this regard, functional characteristics of the BzATP-induced current have been examined in osteoclasts isolated from wild-type (WT) and $P2X_7$ receptor knockout (KO) mice.[2] BzATP (300 μM) or a high concentration of ATP (1 mM) caused development of slowly deactivating inward current in WT, but not $P2X_7$ receptor KO osteoclasts (Figure 2.5A), providing strong evidence that this inward current is mediated by $P2X_7$ receptors. The current–voltage relationship of the BzATP- and ATP-activated currents in WT osteoclasts showed

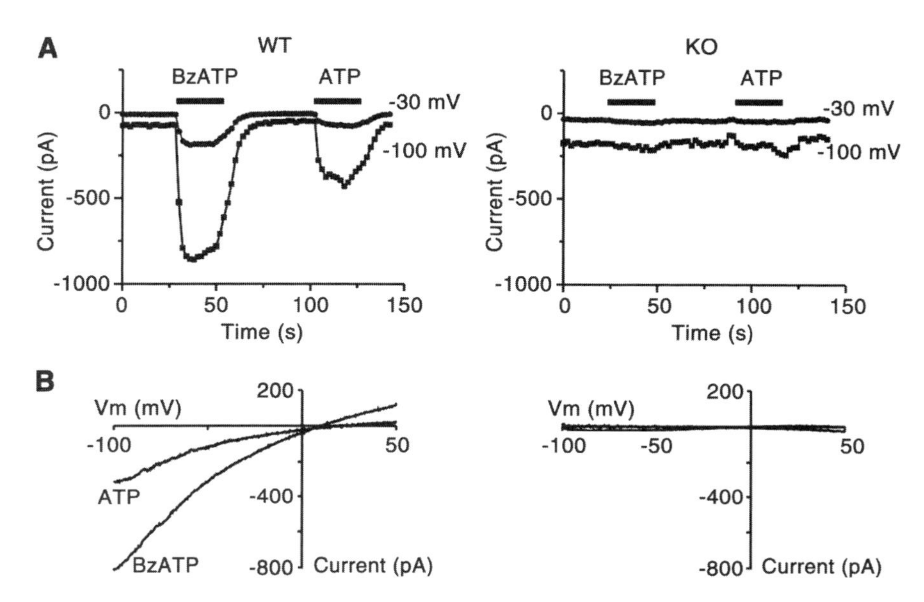

FIGURE 2.5 Nucleotide-induced currents compared in osteoclasts from wild type (WT) and $P2X_7$ receptor knockout (KO) mice. Whole-cell currents were recorded in osteoclasts isolated from WT and $P2X_7$ receptor KO mice. (A) BzATP (300 μM) or ATP (1 mM) caused development of slowly deactivating inward current in WT osteoclasts (at left). Bars above current trace represent duration of agonist application. Current values are shown for –30 and –100 mV, as indicated. In contrast, neither BzATP nor ATP elicited inward current in osteoclasts from KO mice (at right). (B) I-V relationship of the activated current was examined using voltage ramp commands. The activated current, determined by subtraction of control current, is plotted for an osteoclast from a WT mouse (at left) and a KO mouse (at right). The BzATP- and ATP-activated currents showed slight inward rectification and reversed direction close to 0 mV, consistent with nonselective cation current. (Reprinted from Ke, H.Z. et al., Deletion of the $P2X_7$ nucleotide receptor reveals its regulatory roles in bone formation and resorption, *Mol. Endocrinol.*, 17, 1356, 2003. With permission. Copyright 2003, The Endocrine Society.)

slight inward rectification and a reversal potential close to 0 mV (Figure 2.5B), consistent with properties reported for the rabbit $P2X_7$ receptor.[61]

Osteoclasts show no evidence for voltage-gated Ca^{2+} channels;[74] moreover, there is no detectable Ca^{2+} influx through $P2X_4$ receptors on osteoclasts.[68] Thus, other pathways, such as store-operated Ca^{2+} channels, likely mediate influx of extracellular Ca^{2+} into osteoclasts.[68,75] To examine whether the $P2X_7$ receptor mediates Ca^{2+} influx in osteoclasts, $[Ca^{2+}]_i$ changes in response to BzATP were examined using combined patch-clamp and fluorescence techniques.[61] BzATP induced progressively increasing elevations of $[Ca^{2+}]_i$ that were closely associated with the activity-dependent current (Figure 2.6). Moreover, removal of extracellular Ca^{2+} abolished the rise in $[Ca^{2+}]_i$, indicating that $P2X_7$ receptors mediate influx of Ca^{2+}. Others have shown that BzATP induces acute Ca^{2+} responses in *in vitro*–derived human osteoclasts.[73] These responses were inhibited by oxidized ATP and were attributed to activation of $P2X_7$ receptors. However, in rabbit osteoclasts, acute elevation of $[Ca^{2+}]_i$ in response to BzATP is caused by P2Y receptor–mediated release of Ca^{2+} from intracellular stores.[76]

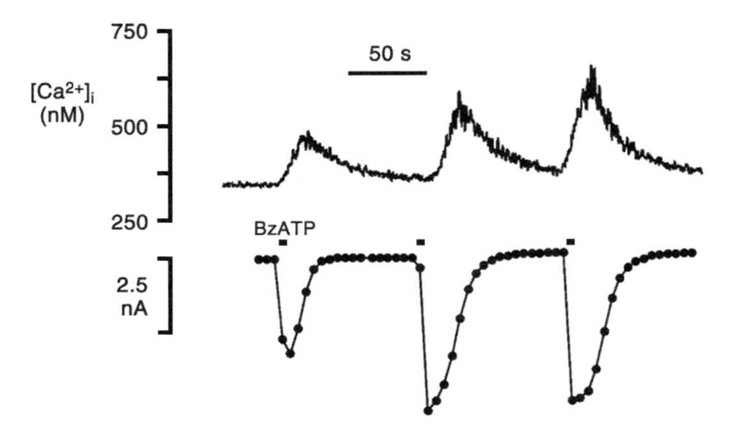

FIGURE 2.6 Activation of $P2X_7$ receptors allows Ca^{2+} influx. Whole-cell currents and $[Ca^{2+}]_i$ were recorded simultaneously in a fura-2-loaded rabbit osteoclast. Bars above the current trace represent the length of BzATP applications. BzATP (100 μM) activated an inward current (bottom) that increased in amplitude with successive stimulations, accompanied by BzATP (100μm) activated an inward current (bottom) that increased in amplitude with successive stimulations, accompanied by Ca^{2+} transients of increasing amplitude (upper trace). (Modified from Naemsch, L.N. et al., Activity-dependent development of $P2X_7$ current and Ca2+ entry in rabbit oseoclasts. *J. Biol. Chem.,* 276, 39107, 2001. With permission.)

2.4.3.3 $P2X_7$ Receptor–Mediated Signal Transduction

In other cell types, the $P2X_7$ receptor has been shown to activate multiple signaling pathways. In osteoclasts, the $P2X_7$ receptor mediates a nonselective cation current and Ca^{2+} influx and, in the absence of extracellular Ca^{2+} and Mg^{2+}, it can induce the formation of membrane pores.[54,77,78] Murine macrophage RAW 264.7 cells have been used extensively as a model system for studying osteoclasts, since in the presence of RANK ligand they differentiate into osteoclastlike cells *in vitro*. In undifferentiated (mononuclear) RAW 264.7 cells, $P2X_7$ receptors have been shown to modulate signaling through ERK1/2 induced by lipopolysaccharide.[79] In the same cell type, the $P2X_7$ receptor was shown to induce actin reorganization and membrane blebbing via the mitogen-activated protein kinase p38 and Rho.[80] However, it is unknown whether these pathways are activated in differentiated RAW 264.7 cells or in authentic osteoclasts.

NF-κB is a transcription factor essential for osteoclast development.[29,30] Inactive NF-κB resides in the cytoplasm and, upon activation, translocates from the cytosol to the nuclei. Thus, activation can be assessed by examining localization of NF-κB. The $P2X_7$ receptor agonist BzATP activates NF-κB in rabbit osteoclasts.[76] To directly examine the role of the $P2X_7$ receptor in NF-κB activation, osteoclasts isolated from WT and $P2X_7$ receptor KO mice were used. Treatment of WT osteoclasts with BzATP, but not vehicle, increased nuclear localization of NF-κB (Figure 2.7A and Figure 2.7B). Importantly, KO osteoclasts treated with BzATP did not show activation of NF-κB (Figure 2.7C and Figure 2.7E). Time course studies revealed that NF-κB activation in WT osteoclasts was transient and reached a maximum 30 min following exposure to BzATP (Figure 2.7D). These data established that activation of the $P2X_7$ receptor causes NF-κB translocation in osteoclasts.

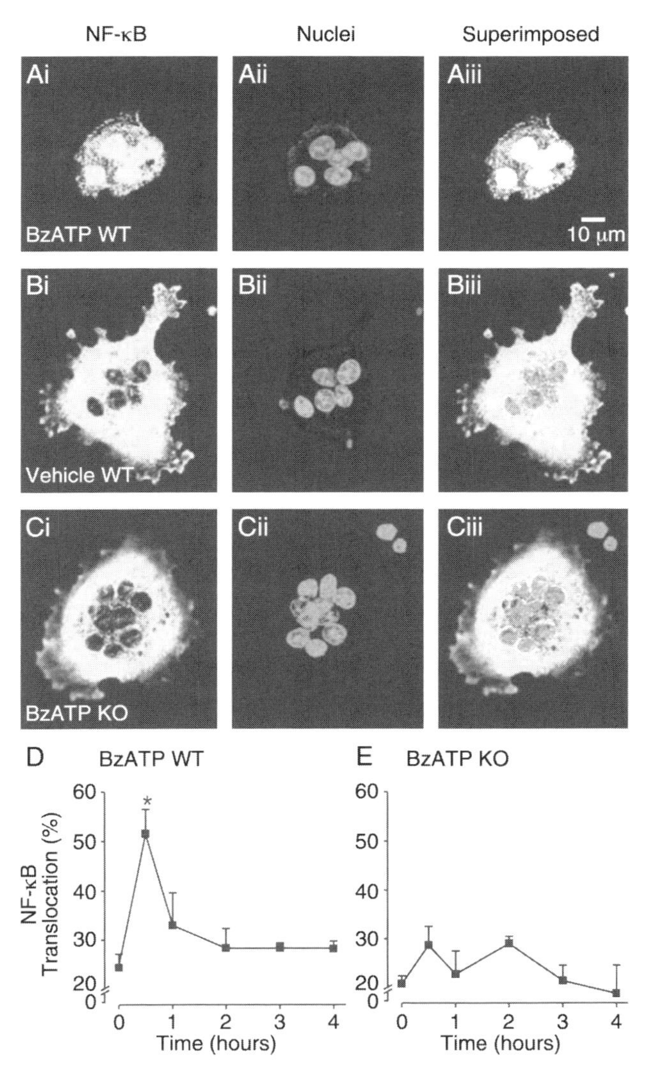

FIGURE 2.7 (See color insert following page 80) BzATP-induced nuclear translocation of NF-κB is mediated by the $P2X_7$ receptor in murine osteoclasts. Osteoclasts isolated from WT and $P2X_7$ receptor KO mice were treated with BzATP (300 μM) or vehicle for 0 to 4 h. The p65 subunit of NF-κB was visualized by immunofluorescence (green, left). Nuclei were stained with TOTO-3 (red, middle) and superimposed images are shown at right. A) BzATP-treated WT osteoclast exhibited nuclear localization of NF-κB at 30 min. B) and C) In contrast, vehicle-treated WT and BzATP-treated KO osteoclasts showed cytoplasmic localization of NF-κB. D) WT osteoclasts treated with BzATP exhibited a significant increase in nuclear translocation of NF-κB at 30 min, compared to time 0 (* $p < 0.05$). E) In contrast, KO osteoclasts did not show a significant change in NF-κB translocation at any time point following BzATP treatment. Data are the percentage of osteoclasts with nuclear localization of NF-κB. (Reproduced from Korcok, J. et al., Extracellular nucleotides act through $P2X_7$ receptors to activate NF-κB in osteoclasts, *J. Bone Miner. Res.*, 19, 642, 2004. With permission of the American Society for Bone and Mineral Research.)

Cells isolated from rabbit long bones contain marrow stromal cells and cells of the osteoblast lineage in addition to osteoclasts. Both osteoclasts and osteoblasts express functional P2 receptors.[4,9] Thus, extracellular nucleotides may act on osteoblasts to induce expression of RANK ligand,[62] which can then bind to its receptor RANK on osteoclasts to stimulate resorption. To examine whether BzATP activates NF-κB in osteoclasts indirectly by increasing expression of RANK ligand, isolated cells were treated with the decoy receptor for RANK ligand, osteoprotegerin. Osteoprotegerin did not inhibit translocation of NF-κB,[76] indicating that BzATP-induced activation of NF-κB is independent of RANK ligand.

2.4.3.4 $P2X_7$ Receptor Function in Osteoclasts

In other cells of hematopoietic origin, the $P2X_7$ receptor plays a role in a variety of processes such as giant cell formation, posttranslational processing of interleukin (IL)-1, and cell death.[81–84] Recent evidence suggests that activation of the $P2X_7$ receptor also permits ATP efflux,[85] presumably through formation of membrane pores. The released ATP could then act through positive feedback to stimulate additional $P2X_7$ receptors or other P2 receptors in the vicinity. The finding that the $P2X_7$ receptor promotes giant cell formation by macrophages[81] led to the proposal that the $P2X_7$ receptor plays a role in the fusion of osteoclast precursors. Supporting data came from studies by Gartland and coworkers, who reported that formation of human osteoclasts from peripheral blood monocytes was inhibited by a blocking antibody directed against the $P2X_7$ receptor or by oxidized ATP.[72] Moreover, RAW 264.7 cells lacking functional $P2X_7$ receptors failed to form multinucleated osteoclast-like cells in response to RANK ligand,[86] contrary to an earlier report from the same group.[87] Critical insights into the role of the $P2X_7$ receptor in osteoclast formation have come from examination of $P2X_7$ receptor KO mice.[2,78] $P2X_7$ receptor KO mice developed by two independent groups showed the presence of multinucleated osteoclasts, providing compelling evidence that the $P2X_7$ receptor is not essential for the fusion of osteoclast precursors and the differentiation of osteoclasts.[2,78] Furthermore, osteoclasts could be generated *in vitro*, and the rate of osteoclastogenesis did not differ in cultures of bone marrow and spleen cells isolated from WT and $P2X_7$ receptor KO mice.

An important role for the $P2X_7$ receptor in bone remodeling was demonstrated by Ke and coworkers using $P2X_7$ receptor KO mice.[2] Radiographs of femurs from $P2X_7$ receptor KO mice show similar length but smaller diameter compared with respective age- and gender-matched WT controls; the effect was most pronounced in adult male animals (Figure 2.8A). Peripheral quantitative computed tomography (pQCT) analysis performed on the distal femoral metaphysis, a site containing both trabecular and cortical bone, revealed smaller bone diameters in KO mice than WT controls (Figure 2.8B). Morphometric analysis revealed significantly lower cortical bone mass in $P2X_7$ receptor KO mice compared to WT littermates, accompanied by a striking reduction in the rate of periosteal bone formation in the knockout. The second important feature observed in $P2X_7$ receptor KO mice was significantly lower trabecular bone mass accompanied by increased osteoclast surface and number.[2]

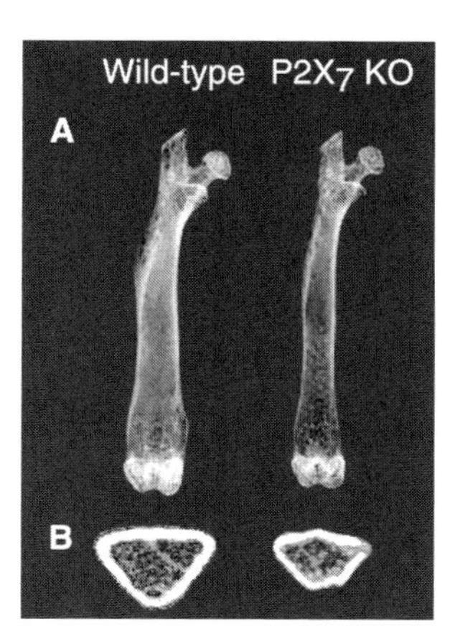

FIGURE 2.8 $P2X_7$ receptor knockout (KO) mice have femurs of smaller diameter than their wild-type (WT) littermates. (A) Radiographs of femurs from adult (9 months) male WT and $P2X_7$ receptor KO mice. $P2X_7$ receptor KO mice had femurs of smaller diameter but similar length compared with WT controls, indicating a role for $P2X_7$ receptor in regulating periosteal bone formation. (B) pQCT images of a distal femoral metaphysis from adult male WT and $P2X_7$ receptor KO mice. (Modified from Ke, H.Z. et al., Deletion of the $P2X_7$ nucleotide receptor reveals its regulatory roles in bone formation and resorption, *Mol. Endocrinol.*, 17, 1356, 2003. With permission. Copyright 2003, The Endocrine Society.)

On the other hand, $P2X_7$ receptor KO mice described by Gartland and colleagues showed no overt skeletal phenotype with the exception of thicker cortical bone when compared to WT mice.[78] Increased cortical thickness in $P2X_7$ receptor KO mice is in contrast to observations made by Ke and coworkers.[2] This difference in the phenotype of the two $P2X_7$ receptor KO lines could be due to different strategies used in their generation or possibly to sex differences; it is noteworthy that the decrease in bone density observed by Ke and coworkers was more pronounced in adult male mice.[2] Mouse strain differences could also have contributed to the difference in phenotype. The $P2X_7$ receptor KO mice developed by Ke and coworkers were from a mixed genetic background of 129/Ola × C57BL/6 × DBA/2, whereas KO mice developed by Gartland and coworkers were from a C57BL/6 × 129 background.

A natural mutation in the cytoplasmic domain of the $P2X_7$ receptor has been reported in some commonly used strains of mice. C57BL/6 and DBA/2 strains have an allelic mutation (Pro451Leu) that reduces the sensitivity of the $P2X_7$ receptor to ATP.[88] Therefore, the reduced sensitivity of the $P2X_7$ receptor in C57BL/6 and DBA/2 strains might mean that the skeletal phenotype of the knockout in these strains is reduced compared to what would be expected in other strains and species.

$P2X_7$ receptor KO mice described by Ke and coworkers show decreased trabecular bone volume accompanied by increased numbers of osteoclasts on the trabecular bone surface.[2] However, there was no difference in osteoclastogenesis in cultures of bone marrow cells from WT and $P2X_7$ receptor KO mice, suggesting that the $P2X_7$ receptor does not play a role in osteoclast formation. This led to the proposal that osteoclast survival is regulated by the $P2X_7$ receptor.[77] By examining authentic osteoclasts, it was found that the $P2X_7$ receptor can regulate osteoclast survival by two mechanisms—induction of acute cytolysis or apoptosis.[77] Osteoclast cytolysis results from activation of $P2X_7$ receptors by high concentrations of exogenous nucleotides in the presence of low concentrations of extracellular Ca^{2+} and Mg^{2+}. However, it is unlikely that these conditions occur *in vivo*; thus, regulation of osteoclast survival was examined under more physiological conditions (in the presence of divalent cations). Under these conditions, WT osteoclasts were more susceptible to apoptosis and showed lower survival rates than did osteoclasts isolated from $P2X_7$ receptor KO mice.

It is interesting that stimulation of the $P2X_7$ receptor can lead to diverse effects such as activation of NF-κB, an anti-apoptotic signaling factor, as well as apoptosis. However, it should be noted that these two effects are achieved at different levels of $P2X_7$ receptor stimulation. The presence of $P2X_7$ receptors enhances osteoclast apoptosis even in the absence of exogenous nucleotides—a condition under which NF-κB is not activated. In contrast, NF-κB is activated by exogenous BzATP, with half-maximal effects observed at ~100 μM.[76] Thus, extracellular nucleotides can act through $P2X_7$ receptors on osteoclasts to activate multiple signaling pathways and induce apoptosis, consistent with the increase in osteoclast number observed in $P2X_7$ receptor knockout mice.[2]

2.4.3.5 Human Studies and Genetic Polymorphisms of the $P2X_7$ Receptor

$P2X_7$ is among the most polymorphic of P2 receptors.[89] In recent years, several polymorphisms that affect its function have been identified within the coding region of the $P2X_7$ receptor. These polymorphisms have a wide range of outcomes, such as gain of function,[90] loss of function,[91,92] impairment of ATP-induced IL-1 and IL-18 release from monocytes,[93,94] and altered cell death.[95] The relatively common Glu496Ala polymorphism results in reduced pore-forming ability[91] without altering channel function[96] and is associated with some forms of chronic lymphocytic leukaemia.[92] This polymorphism has an anti-apoptotic effect that leads to an increase in B-cell numbers and could thus contribute to the pathogenesis of leukemia. Evidence from $P2X_7$ receptor KO mice and *in vitro* experiments suggests that the $P2X_7$ receptor on osteoclasts promotes apoptosis.[2,77] Thus, it is possible that polymorphisms leading to loss of function of the human $P2X_7$ receptor would render osteoclasts less susceptible to apoptosis, thereby increasing bone resorption *in vivo*. In this regard, there is a preliminary report that the Glu496Ala polymorphism in the human $P2X_7$ receptor is associated with decreased susceptibility to ATP-induced death of osteoclast-like cells derived from peripheral blood monocytes.[97] It will be of interest to examine the association of $P2X_7$ receptor polymorphisms with bone mineral density and susceptibility to osteoporosis in humans.

2.5 P2Y RECEPTOR SUBTYPES IN OSTEOCLASTS: EXPRESSION, SIGNALING, AND FUNCTION

P2Y receptors are G protein–coupled receptors that, in many cases, signal through PLCβ, leading to formation of inositol 1,4,5-trisphosphate (IP_3) and subsequent release of Ca^{2+} from intracellular stores.[58] To date, at least eight P2Y receptors, $P2Y_{1, 2, 4, 6, 11-14}$ have been identified in mammalian cells.

Functional studies first established the presence of P2Y receptors on osteoclasts.[52,53] Later studies using RT-PCR demonstrated that osteoclast-like cells derived *in vitro* from human peripheral blood monocytes express a wide range of P2Y receptors ($P2Y_{1, 2, 4, 6, 11}$).[62] However, since the preparation used in this study contained some monocytes, it is unclear whether these receptors were on monocytes or osteoclast-like cells. More recently, RT-PCR was performed on authentic osteoclasts purified by micromanipulation (Figure 2.9).[98] Rabbit bone marrow cells, containing cells of both osteoblast and osteoclast lineages (bone cells), possessed transcripts for $P2Y_1$, $P2Y_2$, $P2Y_6$, and $P2Y_{11}$ receptors as well as alkaline phosphatase, an osteoblast marker (Figure 2.9Ai). In contrast, homogeneous populations of rabbit osteoclasts purified by micromanipulation expressed $P2Y_1$, $P2Y_2$, and $P2Y_6$ receptors, whereas the $P2Y_{11}$ receptor and alkaline phosphatase were absent (Figure 2.9Aii). Future studies should examine the possibility that $P2Y_{12}$, $P2Y_{13}$, or $P2Y_{14}$ receptor subtypes are expressed on osteoclasts or their precursors.

Patch-clamp electrophysiology and Ca^{2+} fluorescence techniques have been used to examine whether P2Y receptors on osteoclasts are functional. Agonists at $P2Y_1$, $P2Y_2$, and $P2Y_6$ receptors induced transient increase of $[Ca^{2+}]_i$ in rabbit osteoclasts (Figure 2.9B). Electrophysiological studies in osteoclasts showed that ATP induces activation of inward current, attributed to P2X channels, followed by P2Y-mediated outward current (Figure 2.3B).[56,60] The P2Y receptor–mediated outward current was blocked by Cs^+ in the electrode solution, suggesting that it was due to K^+ channels.[56] To examine the voltage dependence of this current, K^+ electrode solution and voltage ramps were used. In an osteoclast in which P2X current was not apparent, ATP induced a transient current that peaked and then declined (Figure 2.10A). This current was outward at potentials positive to the K^+ equilibrium potential (E_K) and inward at potentials negative to E_K (Figure 2.10A), providing evidence for K^+ selectivity. The I-V relationship of the ATP-induced current was linear and reversed direction close to E_K (Figure 2.10B). The latency in activation of the outward K^+ current compared to the more rapidly activating inward $P2X_4$ current (see Figure 2.3B) suggested the involvement of a second messenger.[56] To examine the role of cytosolic Ca^{2+} in activating this K^+ current, a calcium ionophore was used to elevate $[Ca^{2+}]_i$ in the absence of nucleotides. Using simultaneous patch-clamp and fluorescence recordings, it was demonstrated that the increase in $[Ca^{2+}]_i$ was accompanied by activation of outward current (Figure 2.10C), displayed a linear I-V relationship and reversed direction at negative potential, characteristics shared by the ATP-induced current. It is notable that elevation of $[Ca^{2+}]_i$ preceded activation of K^+ current (Figure 2.10C), consistent with Ca^{2+} activating the K^+ current. Plotting I_K as a function of $[Ca^{2+}]_i$ showed the Ca^{2+} sensitivity of K^+ current (Figure 2.10D). Current was activated at $[Ca^{2+}]_i$ levels typically achieved following stimulation of osteoclasts with nucleotides. The Hill coefficient of ~5 is consistent with multiple Ca^{2+}-binding sites being involved in channel gating.[56]

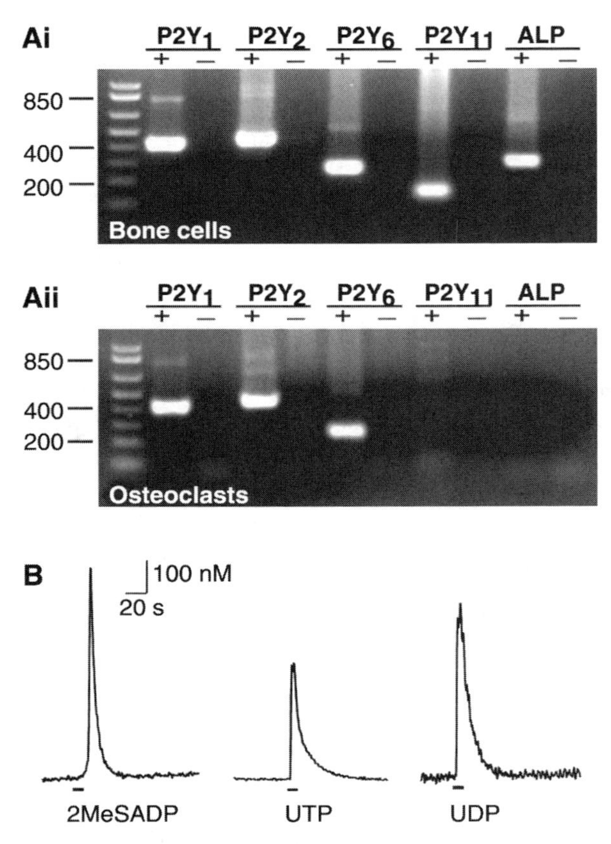

FIGURE 2.9 Osteoclasts express $P2Y_1$, $P2Y_2$, and $P2Y_6$ receptors, which induce transient elevation of $[Ca^{2+}]_i$. (A) Cells were isolated from the long bones of neonatal rabbits and cultured for 2–4 days on glass coverslips. Some samples were then washed and RNA was isolated (bone cells). In other samples, osteoclasts were purified using pronase E, followed by removal of nonosteoclasts using micropipettes prior to RNA isolation (osteoclasts). Samples were divided into those that did (+) or did not (–) undergo reverse transcription. PCR was then carried out to detect mRNA transcripts for P2Y receptors or alkaline phosphatase (ALP), an osteoblast marker. $P2Y_1$, $P2Y_2$, and $P2Y_6$ receptor transcripts were detected in preparations of both bone cells and purified osteoclasts. In contrast, $P2Y_{11}$ receptor and ALP were detected only in bone cell preparations. (B) Rabbit osteoclasts were loaded with fura-2 to monitor $[Ca^{2+}]_i$. 2MeSADP (10 μM), UTP (10 μM), or UDP (10 μM) was applied to single osteoclasts for 10 s by pressure ejection from micropipettes, as indicated by bars below the traces. All nucleotides tested induced transient elevation of $[Ca^{2+}]_i$. (Reproduced from Korcok, J. et al., $P2Y_6$ nucleotide receptors activate NF-κB and increase survival of osteoclasts, *J. Biol. Chem.*, 280, 16909, 2005. With permission.)

In contrast to P2X receptors, some P2Y receptor subtypes are activated by uridine nucleotides. $P2Y_1$ receptors are activated selectively by ADP,[99] the $P2Y_2$ receptor by ATP and UTP with approximately equal potency, and the human $P2Y_4$ receptor selectively by UTP, whereas at the rat $P2Y_4$ receptor ATP and UTP are equipotent. The $P2Y_6$ receptor is activated selectively by UDP.[58] Using a variety of P2 receptor agonists,

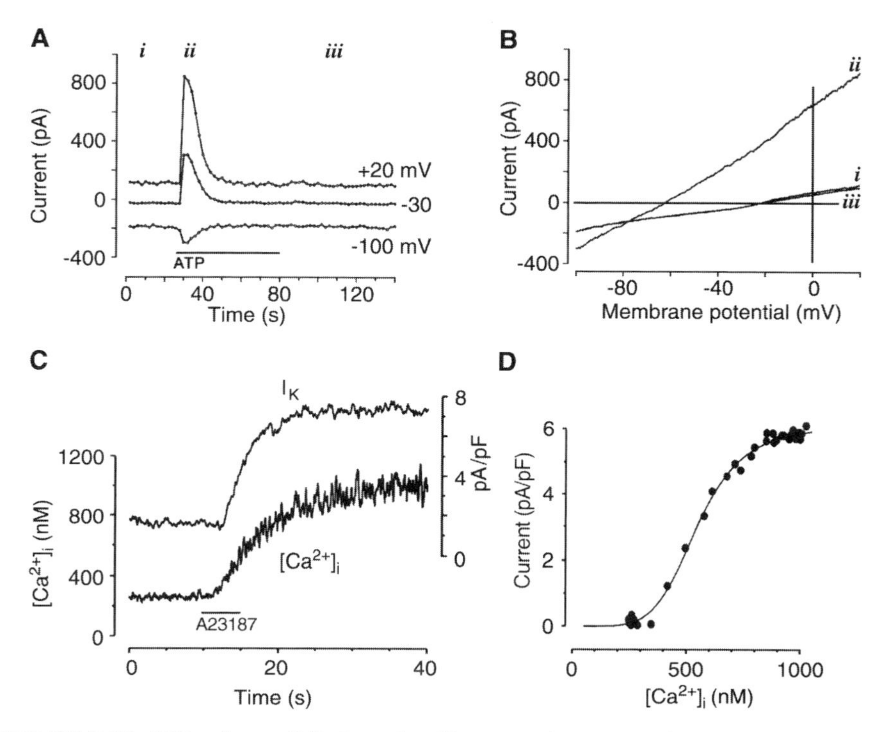

FIGURE 2.10 ATP activates Ca^{2+}-dependent K^+ current in rat osteoclasts. (A) ATP (10 μM, applied for time indicated by the bar) transiently activated current in an osteoclast (no P2X current was apparent in this cell). Whole-cell currents were measured at various membrane potentials (indicated at the right) during voltage ramp commands, revealing reversal of the ATP-activated current near equilibrium potential for K^+ (E_K). (B) Current traces from the cell in A showing I-V relationships obtained under control conditions (i), during ATP application (ii), and after recovery (iii). ATP elicited a linear current that reversed direction near E_K. (C) Whole-cell current and $[Ca^{2+}]_i$ were recorded simultaneously in a rat osteoclast loaded with fura-2. Local superfusion with the Ca^{2+} ionophore 4Br-A23187 (5 μM, applied for time indicated by the bar) induced a slow increase of $[Ca^{2+}]_i$ (bottom trace) accompanied by an increase of K^+ current (I_K) (top trace, recorded at 0 mV, normalized per cell capacitance). (D) Ca^{2+} sensitivity was determined from the amplitude of I_K above baseline at 0 mV as a function of $[Ca^{2+}]_i$ for the response in part C. Data were fitted using a sigmoidal function (continuous line) revealing half-maximal activation at 550 nM $[Ca^{2+}]_i$ and a Hill coefficient of 5.4. (Modified from Weidema, A.F. et al., Extracellular nucleotides activate non-selective cation and Ca^{2+}-dependent K^+ channels in rat osteoclasts, *J. Physiol.*, 503, 303, 1997. With permission.)

it was shown that most nucleotide-induced Ca^{2+} responses in osteoclasts are mediated by P2Y receptors that cause release of Ca^{2+} from intracellular stores.[68] Ca^{2+} responses induced by low concentrations of ATP were abolished by Cibacron blue, a P2 receptor antagonist, but were restored following washout, indicating that the lack of response was not due to desensitization (Figure 2.11A). Ectonucleotidases, present in the extracellular environment, can rapidly hydrolyze ATP to ADP, AMP, and adenosine.[100] The effect of ATP was not mediated through ADP, since ATPγS (a poorly hydrolyzable analogue of ATP) induced increase in $[Ca^{2+}]_i$ with characteristics similar to that of ATP

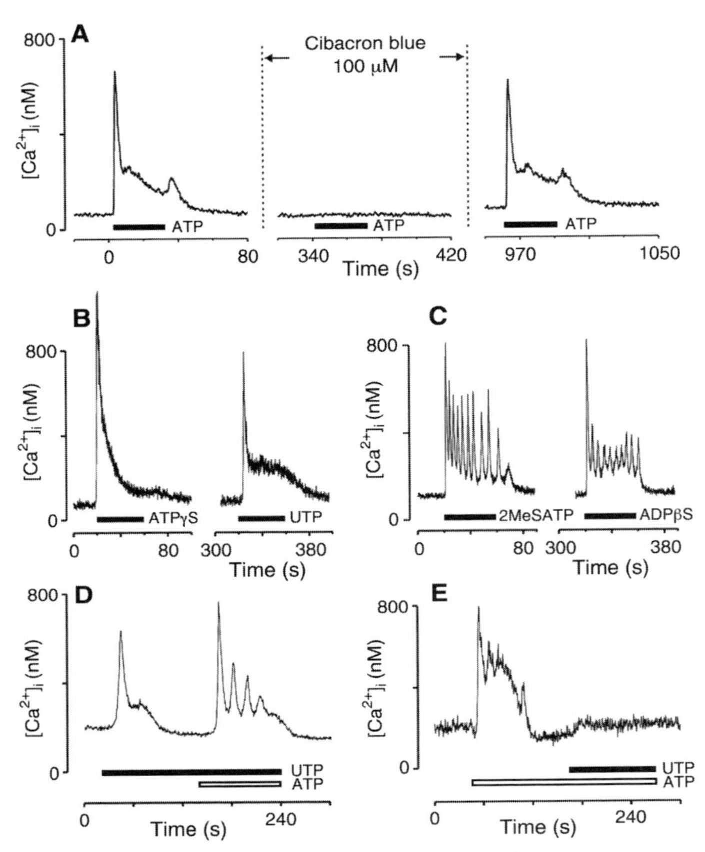

FIGURE 2.11 Multiple subtypes of P2Y receptors on mammalian osteoclasts. Nucleotides were applied to single osteoclasts for the times indicated by the bars below the Ca^{2+} traces. (A) ATP (20 μM) applied to an isolated osteoclast caused rapid rise of $[Ca^{2+}]_i$ with transient and sustained components. Following 5 min for recovery, the P2 receptor antagonist Cibacron blue (100 μM) was applied by bath superfusion, blocking the subsequent response to ATP (middle panel). Following washout and recovery, the osteoclast again responded to ATP (right trace). (B) The poorly hydrolyzable analogue ATPγS (10 μM) induced an increase in $[Ca^{2+}]_i$ with transient and sustained phases. After 3 min recovery, UTP (10 μM, an agonist at $P2Y_2$ and $P2Y_4$ receptors), elicited similar elevation of $[Ca^{2+}]_i$ in the same cell. (C) Nucleotides also elicited oscillatory changes of $[Ca^{2+}]_i$ in some osteoclasts. 2-Methylthio ATP (2-MeSATP, 1 μM) elicited a rise of $[Ca^{2+}]_i$ followed by oscillations. Following 3 min recovery, a similar response was elicited by ADPβS (1 μM). (D) and (E) Evidence for multiple nucleotide receptors in individual cells. Prolonged application of UTP (100 μM) induced a transient elevation of $[Ca^{2+}]_i$ that desensitized after ~50 s. Subsequent stimulation with ATP (100 μM) in the continued presence of UTP elicited a rise of $[Ca^{2+}]_i$. When the agonists were applied in the reverse order to another osteoclast, bath application of ATP elicited the typical rise of $[Ca^{2+}]_i$ but subsequently UTP elicited little response (panel E). Responses in A, B, and C were recorded from rat osteoclasts, and in D and E from rabbit osteoclasts. All responses were observed in osteoclasts from both species. (Reproduced from Weidema, A.F., Dixon, S.J., and Sims, S.M., Activation of P2Y but not $P2X_4$ nucleotide receptors causes elevation of $[Ca^{2+}]_i$ in mammalian osteoclasts, *Am. J. Physiol.*, 280, C1531, 2001. With permission.)

(Figure 2.11B). Furthermore, osteoclasts respond to ADP, but not AMP and adenosine, with increase in $[Ca^{2+}]_i$,[68] indicating that P2, but not P1, receptors are coupled to changes in $[Ca^{2+}]_i$. Following recovery, the same osteoclast that responded to ATPγS responded to UTP, consistent with activation of $P2Y_2$ and/or $P2Y_4$ receptors at which UTP is an agonist (Figure 2.11B). Interestingly, application of 2-MeSATP or ADPβS (agonist at the $P2Y_1$ receptor) resulted in oscillatory changes of $[Ca^{2+}]_i$ in some cells (Figure 2.11C). When responses to UTP were desensitized by the continuous application of UTP, ATP still elicited a rise of $[Ca^{2+}]_i$ (Figure 2.11D), indicating that other receptors in addition to $P2Y_2$ and $P2Y_4$ are expressed on osteoclasts. On the other hand, prolonged application of ATP elicited an initial rise in $[Ca^{2+}]_i$; however, subsequent exposure to UTP did not elicit a response (Figure 2.11E). These findings are consistent with the presence of at least two P2Y subtypes in osteoclasts—one activated by UTP or ATP, and the second activated by ATP or its breakdown product ADP.

2.5.1 $P2Y_1$ Receptors

$P2Y_1$ receptor expression was revealed by immunocytochemistry (Figure 2.12A) and *in situ* hybridization in isolated rat osteoclasts[66] and by RT-PCR in microisolated rabbit osteoclasts.[98] The $P2Y_1$ receptor was also evident in cultured rat calvarial osteoblasts and in osteoblasts in histological sections of rat long bones.[66]

Like several other P2Y receptors, activation of the $P2Y_1$ receptor leads to PLC-mediated IP_3 formation and increase in $[Ca^{2+}]_i$.[58] Ca^{2+} responses to the $P2Y_1$ agonists ADP, its more stable analogue ADPβS, and 2-MeSADP have been shown in isolated rabbit and rat osteoclasts (Figure 2.9B).[68,76,98] Furthermore, stimulation of rabbit osteoclasts with ADP, but not AMP, leads to a decrease in cytosolic pH.[101] This effect was not dependent on changes of Ca^{2+} and was suggested to be due to increased activity of the Cl^-/HCO_3^- exchanger.[101] Patch-clamp electrophysiology has shown that ADP and ADPβS induce outward current, consistent with activation

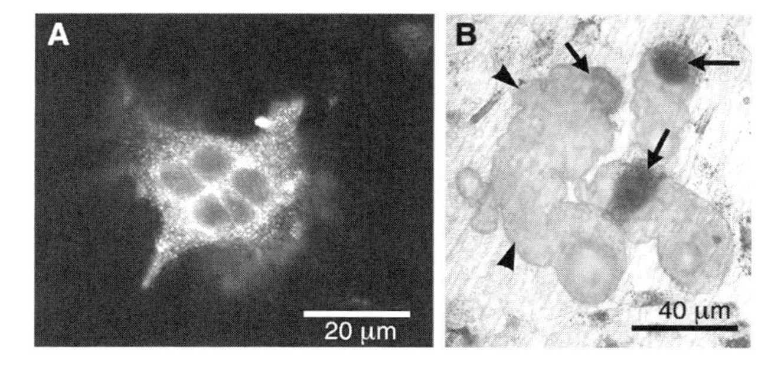

FIGURE 2.12 (See color insert following page 80) Activation of $P2Y_1$ receptor on rat osteoclasts stimulates bone resorption. (A) Immunofluorescence shows localization of $P2Y_1$ receptor on an isolated rat osteoclast. (B) A representative photomicrograph of a 26-h culture of rat long bone cells on dentin discs shows ADP-stimulated, TRAP-stained multinucleated osteoclasts (arrows) with corresponding resorption pits (arrowheads), visualized by reflective light microscopy. (Modified from Hoebertz, A. et al., Extracellular ADP is a powerful osteolytic agent: evidence for signaling through the $P2Y_1$ receptor on bone cells, *FASEB J.*, 15, 1139, 2001. With permission.)

of Ca^{2+}-dependent K^+ channels,[60] effects that are most likely mediated by $P2Y_1$ receptors.

ATP has been shown to have a potent stimulatory effect on osteoclastic resorption.[57] More recent studies suggest that this effect is mediated by $P2Y_1$ receptors, since ADP and 2-MeSADP showed potent stimulatory effects on resorption by mature rat osteoclasts and osteoclasts derived from mouse marrow cultures.[66] A representative photomicrograph shows ADP-stimulated, TRAP-positive osteoclasts on a dentin disc (Figure 2.12B). Stimulatory effects were observed at ADP concentrations as low as 20 nM, and these effects were further enhanced by acidification of the extracellular medium. In contrast, the degradation products AMP and adenosine had no effect. ADP also showed potent stimulatory effects on osteoclast formation in mouse marrow cultures,[66] suggesting a dual effect of $P2Y_1$ receptor activation on osteoclasts and their precursors. The mechanism of $P2Y_1$ receptor action on osteoclast function is not entirely understood; however, the cyclooxygenase inhibitor indomethacin blocked ADP-induced Ca^{2+} release from cultured mouse calvariae, indicating a requirement for prostaglandin synthesis.[66] It remains to be determined whether ADP stimulates osteoclast formation and activation directly, by acting through $P2Y_1$ receptors on osteoclasts, or indirectly, by acting through $P2Y_1$ receptors on other cell types that subsequently stimulate osteoclastic resorption.

2.5.2 $P2Y_2$ Receptors

The $P2Y_2$ receptor was cloned from a human osteoclastoma cDNA library, and transcripts were found to be expressed by osteoclast-like cells derived from a giant cell tumor of bone.[55] These observations are consistent with identification of the $P2Y_2$ receptor in rat osteoclasts using *in situ* hybridization[59] and in microisolated rabbit osteoclasts using RT-PCR.[98]

The $P2Y_2$ receptor is activated by ATP and UTP, with approximately equal potency. Despite the presence of the $P2Y_2$ receptor in osteoclasts, several authors have reported that UTP fails to elicit elevation of $[Ca^{2+}]_i$ in these cells.[53,73,102] In contrast, others have reported that both ATP and UTP induce transient rise of $[Ca^{2+}]_i$ in isolated osteoclasts.[68,98,103] The PLC inhibitor U-73122 abolished Ca^{2+} responses induced by low concentrations of ATP, indicating that Ca^{2+} release is dependent on PLC.[68] Furthermore, Ca^{2+} responses were still observed in the absence of extracellular Ca^{2+}. Thus, it is likely that the Ca^{2+} responses induced by UTP are mediated by $P2Y_2$ receptors and possibly $P2Y_4$ receptors.

The physiological role of $P2Y_2$ receptors remains unknown in osteoclasts. However, the $P2Y_2$ receptor has been shown to inhibit bone formation by osteoblasts.[104] In this study, ATP and UTP, but not ADP, inhibited bone formation at concentrations of 1 to 10 μM. The results of studies on bone resorption indicate that the stimulatory actions of ATP on osteoclastic bone resorption are mediated by ADP acting on the $P2Y_1$ receptor.[66] Thus, ATP released into the extracellular environment may act to inhibit bone formation by osteoblasts. ATP would then be degraded by ectonucleotidases to ADP, which in turn would stimulate bone resorption by osteoclasts, initiating a cycle of bone remodeling.

2.5.3 $P2Y_6$ Receptors

The $P2Y_6$ receptor was cloned from a human cDNA library and was found to be expressed in human bone and some osteoblastic cell lines.[105] Recently, the $P2Y_6$ receptor was also shown to be expressed by authentic osteoclasts (Figure 2.9A_{ii}).[98] In other cell types, $P2Y_6$ receptors signal through pertussis toxin-insensitive G_q to activate PLCβ, leading to formation of IP_3 and subsequent release of Ca^{2+} from intracellular stores.[58] Stimulation of rat and rabbit osteoclasts with the $P2Y_6$ receptor agonists UDP or INS48823 induced transient rise of $[Ca^{2+}]_i$, demonstrating the presence of functional receptors (Figure 2.9B and Figure 2.13A). INS48823 (P1-((2-benzyl-1,3-dioxolo-4-yl)uridine 5′) P3-(uridine 5′–) triphosphate), a stable analogue of diuridine 5′–triphosphate, is a selective P2Y6 receptor agonist.[98]

Surprisingly, $P2Y_6$ receptor agonists UDP and INS48823, but not agonists at other P2Y receptors, transiently activated NF-κB in rabbit osteoclasts.[98] G protein–coupled receptors can couple to NF-κB through the sequential activation of conventional protein kinase C (PKC) isoforms and IκB kinase, leading to IκB degradation by the proteasome.[106] NF-κB activation through the $P2Y_6$ receptor in osteoclasts is a proteasome-dependent process; however, the exact mechanism of activation remains to be determined. Activation of $P2Y_1$, $P2Y_2$, and $P2Y_6$ receptors induces transient rise in $[Ca^{2+}]_i$, while only the $P2Y_6$ receptor couples to NF-κB activation, indicating that the rise in $[Ca^{2+}]_i$ alone is not sufficient to activate NF-κB.[98] ATP can activate osteoclasts indirectly by inducing RANK ligand expression on osteoblasts.[62] However, osteoprotegerin (a decoy receptor for RANK ligand) had no effect on UDP-induced NF-κB activation in osteoclasts, suggesting that NF-κB activation is mediated directly through $P2Y_6$ receptors on osteoclasts.

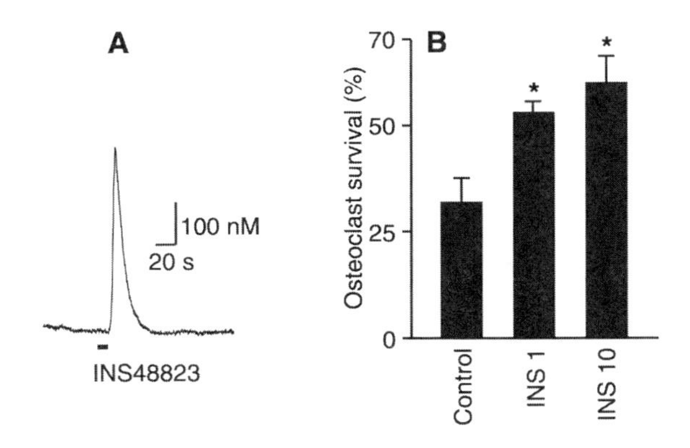

FIGURE 2.13 $P2Y_6$-selective agonist, INS48823, increases $[Ca^{2+}]_i$ and enhances osteoclast survival. (A) Rabbit and rat osteoclasts were loaded with fura-2 to monitor $[Ca^{2+}]_i$. INS48823 (10 μM) was applied for 10 s using pressure ejection from a micropipette, as indicated by bar below the trace. Shown is the response of a single rabbit osteoclast (representative of 4 out of 4 rabbit and 3 out of 4 rat osteoclasts tested). (B) INS48823 (INS) enhances survival of rat osteoclasts. Rat osteoclasts were counted prior to and after 18-h incubation with or without INS48823 (1 and 10 μM). INS48823 significantly enhanced osteoclast survival when compared to control ($*p < 0.05$). (Modified from Korcok, J. et al., $P2Y_6$ nucleotide receptors activate NF-κB and increase survival of osteoclasts, *J. Biol. Chem.*, 280, 16909, 2005. With permission.)

Activation of the $P2Y_6$ receptor by UDP or INS48823 enhances the survival of rat osteoclasts *in vitro* (Figure 2.13B).[98] Since the predominant form of cell death for osteoclasts *in vitro* and *in vivo* is apoptosis, it is likely that activation of $P2Y_6$ receptors initiates anti-apoptotic signals. In this regard, the NF-κB inhibitor SN50 blocked NF-κB activation and prevented the increase in survival induced by UDP and INS48823, thus showing that the $P2Y_6$ receptor acts through NF-κB to enhance osteoclast survival. Increasing the life span of osteoclasts is likely an important mechanism for regulating bone resorption *in vivo*.[41] Thus, by acting through $P2Y_6$ receptors, nucleotides could stimulate bone resorption.

2.6 P2 RECEPTORS AS POTENTIAL TARGETS FOR ANTIRESORPTIVE DRUGS

Osteoclasts express functional P2X ($P2X_4$ and $P2X_7$) and P2Y ($P2Y_1$, $P2Y_2$, and $P2Y_6$) receptors (Figure 2.14). Activation of $P2X_4$ and $P2X_7$ receptors on osteoclasts induces inward current, while the $P2X_7$ receptor also mediates influx of Ca^{2+} and signals through NF-κB.[60,61,76] $P2Y_1$, $P2Y_2$, and $P2Y_6$ receptors couple to release Ca^{2+} from intracellular stores, whereas only the $P2Y_6$ receptor activates NF-κB.[98]

Novel functional roles have been described for P2 receptors on osteoclasts and osteoblasts. Based on evidence from knockout mice, $P2X_7$ receptors stimulate bone formation and inhibit bone resorption.[2] ADP stimulates $P2Y_1$ receptors to increase resorption by enhancing the differentiation and activity of osteoclasts.[66] Moreover, activation of the $P2Y_6$ receptor by UDP enhances osteoclast survival,[98] thus prolonging osteoclast-mediated bone resorption. In osteoblasts, ATP and UTP activate $P2Y_2$ receptors to inhibit bone formation.[104] Nucleotides such as ATP and UTP are released into the extracellular fluid of bone during inflammation or in response to mechanical

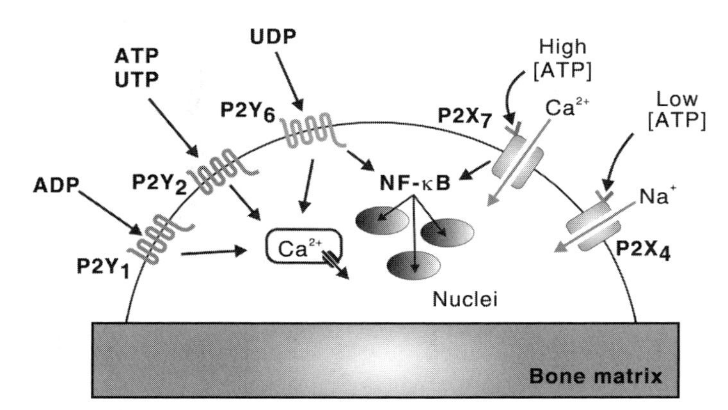

FIGURE 2.14 Summary of P2 receptors on osteoclasts. Osteoclasts express functional $P2Y_1$, $P2Y_2$, $P2Y_6$, $P2X_4$, and $P2X_7$ receptors. Activation of all P2Y receptor subtypes expressed on osteoclasts results in transient rise of $[Ca^{2+}]_i$; however, only the $P2Y_6$ receptor activates NF-κB to enhance osteoclast survival. $P2X_4$ and $P2X_7$ receptors are nonselective cation channels. Activation of the $P2X_7$ receptor by high concentrations of ATP also permits Ca^{2+} influx and activates NF-κB signaling.

stimulation. Initially, bone resorption would be inhibited via ATP acting on $P2X_7$ receptors on osteoclasts, and bone formation would be inhibited via ATP and UTP acting on $P2Y_2$ receptors on osteoblasts. Subsequent hydrolysis of ATP and UTP by ectonucleotidases (at a later time or at a distance from the site of release) would yield ADP and UDP, which could then initiate bone resorption by acting on $P2Y_1$ and $P2Y_6$ receptors. Thus, the rapidly expanding field of P2 receptor biology suggests new potential drug targets for the treatment of bone diseases. However, several important issues need to be considered when developing agents targeting P2 receptors, namely lack of selective and stable agonists and antagonists for these receptors and their wide distribution not only in bone but in other cell types as well.

Endogenous nucleotides are unstable compounds that are hydrolyzed in the extracellular environment by ectonucleotidases. Several more selective and stable nucleotide analogues have been identified as P2 receptor agonists. ATP analogues such as α,β–meATP and β,γ–meATP are relatively stable and selective for $P2X_1$ and $P2X_3$ receptors, whereas the ADP analogue ADPβS is a potent $P2Y_1$ receptor agonist.[58] UTP, UDP, and 2-MeSADP are relatively selective for P2Y receptors, but their stability in the presence of ectonucleotidases is poor. BzATP is a more potent agonist than ATP at $P2X_7$ receptors; however, this analogue is labile and can activate a number of P2 receptors in addition to $P2X_7$.[107] Activation of the $P2X_7$ receptor requires high ATP concentrations (>1 mM). However, recent evidence from our laboratory suggests that osteoclast apoptosis can occur due to activation of the $P2X_7$ receptor in the absence of exogenous nucleotides.[77] Thus, the concentration of endogenous ATP in the extracellular environment may be sufficiently high to activate this receptor or the $P2X_7$ receptor may be activated by other mechanisms such as ADP ribosylation.[69]

Another approach taken in developing stable P2 receptor agonists has been the synthesis of dinucleotide derivatives. Several pyrimidine dinucleotides Up_nU ($n = 2–7$) have been synthesized and their potencies at $P2Y_1$, $P2Y_2$, $P2Y_4$, and $P2Y_6$ receptors have been examined.[108] Up_4U (INS365) has potency comparable to UTP at $P2Y_2$ and $P2Y_4$ receptors and is in clinical development of the treatment of dry eye[109] and chronic lung diseases, such as cystic fibrosis.[110] In our studies, we have used INS48823, a stable analogue of diuridine 5′-triphosphate, as an agonist at $P2Y_6$ receptors.[98] In isolated osteoclasts, INS48823 induced a transient rise in $[Ca^{2+}]_i$, activated NF-κB, and enhanced cell survival with greater potency than UDP. The higher potency of INS48823 is likely due to its greater stability.

Activation of the $P2Y_1$ receptor has been shown to stimulate bone resorption by enhancing osteoclast differentiation and activity,[66] whereas activation of the $P2Y_6$ receptor enhances osteoclast survival.[98] Thus, stable and selective antagonists at these receptors might be of particular interest as potential antiresorptive drugs. A variety of P2 receptor antagonists are available; however, many are nonspecific. Some antagonists do not discriminate between P2X and P2Y receptors and are even less effective in discriminating between individual subtypes of P2X and P2Y receptors. For example, the trypanoside suramin is an antagonist at P2 receptors but does not discriminate P2Y and P2X subtypes.[58] Suramin inhibits bone resorption *in vitro* and *in vivo*,[57,111,112] an effect that could at least in part be due to blockade of P2 receptors. However, suramin inhibits ectonucleotidases and has other actions that complicate the interpretation of results obtained using this antagonist. Other nonspecific antagonists for P2 receptors include pyridoxal-phosphate-6-azophenyl-2′,4′–disulphonic acid (PPADS), which was originally thought to be P2X-selective, and the anthraquinone-sulphonic acid derivative, Cibacron blue.[58] In the search for more selective compounds, oxidized ATP was found to be an irreversible antagonist at the $P2X_7$ receptor.[113] However, oxidized ATP has been shown to have effects independent of its actions on P2 receptors.[114,115] Some specific P2 antagonists have been developed, such as the selective $P2Y_1$ receptor antagonist MRS2279.[116] Because of the proresorptive effects of this receptor, $P2Y_1$ antagonists may have therapeutic potential in the prevention of excessive bone resorption. In addition, $P2X_7$ receptor antagonists based on KN-62, cyclic imides, and adamantane amides are currently under development.[117] As more selective antagonists for individual P2 receptors become available, their effects on osteoclast functions should be assessed.

The wide distribution of P2 receptors in the body needs to be considered when developing drugs targeting these receptors. Bisphosphonates are synthetic analogues of pyrophosphate used in the prevention and treatment of osteoporosis.[118] Administered orally, these drugs selectively localize to mineral surfaces of the bone, from which they are released during the process of bone resorption. Once in the extracellular environment, bisphosphonates can act through one of the two distinct mechanisms to block osteoclastic resorption. Smaller bisphosphonates, such as clodronate and etidronate, are metabolized to poorly hydrolyzable ATP analogues that cause osteoclast cell death.[119] Like bisphosphonates, nucleotide analogues may accumulate in bone via interactions of negatively charged phosphate groups with hydroxyapatite. Another option is to develop bisphosphonates that, once released into the extracellular environment by the action of osteoclasts, are metabolized to potent P2 agonists or antagonists.[7] These compounds could then act locally on P2 receptors to therapeutically modulate bone resorption.

2.7 PROSPECTS AND CONCLUSIONS

Nucleotides can be released into the extracellular environment of the bone by a variety of stimuli, such as inflammation and mechanical stimulation. Once in the extracellular environment, nucleotides act on multiple P2 receptors on osteoclasts and osteoblasts to modulate cellular activity and interactions between these two cell types. Recent evidence has provided insight into the importance of at least some of these receptors in osteoclast function. The $P2X_7$ receptor activates multiple signaling pathways and exerts different effects on osteoclasts, depending on the duration and degree of activation. Under physiological conditions, the $P2X_7$ receptor promotes osteoclast apoptosis.[77] Several polymorphisms of the human $P2X_7$ receptor have been identified and it is possible that these could affect skeletal remodeling. Further studies are needed to examine the association of these polymorphisms with bone density and susceptibility to osteoporosis.

To date, at least eight P2Y receptors, $P2Y_{1,\,2,\,4,\,6,\,11\text{–}14}$, have been identified in mammalian cells. Using multiple approaches, $P2Y_{1,\,2,\,6}$ receptors have been shown to be expressed in osteoclasts.[55,59,66,98,102] Moreover, agonists at these receptors exert potent effects on osteoclast and osteoblast function. Additional P2 receptors were shown by RT-PCR to be expressed in osteoclast-like cells.[62] However, further studies are needed to examine whether these receptors are also expressed in authentic osteoclasts, whether receptor expression is regulated, and how these P2 receptors contribute to the control of osteoclast function. It will also be of considerable interest to examine cross talk among signaling pathways activated by nucleotides and those activated by other soluble ligands, as well as cell–cell and cell–matrix interactions.

Nucleotides are released from many cell types in response to mechanical stimulation and have been suggested to mediate mechanotransduction in bone.[3,4] In this regard, the pattern of skeletal changes observed in the $P2X_7$ receptor KO mouse by Ke and coworkers is consistent with decreased sensitivity to mechanical loading.[2] Indeed, there is a recent report that skeletal responses to mechanical loading are greatly attenuated in $P2X_7$ receptor KO mice.[120] Thus, the role of P2 receptors in the sensing of mechanical stimuli by osteoclasts and other skeletal cells is likely to be a fertile ground for future research.

Nucleotides can induce chemotaxis in several different cell types. In bone marrow hematopoietic stem cells, microglia, and dendritic cells, nucleotides act through P2Y receptors to induce chemotaxis.[121–123] Chemotaxis is involved in the recruitment and migration of osteoclasts to new sites of bone resorption, and it would be interesting to examine whether nucleotides induce chemotaxis of osteoclasts. Furthermore, identification of P2 receptors involved in osteoclast chemotaxis could provide therapeutic targets for inhibiting pathological bone resorption. However, the lack of selective agonists and antagonists is still a limitation when studying P2 receptors. A possible solution for this limitation is the use of genetically modified mice, such as the $P2X_7$ receptor KO mouse, which has already provided novel insights into the roles played by P2 receptors in osteoclasts. Data obtained from these studies may assist in the development of new therapies for metabolic, inflammatory, and neoplastic diseases of the skeleton.

ACKNOWLEDGEMENTS

We thank researchers in this rapidly advancing field for providing their latest findings and for permission to reproduce illustrative material from their previously published work. We also apologize to those investigators whose work was omitted due to space constraints or to oversight. Studies from the authors' laboratories that were reviewed in this chapter were supported by the Canadian Institutes of Health Research (CIHR), the CIHR Institute of Musculoskeletal Health and Arthritis, The Arthritis Society, and the Canadian Arthritis Network. Jasminka Korcok was supported by a CIHR Doctoral Research Award.

REFERENCES

1. Rodan, G.A. and Martin, T.J., Therapeutic approaches to bone diseases, *Science*, 289, 1508, 2000.
2. Ke, H.Z. et al., Deletion of the $P2X_7$ nucleotide receptor reveals its regulatory roles in bone formation and resorption, *Mol. Endocrinol.*, 17, 1356, 2003.

3. Dixon, S.J. et al., P2 purinoceptors in skeletal cells: receptor subtypes, signalling pathways, and possible role in mechanotransduction, in *Biological Mechanisms of Tooth Eruption, Resorption and Replacement by Implants*, Davidovitch, Z. and Mah, J., Eds., Harvard Society for the Advancement of Orthodontics, Boston, 1998, p. 301.
4. Dixon, S.J. and Sims, S.M., P2 purinergic receptors on osteoblasts and osteoclasts: potential targets for drug development, *Drug Dev. Res.*, 49, 187, 2000.
5. Bowler, W.B. et al., Extracellular nucleotide signaling: a mechanism for integrating local and systemic responses in the activation of bone remodeling, *Bone*, 28, 507, 2001.
6. Komarova, S.V., Dixon, S.J., and Sims, S.M., Osteoclast ion channels: potential targets for antiresorptive drugs, *Curr. Pharm. Des.*, 7, 637, 2001.
7. Naemsch, L.N. et al., P2 nucleotide receptors in osteoclasts, *Drug. Dev. Res.*, 53, 130, 2001.
8. Gartland, A. et al., P2 receptors in bone—modulation of osteoclast formation and activity via $P2X_7$ activation, *Crit. Rev. Eukaryot. Gene Expr.*, 13, 237, 2003.
9. Hoebertz, A., Arnett, T.R., and Burnstock, G., Regulation of bone resorption and formation by purines and pyrimidines, *Trends Pharmacol. Sci.*, 24, 290, 2003.
10. Gallagher, J.A., ATP P2 receptors and regulation of bone effector cells, *J. Musculoskelet. Neuronal Interact.*, 4, 125, 2004.
11. Roodman, G.D., Advances in bone biology: the osteoclast, *Endocr. Rev.*, 17, 308, 1996.
12. Teitelbaum, S.L., Bone resorption by osteoclasts, *Science*, 289, 1504, 2000.
13. Kong, Y.Y. et al., Activated T cells regulate bone loss and joint destruction in adjuvant arthritis through osteoprotegerin ligand, *Nature*, 402, 304, 1999.
14. Roodman, G.D., Cell biology of the osteoclast, *Exp. Hematol.*, 27, 1229, 1999.
15. Boyle, W.J., Simonet, W.S., and Lacey, D.L., Osteoclast differentiation and activation, *Nature*, 423, 337, 2003.
16. Teitelbaum, S.L. and Ross, F.P., Genetic regulation of osteoclast development and function, *Nat. Rev. Genet.*, 4, 638, 2003.
17. Komarova, S.V. et al., Convergent signaling by acidosis and receptor activator of NF-κB ligand (RANKL) on the calcium/calcineurin/NFAT pathway in osteoclasts, *Proc. Natl. Acad. Sci. U.S.A.*, 102, 2643, 2005.
18. Yoshida, H. et al., The murine mutation osteopetrosis is in the coding region of the macrophage colony stimulating factor gene, *Nature*, 345, 442, 1990.
19. Lacey, D.L. et al., Osteoprotegerin ligand is a cytokine that regulates osteoclast differentiation and activation, *Cell*, 93, 165, 1998.
20. Klemsz, M.J. et al., The macrophage and B cell–specific transcription factor PU.1 is related to the *ets* oncogene, *Cell*, 61, 113, 1990.
21. Zhang, D.E. et al., Identification of a region which directs the monocytic activity of the colony-stimulating factor 1 (macrophage colony-stimulating factor) receptor promoter and binds PEBP2/CBF (AML1), *Mol. Cell. Biol.*, 14, 8085, 1994.
22. Matsuo, K. et al., Nuclear factor of activated T-cells (NFAT) rescues osteoclastogenesis in precursors lacking c-Fos, *J. Biol. Chem.*, 279, 26475, 2004.
23. Ishida, N. et al., Large scale gene expression analysis of osteoclastogenesis *in vitro* and elucidation of NFAT2 as a key regulator, *J. Biol. Chem.*, 277, 41147, 2002.
24. Takayanagi, H. et al., Induction and activation of the transcription factor NFATc1 (NFAT2) integrate RANKL signaling in terminal differentiation of osteoclasts, *Dev. Cell*, 3, 889, 2002.
25. Takayanagi, H., Mechanistic insight into osteoclast differentiation in osteoimmunology, *J. Mol. Med.*, 83, 170, 2005.

26. Komarova, S.V. et al., RANK ligand-induced elevation of cytosolic Ca^{2+} accelerates nuclear translocation of nuclear factor κB in osteoclasts, *J. Biol. Chem.*, 278, 8286, 2003.
27. Senftleben, U. and Karin, M., The IKK/NF-κB pathway, *Crit. Care Med.*, 30, S18, 2002.
28. Karin, M., Yamamoto, Y., and Wang, Q.M., The IKK NF-κB system: a treasure trove for drug development, *Nat. Rev. Drug Discov.*, 3, 17, 2004.
29. Franzoso, G. et al., Requirement for NF-κB in osteoclast and B-cell development, *Genes Dev.*, 11, 3482, 1997.
30. Iotsova, V. et al., Osteopetrosis in mice lacking NF-κB1 and NF-κB2, *Nat. Med.*, 3, 1285, 1997.
31. Väänänen, H.K. et al., The cell biology of osteoclast function, *J. Cell Sci.*, 113, 377, 2000.
32. Duong, L.T. et al., Integrins and signaling in osteoclast function, *Matrix Biol.*, 19, 97, 2000.
33. Horton, M.A., The αvβ3 integrin "vitronectin receptor," *Int. J. Biochem. Cell Biol.*, 29, 721, 1997.
34. Baron, R., Anatomy and ultrastructure of bone, in *Primer on the Metabolic Bone Diseases and Disorders of Mineral Metabolism*, Favus, M.J., Ed., Lippincot Williams & Wilkins, New York, 1999, p. 3.
35. Blair, H.C., How the osteoclast degrades bone, *Bioessays*, 20, 837, 1998.
36. Drake, F.H. et al., Cathepsin K, but not cathepsins B, L, or S, is abundantly expressed in human osteoclasts, *J. Biol. Chem.*, 271, 12511, 1996.
37. Delaisse, J.M. et al., Matrix metalloproteinases (MMP) and cathepsin K contribute differently to osteoclastic activities, *Microsc. Res. Tech.*, 61, 504, 2003.
38. Hayman, A.R. et al., Mice lacking tartrate-resistant acid phosphatase (Acp 5) have disrupted endochondral ossification and mild osteopetrosis, *Development*, 122, 3151, 1996.
39. Nesbitt, S.A. and Horton, M.A., Trafficking of matrix collagens through bone-resorbing osteoclasts, *Science*, 276, 266, 1997.
40. Salo, J. et al., Removal of osteoclast bone resorption products by transcytosis, *Science*, 276, 270, 1997.
41. Manolagas, S.C., Birth and death of bone cells: basic regulatory mechanisms and implications for the pathogenesis and treatment of osteoporosis, *Endocr. Rev.*, 21, 115, 2000.
42. Fuller, K. et al., Macrophage colony-stimulating factor stimulates survival and chemotactic behavior in isolated osteoclasts, *J. Exp. Med.*, 178, 1733, 1993.
43. Pilkington, M.F., Sims, S.M., and Dixon, S.J., Wortmannin inhibits spreading and chemotaxis of rat osteoclasts *in vitro*, *J. Bone Miner. Res.*, 13, 688, 1998.
44. Pilkington, M.F., Sims, S.M., and Dixon, S.J., Transforming growth factor–β induces osteoclast ruffling and chemotaxis: potential role in osteoclast recruitment, *J. Bone Miner. Res.*, 16, 1237, 2001.
45. Buccione, R., Orth, J.D., and McNiven, M.A., Foot and mouth: podosomes, invadopodia and circular dorsal ruffles, *Nat. Rev. Mol. Cell Biol.*, 5, 647, 2004.
46. Weinstein, R.S. and Manolagas, S.C., Apoptosis and osteoporosis, *Am. J. Med.*, 108, 153, 2000.
47. Boyce, B. et al., Apoptosis in bone cells, in *Principles of Bone Biology*, Bilezikian, J.P., Raisz, L.G., and Rodan, G.A., Eds., Academic Press, San Diego, 2002, p. 151.
48. Tanaka, S. et al., Signal transduction pathways regulating osteoclast differentiation and function, *J. Bone Miner. Metab.*, 21, 123, 2003.

49. Parfitt, A.M. et al., A new model for the regulation of bone resorption, with particular reference to the effects of bisphosphonates, *J. Bone Miner. Res.*, 11, 150, 1996.
50. Boyce, B.F. et al., Recent advances in bone biology provide insight into the pathogenesis of bone diseases, *Lab. Invest.*, 79, 83, 1999.
51. Hughes, D.E. et al., Bisphosphonates promote apoptosis in murine osteoclasts *in vitro* and *in vivo*, *J. Bone Miner. Res.*, 10, 1478, 1995.
52. Yu, H. and Ferrier, J., ATP induces an intracellular calcium pulse in osteoclasts, *Biochem. Biophys. Res. Commun.*, 191, 357, 1993.
53. Yu, H. and Ferrier, J., Mechanisms of ATP-induced Ca^{2+} signaling in osteoclasts, *Cell. Signal.*, 6, 905, 1994.
54. Modderman, W.E. et al., Permeabilization of cells of hemopoietic origin by extracellular ATP^{4-}: elimination of osteoclasts, macrophages, and their precursors from isolated bone cell populations and fetal bone rudiments, *Calcif. Tissue Int.*, 55, 141, 1994.
55. Bowler, W.B. et al., Identification and cloning of human P2U purinoceptor present in osteoclastoma, bone, and osteoblasts, *J. Bone Miner. Res.*, 10, 1137, 1995.
56. Weidema, A.F. et al., Extracellular nucleotides activate non-selective cation and Ca^{2+}-dependent K^+ channels in rat osteoclasts, *J. Physiol.*, 503, 303, 1997.
57. Morrison, M.S. et al., ATP is a potent stimulator of the activation and formation of rodent osteoclasts, *J. Physiol.*, 511, 495, 1998.
58. Ralevic, V. and Burnstock, G., Receptors for purines and pyrimidines, *Pharmacol. Rev.*, 50, 413, 1998.
59. Hoebertz, A. et al., Expression of P2 receptors in bone and cultured bone cells, *Bone*, 27, 503, 2000.
60. Naemsch, L.N. et al., $P2X_4$ purinoceptors mediate an ATP-activated, non-selective cation current in rabbit osteoclasts, *J. Cell Sci.*, 112, 4425, 1999.
61. Naemsch, L.N., Dixon, S.J., and Sims, S.M., Activity-dependent development of $P2X_7$ current and Ca^{2+} entry in rabbit osteoclasts, *J. Biol. Chem.*, 276, 39107, 2001.
62. Buckley, K.A. et al., Adenosine triphosphate stimulates human osteoclast activity via upregulation of osteoblast-expressed receptor activator of nuclear factor-κB ligand, *Bone*, 31, 582, 2002.
63. King, B.F. et al., Full sensitivity of $P2X_2$ purinoceptor to ATP revealed by changing extracellular pH, *Br. J. Pharmacol.*, 117, 1371, 1996.
64. Wildman, S.S., King, B.F., and Burnstock, G., Potentiation of ATP-responses at a recombinant $P2X_2$ receptor by neurotransmitters and related substances, *Br. J. Pharmacol.*, 120, 221, 1997.
65. Arnett, T.R. and Spowage, M., Modulation of the resorptive activity of rat osteoclasts by small changes in extracellular pH near the physiological range, *Bone*, 18, 277, 1996.
66. Hoebertz, A. et al., Extracellular ADP is a powerful osteolytic agent: evidence for signaling through the $P2Y_1$ receptor on bone cells, *FASEB J.*, 15, 1139, 2001.
67. North, R.A., Molecular physiology of P2X receptors, *Physiol. Rev.*, 82, 1013, 2002.
68. Weidema, A.F., Dixon, S.J., and Sims, S.M., Activation of P2Y but not $P2X_4$ nucleotide receptors causes elevation of $[Ca^{2+}]_i$ in mammalian osteoclasts, *Am. J. Physiol.*, 280, C1531, 2001.
69. Seman, M. et al., NAD-induced T cell death: ADP-ribosylation of cell surface proteins by ART2 activates the cytolytic $P2X_7$ purinoceptor, *Immunity*, 19, 571, 2003.
70. Surprenant, A. et al., The cytolytic P2Z receptor for extracellular ATP identified as a P2X receptor ($P2X_7$), *Science*, 272, 735, 1996.
71. Virginio, C. et al., Kinetics of cell lysis, dye uptake and permeability changes in cells expressing the rat $P2X_7$ receptor, *J. Physiol.*, 519, 335, 1999.

72. Gartland, A. et al., Blockade of the pore-forming $P2X_7$ receptor inhibits formation of multinucleated human osteoclasts *in vitro*, *Calcif. Tissue Int.*, 73, 361, 2003.
73. Jorgensen, N.R. et al., Intercellular calcium signaling occurs between human osteoblasts and osteoclasts and requires activation of osteoclast $P2X_7$ receptors, *J. Biol. Chem.*, 277, 7574, 2002.
74. Sims, S.M., Kelly, M.E., and Dixon, S.J., K^+ and Cl^- currents in freshly isolated rat osteoclasts, *Pflügers Arch.*, 419, 358, 1991.
75. Zaidi, M. et al., Linkage of extracellular and intracellular control of cytosolic Ca^{2+} in rat osteoclasts in the presence of thapsigargin, *J. Bone Miner. Res.*, 8, 961, 1993.
76. Korcok, J. et al., Extracellular nucleotides act through $P2X_7$ receptors to activate NF-κB in osteoclasts, *J. Bone Miner. Res.*, 19, 642, 2004.
77. Korcok, J., Sims, S.M., and Dixon, S.J., $P2X_7$ nucleotide receptors act through two distinct mechanisms to regulate osteoclast survival, *J. Bone Miner. Res.*, 19, S418, 2004 (abstract).
78. Gartland, A. et al., $P2X_7$ receptor–deficient mice maintain the ability to form multinucleated osteoclasts *in vivo* and *in vitro*, *Crit. Rev. Eukaryot. Gene Expr.*, 13, 243, 2003.
79. Hu, Y. et al., Purinergic receptor modulation of lipopolysaccharide signaling and inducible nitric-oxide synthase expression in RAW 264.7 macrophages, *J. Biol. Chem.*, 273, 27170, 1998.
80. Pfeiffer, Z.A. et al., The nucleotide receptor $P2X_7$ mediates actin reorganization and membrane blebbing in RAW 264.7 macrophages via p38 MAP kinase and Rho, *J. Leukoc. Biol.*, 75, 1173, 2004.
81. Chiozzi, P. et al., Spontaneous cell fusion in macrophage cultures expressing high levels of the P2Z/$P2X_7$ receptor, *J. Cell Biol.*, 138, 697, 1997.
82. Di Virgilio, F. et al., The P2Z/$P2X_7$ receptor of microglial cells: a novel immunomodulatory receptor, *Prog. Brain Res.*, 120, 355, 1999.
83. Humphreys, B.D. et al., Stress-activated protein kinase/JNK activation and apoptotic induction by the macrophage $P2X_7$ nucleotide receptor, *J. Biol. Chem.*, 275, 26792, 2000.
84. Solle, M. et al., Altered cytokine production in mice lacking $P2X_7$ receptors, *J. Biol. Chem.*, 276, 125, 2001.
85. Pellegatti, P. et al., A novel recombinant plasma membrane–targeted luciferase reveals a new pathway for ATP secretion, *Mol. Biol. Cell*, 16, 3659, 2005.
86. Hiken, J.F. and Steinberg, T.H., ATP downregulates $P2X_7$ and inhibits osteoclast formation in RAW cells, *Am. J. Physiol.*, 287, C403, 2004.
87. Steinberg, T.H. et al., P2-mediated responses in osteoclasts and osteoclast-like cells, *Drug Dev. Res.*, 53, 126, 2001.
88. Adriouch, S. et al., Cutting edge: a natural P451L mutation in the cytoplasmic domain impairs the function of the mouse $P2X_7$ receptor, *J. Immunol.*, 169, 4108, 2002.
89. Di Virgilio, F. and Wiley, J.S., The $P2X_7$ receptor of CLL lymphocytes—a molecule with a split personality, *Lancet*, 360, 1898, 2002.
90. Cabrini, G. et al., A His-155 to Tyr polymorphism confers gain-of-function to the human $P2X_7$ receptor of human leukemic lymphocytes, *J. Immunol.*, 175, 82, 2005.
91. Gu, B.J. et al., A Glu-496 to Ala polymorphism leads to loss of function of the human $P2X_7$ receptor, *J. Biol. Chem.*, 276, 11135, 2001.
92. Wiley, J.S. et al., A loss-of-function polymorphic mutation in the cytolytic $P2X_7$ receptor gene and chronic lymphocytic leukaemia: a molecular study, *Lancet*, 359, 1114, 2002.

93. Sluyter, R., Shemon, A.N., and Wiley, J.S., Glu496 to Ala polymorphism in the $P2X_7$ receptor impairs ATP-induced IL-1β release from human monocytes, *J. Immunol.*, 172, 3399, 2004.
94. Sluyter, R., Dalitz, J.G., and Wiley, J.S., $P2X_7$ receptor polymorphism impairs extracellular adenosine 5′–triphosphate-induced interleukin-18 release from human monocytes, *Genes Immun.*, 5, 588, 2004.
95. Le Stunff, H. et al., The Pro-451 to Leu polymorphism within the C-terminal tail of $P2X_7$ receptor impairs cell death but not phospholipase D activation in murine thymocytes, *J. Biol. Chem.*, 279, 16918, 2004.
96. Boldt, W. et al., Glu496Ala polymorphism of human $P2X_7$ receptor does not affect its electrophysiological phenotype, *Am. J. Physiol.*, 284, C749, 2003.
97. Ohlendorff, S. et al., Effect of the $P2X_7$ Glu496Ala polymorphism in osteoclasts and association to bone mass, *J. Bone Miner. Res.*, 19, S248, 2004 (abstract).
98. Korcok, J. et al., $P2Y_6$ nucleotide receptors activate NF-κB and increase survival of osteoclasts, *J. Biol. Chem.*, 280, 16909, 2005.
99. Jacobson, K.A., Jarvis, M.F., and Williams, M., Purine and pyrimidine (P2) receptors as drug targets, *J. Med. Chem.*, 45, 4057, 2002.
100. Zimmermann, H., Extracellular metabolism of ATP and other nucleotides, *N. Schmied. Arch. Pharmacol.*, 362, 299, 2000.
101. Yu, H. and Ferrier, J., Osteoclast ATP receptor activation leads to a transient decrease in intracellular pH, *J. Cell. Sci.*, 108, 3051, 1995.
102. Bowler, W.B. et al., $P2Y_2$ receptors are expressed by human osteoclasts of giant cell tumor but do not mediate ATP-induced bone resorption, *Bone*, 22, 195, 1998.
103. Wiebe, S.H., Sims, S.M., and Dixon, S.J., Calcium signalling via multiple P2 purinoceptor subtypes in rat osteoclasts, *Cell Physiol. Biochem.*, 9, 323, 1999.
104. Hoebertz, A. et al., ATP and UTP at low concentrations strongly inhibit bone formation by osteoblasts: a novel role for the $P2Y_2$ receptor in bone remodeling, *J. Cell. Biochem.*, 86, 413, 2002.
105. Maier, R. et al., Cloning of $P2Y_6$ cDNAs and identification of a pseudogene: comparison of P2Y receptor subtype expression in bone and brain tissues, *Biochem. Biophys. Res. Commun.*, 240, 298, 1997.
106. Ye, R.D., Regulation of nuclear factor κB activation by G-protein–coupled receptors, *J. Leukoc. Biol.*, 70, 839, 2001.
107. North, R.A. and Surprenant, A., Pharmacology of cloned P2X receptors, *Ann. Rev. Pharmacol. Toxicol.*, 40, 563, 2000.
108. Pendergast, W. et al., Synthesis and P2Y receptor activity of a series of uridine dinucleoside 5′–polyphosphates, *Bioorg. Med. Chem. Lett.*, 11, 157, 2001.
109. Mundasad, M.V. et al., Ocular safety of INS365 ophthalmic solution: a $P2Y_2$ agonist in healthy subjects, *J. Ocul. Pharmacol. Ther.*, 17, 173, 2001.
110. Noone, P.G. et al., Safety of aerosolized INS 365 in patients with mild to moderate cystic fibrosis: results of a phase I multi-center study, *Pediatr. Pulmonol.*, 32, 122, 2001.
111. Walther, M.M. et al., Suramin inhibits bone resorption and reduces osteoblast number in a neonatal mouse calvarial bone resorption assay, *Endocrinology*, 131, 2263, 1992.
112. Farsoudi, K.H. et al., Suramin is a potent inhibitor of calcemic hormone- and growth factor–induced bone resorption *in vitro*, *J. Pharmacol. Exp. Ther.*, 264, 579, 1993.
113. Murgia, M. et al., Oxidized ATP. An irreversible inhibitor of the macrophage purinergic P2Z receptor, *J. Biol. Chem.*, 268, 8199, 1993.
114. Beigi, R.D. et al., Oxidized ATP (oATP) attenuates proinflammatory signaling via P2 receptor–independent mechanisms, *Br. J. Pharmacol.*, 140, 507, 2003.

115. Di Virgilio, F., Novel data point to a broader mechanism of action of oxidized ATP: the $P2X_7$ receptor is not the only target, *Br. J. Pharmacol.*, 140, 441, 2003.
116. Boyer, J.L. et al., 2-Chloro N6-methyl-(N)-methanocarba-2′–deoxyadenosine-3′,5′–bisphosphate is a selective high affinity $P2Y_1$ receptor antagonist, *Br. J. Pharmacol.*, 135, 2004, 2002.
117. Baraldi, P.G., Di Virgilio, F., and Romagnoli, R., Agonists and antagonists acting at $P2X_7$ receptor, *Curr. Top. Med. Chem.*, 4, 1707, 2004.
118. Russell, R.G. and Rogers, M.J., Bisphosphonates: from the laboratory to the clinic and back again, *Bone*, 25, 97, 1999.
119. Frith, J.C. et al., Clodronate and liposome-encapsulated clodronate are metabolized to a toxic ATP analog, adenosine 5′–(β,γ–dichloromethylene) triphosphate, by mammalian cells *in vitro*, *J. Bone Miner. Res.*, 12, 1358, 1997.
120. Li, J. et al., The $P2X_7$ nucleotide receptor mediates skeletal mechanotransduction, *J. Biol. Chem.*, 280, 42952, 2005.
121. Lee, B.C. et al., P2Y-like receptor, GPR105 ($P2Y_{14}$), identifies and mediates chemotaxis of bone-marrow hematopoietic stem cells, *Genes Dev.*, 17, 1592, 2003.
122. Idzko, M. et al., Nucleotides induce chemotaxis and actin polymerization in immature but not mature human dendritic cells via activation of pertussis toxin–sensitive P2y receptors, *Blood*, 100, 925, 2002.
123. Honda, S. et al., Extracellular ATP or ADP induce chemotaxis of cultured microglia through $G_{i/o}$-coupled P2Y receptors, *J. Neurosci.*, 21, 1975, 2001.
124. Weidema, A.F., Dixon, S.J., and Sims, S.M., Electrophysiological characterization of ion channels in osteoclasts isolated from human deciduous teeth, *Bone*, 27, 5, 2000.

3 The Role of Purinergic Signaling in the Interactions between Skeletal Cells

Katherine A Buckley, Alison Gartland, and James A Gallagher

CONTENTS

3.1 INTRODUCTION

All living cells and tissues, from unicellular organisms to the connective tissues of large mammals, are subjected to physical stresses in response to mechanical loading. The magnitudes and types of stresses to which cells are exposed vary greatly depending on the physical environment in which they live. Tissue specialization in higher organisms allows for some tissues to be largely protected from physical stress, whereas other tissues are primarily designed to withstand mechanical loading. The latter group of tissues, of which bone is a good example, have had their architecture and composition shaped by evolution to perform this function.

Specialization of tissues to resist mechanical loading is not cost free. Formation, maintenance, and movement of specialized tissues have nutritional and metabolic costs, and so in the case of tissues specialized to resist mechanical loading, it is important that the amount of tissue is appropriate to the applied load. Too little would lead to frequent wounding, severe tissue damage, and eventually death, whereas too much would put the individual at survival disadvantage by devoting scarce resources to an unnecessary function. In order to provide the appropriate amount of tissue, complex mechanotransduction mechanisms have evolved to allow tissues to detect and respond to mechanical loading. It is now well established that such mechanisms contribute to the regulation of bone mass and architecture, and their identification has become one of the major areas of research in bone biology.

Several candidates have been suggested as mediators of mechanotranduction, including prostaglandins, nitric oxide, and glutamate. Over the past decade, we have suggested that extracellular nucleotides may be one of the major mediators. ATP is an ideal candidate for a mediator of mechanotranduction. Loss of ATP from cells is an inevitable consequence of cell damage, and ATP released in the extracellular environment provides an appropriate cue to neighboring cells to embark on a wound-healing response. We and others have demonstrated that:

1. ATP is released into the extracellular environment by bone cells.
2. This release is stimulated by mechanical perturbation.
3. Bone cells express a spectrum of P2 receptors.
4. Activation of these receptors brings about changes in bone formation and bone resorption.
5. Extracellular ATP works in concert with systemic hormones such as parathyroid hormone (PTH), providing a mechanism for the integration of local and systemic responses in bone.

Here we review these finding in the context of the interactions between skeletal cells.

3.2 P2 RECEPTORS IN BONE

The first evidence for the expression of P2 receptors in bone was obtained from studies in which osteoblastic cells were exposed to exogenous ATP *in vitro* and elevations in intracellular calcium ($[Ca^{2+}]_i$) were observed. The responses were consistent with the expression of functional P2 receptors by these cells.[1–4] Then, the receptor only existed as a pharmacological entity, but following advances in molecular cloning, Bowler and colleagues confirmed the expression of $P2Y_2$ receptors in primary human osteoblasts.[5] Subsequent studies have reported the expression of multiple subtypes of P2Y and P2X receptors by osteoblastic cells.[6–9] Functional and molecular studies have also revealed the expression of many P2Y and P2X receptor subtypes by osteoclasts and their precursors.[10–15] Heterogeneity of receptor subtype expression within osteoblast and osteoclast populations may serve to modulate purinergic signaling between bone cells, and this heterogeneity may reflect differences in the differentiation status of individual cells.[16] The expression of multiple P2 receptor subtypes in bone is manifested in the numerous processes they influence in the course of bone remodeling, which will be discussed below. The emerging picture of a pivotal role of P2 receptors in the regulation of skeletal homeostasis has resulted in growing interest in these receptors as potential therapeutic targets in bone disease in recent years.[17]

3.3 ATP RELEASE FROM OSTEOBLASTS

The regulation of P2 receptor signaling between bone cells is controlled not only by modulation of P2 receptor expression by osteoblasts and osteoclasts, but also by availability and concentration of nucleotide agonists. ATP is a ubiquitous intracellular molecule, and therefore every cell is a potential source of this nucleotide. Given its relatively large molecular mass and anionic nature, there is little permeation of ATP (or MgATP, the predominant cytosolic form) across the plasma membrane. The most obvious route of ATP into the extracellular space is due to cell lysis at sites of tissue trauma. Although cytosolic ATP is usually at a concentration of around 3 to 5mM, it is easy to conceive that physiologically relevant micromolar levels of ATP could be achieved due to dilution effects and the action of local ecto-nucleotidases. In fact, once in the extracellular environment, nucleotides undergo both catabolic and anabolic processes via numerous ecto-enzymatic processes, which may act to control the duration of nucleotide signal.[18]

Adenine and uridine nucleotides may also be transported into the extracellular environment nonlytically via their inclusion in the exocytotic vesicles of neuronal and nonneuronal cells,[19–21] and it has been suggested that nucleotides can be released actively via intrinsic plasma membrane proteins (Figure 3.1).[22] Osteoblasts have been shown to release ATP constitutively,[18,23] and this finding has been mirrored in other cell types.[24] The exact mechanism of this constitutive release by osteoblasts remains to be identified, although exocytotic release is thought to be the most likely means. Mechanical stimulation increases ATP release,[25,26] (Figure 3.1) indicating that nucleotides may play a role in mechanotransduction in bone, the process by which detection of mechanical deformation of fluid shear by skeletal cells results in remodeling. Ultrasound stimulation of osteoblasts *in vitro* enhances constitutive ATP release from these cells, a process that may contribute to the observed decrease in time to fracture healing *in vivo* when ultrasound is applied to fracture sites.[27]

Modulation of nucleotide release and metabolism by osteoblasts may provide a gap junction–independent mechanism for the transduction of waves of intracellular calcium signals between neighboring osteoblasts.[28,29] This form of communication has also been observed between osteoblasts and osteoclasts.[30] These findings indicate the presence of autocrine and paracrine signaling mechanisms at focal sites in bone whereby osteoblasts and osteoclasts may be able to modulate many aspects of their own activity.

3.4 REGULATION OF OSTEOCLASTOGENESIS BY RANKL/OPG

It was understood for some time that osteoblasts played a vital role in bone resorption, and it was hypothesized that the signal to resorb bone was received by the osteoblasts, which resulted in the production of a molecule or molecules that recruited and

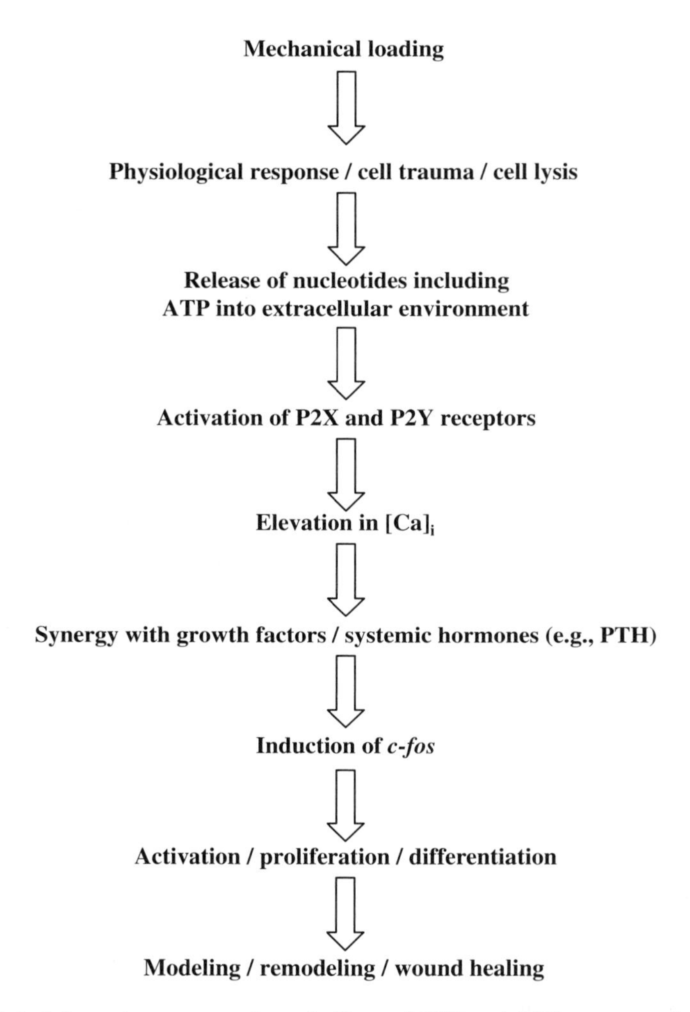

FIGURE 3.1 Schematic representation of effects of P2X and P2Y receptor stimulation on mechanotransduction in bone.

stimulated the activity of osteoclasts. It has only been relatively recently, however, that this molecule was identified as receptor activator of nuclear factor–kB ligand (RANKL).[31] CD14-positive monocytic osteoclast precursor cells differentiate into osteoclasts at bone-resorbing sites, and although this process is also influenced by osteotropic hormones and local factors in the bone microenvironment, RANKL provides the essential signal for osteoclastogenesis to occur. This ligand, presented on the osteoblast cell membrane, interacts with the RANK receptor on monocytic osteoclast precursor cells, inducing fusion of these cells to form multinucleated osteoclasts (Figure 3.2). RANKL also enhances resorption by mature osteoclasts, which retain expression of the RANK receptor.[32]

Osteoprotegerin (OPG) is a soluble decoy receptor to RANKL and is also produced by osteoblasts.[33] Binding of osteoprotegerin to RANKL prevents osteoclastogenesis

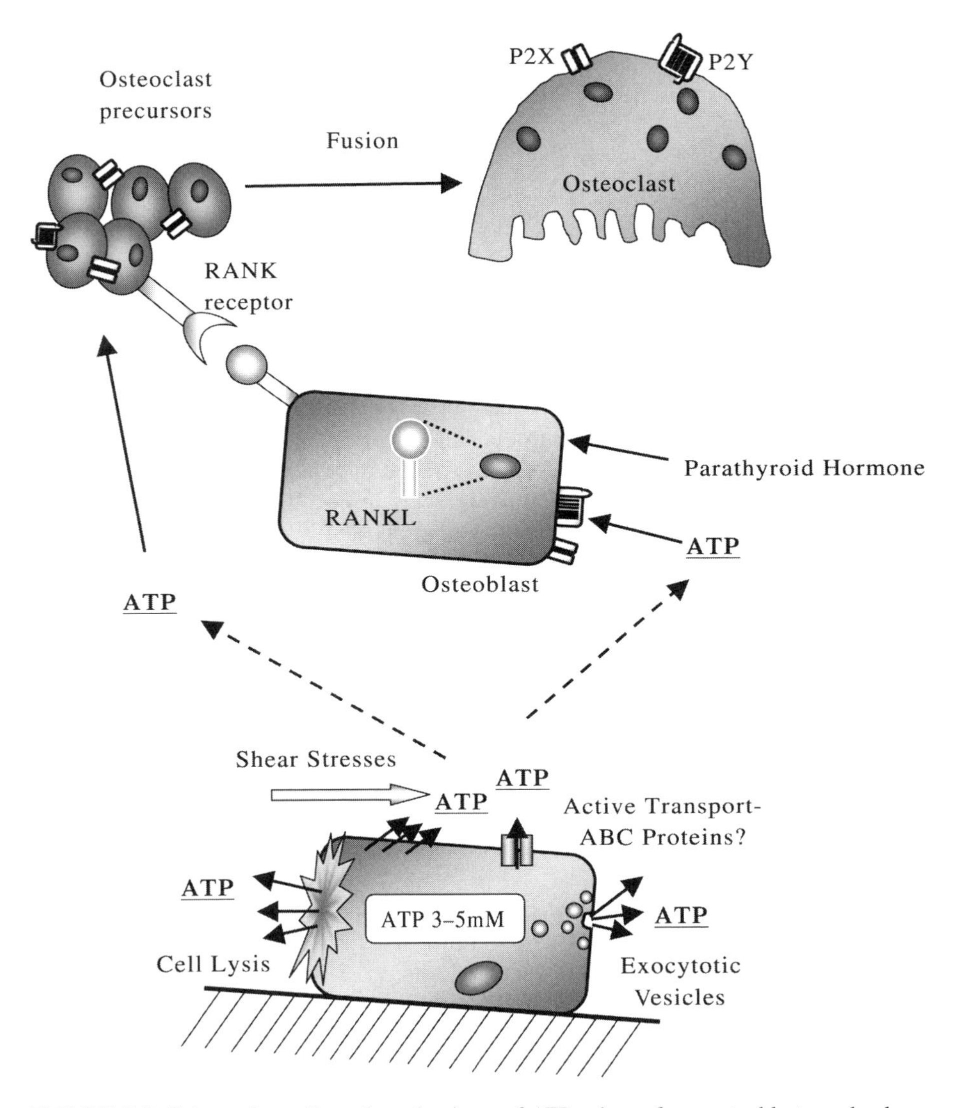

FIGURE 3.2 Schematic outline of mechanisms of ATP release from osteoblasts and subsequent methods of osteoclast formation by ATP. ATP may be released from osteoblasts by the methods shown, and this release may be enhanced by shear stresses. ATP induces osteoclast formation by increasing RANKL expression by osteoblasts or by stimulating osteoclast precursor cell $P2X_7$ receptors. Extracellular nucleotides may subsequently act on P2 receptors expressed by the mature multinucleated osteoclast cell.

from occurring; osteoblasts may therefore alter their levels of expression of both RANKL and OPG to modulate osteoclast formation and activity. Many regulatory factors control the ratio of expression of these two proteins: 1,25 dihydroxyvitamin D_3, PTH and interleukin (IL)-11 induce an increase in RANKL mRNA expression in osteoblasts, while OPG levels are decreased.[34] Transforming growth factor–β has the opposite effect.[35]

3.4.1 Effects of ATP on RANKL/OPG Expression

Morrison and colleagues first reported that low-concentration ATP (0.2 to 2 μM) enhanced both osteoclast formation and resorption in rodent osteoclast-containing bone cell populations.[36] Hoebertz and coworkers subsequently demonstrated that similar concentrations of ADP and 2-methylthio ADP caused increases in osteoclastic bone resorption, again in rodent mixed bone cell populations.[37] The ADP effects were blocked by MRS2179, a $P2Y_1$ receptor antagonist, and evidence for $P2Y_1$ receptor expression on both osteoclasts and osteoblasts was found. The authors therefore concluded that ADP could exert its actions both directly on osteoclasts and indirectly via $P2Y_1$ receptors on osteoblasts. It was more recently shown that ATP at similar low micromolar concentrations enhanced resorption by human osteoclasts produced in cocultures of human peripheral blood monocytes and UMR-106 rat osteoblast-like cells.[15] In contrast, introduction of ATP to peripheral blood monocytes cultured with recombinant RANKL had no effect on subsequent resorption, indicating that the low micromolar ATP-induced effects in cocultures were mediated via UMR-106 P2 receptors. These findings were supported by the discovery that ATP induced increased RANKL expression by these osteoblast-like cells at low and high micromolar concentrations, confirming the involvement of osteoblasts in stimulating osteoclast activity following exposure to ATP (Figure 3.2).[15] Because RANKL modulates both osteoclast formation and resorption, it is likely that the stimulatory effects of ATP affect both of these processes.

More recent studies have revealed that ATP increases RANKL expression by human SaOS-2 osteoblast-like cells and also decreases expression of the decoy receptor OPG (KA Buckley, unpublished observations). Parathyroid hormone is known to induce similar effects on RANKL/OPG osteoblast expression, and PTH and ATP have additive effects on the RANKL/OPG expression ratio by SaOS-2 cells (Figure 3.3).

At fracture healing sites, RANKL expression is strongly induced,[38] and at sites of tissue trauma such as this, larger quantities of ATP are released from damaged cells, in agreement with the findings of elevated osteoblast RANKL expression following ATP stimulation. In healthy bone, tissue sites of active remodeling are distributed at discrete loci in the skeleton where localized release of ATP may activate osteoclast formation and resorption to initiate the remodeling cycle via modulation of RANKL expression.

3.5 $P2X_7$ Receptor Expression and Activation in Bone

The $P2X_7$ receptor is expressed by both osteoblasts[14] and osteoclasts.[9,30] In common with other P2X receptor subtypes, $P2X_7$ is an ATP-gated ion channel, but it has the unique ability among the P2 receptor family to form membrane pores that are permeable to molecules of up to 900 Da when exposed to repeated or prolonged application of nucleotide agonist. The involvement of this membrane pore in the fusion of osteoclast precursors to form multinucleated osteoclasts has recently been described; introduction of a $P2X_7$-specific monoclonal antibody or $P2X_7$ receptor

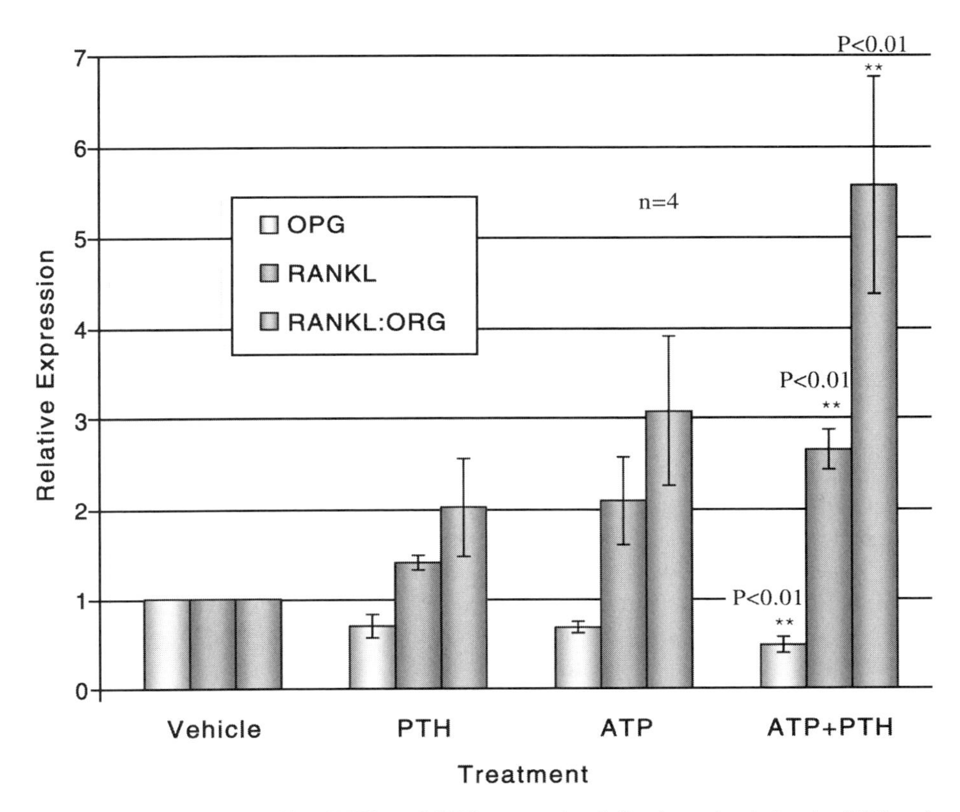

FIGURE 3.3 SaOS-2 Cell RANKL and OPG expression following stimulation by PTH and ATP. Quiescent SaOS-2 cells were induced with 100 ng/mL PTH 1-34 or 10 μM ATP (or both) for 8 h. PCRs using primers specific to RANKL and OPG were performed with cDNA from these cells on a Biorad icycler. The PCR threshold cycle number (C_T) for each test cDNA sample was calculated and relative expression levels of RANKL and OPG were calculated using the C_T values for PCRs performed with β–actin primers. PTH and ATP both increased RANKL and decreased OPG expression. ATP in combination with PTH induced a further increase in RANKL and decrease in OPG expression compared to PTH fragments alone. Statistical analysis was by ANOVA. Data represent means ± SEM. Primer sequences, T_ms, and product sizes are shown in Table 3.1.

antagonist to cultures of mononuclear osteoclast precursors prevented the formation of osteoclasts.[39] In addition, activation of the $P2X_7$ receptor can result in cytokine release[40] and apoptosis,[41] indicating a central role of this receptor in the control of bone resorption by purinergic signaling. While apoptosis of both osteoblasts and osteoclasts may occur where high concentrations of ATP are found extracellularly, such as at sites of inflammation or trauma, osteoclast precursor cell fusion may be induced by transient localized high ATP concentrations at the surface of tightly juxtaposed cells (Figure 3.1). The continuation of the presence of this high extracellular ATP concentration once the multinucleated osteoclast has formed, however, would be likely to induce its apoptosis via $P2X_7$ receptor activation. Dynamic fluctuations in extracellular ATP concentrations are therefore fundamental to osteoclast

TABLE 3.1
PCR primers, Tms, and product sizes

Primer	Sequences	T_m	Product Size	Accession No.
β–Actin	GGACCTGACTGACTACCTC (sense) GCCATCTCTTGCTCGAAG (anti-sense)	53.9	135	NM_001101
RANKL	GCAGCGGCACATTGGACATG (sense) AGGATCTGGTCACTGGGTTTGC (anti-sense)	52.2	142	AB037599
OPG	GCAGCGGCACATTGGACATG (sense) AGGATCTGGTCACTGGGTTTGC (anti-sense)	54.1	135	NM_002546

formation and survival. At lower micromolar ATP concentrations, osteoclast formation is likely to predominate due to increased RANKL expression. At higher millimolar ATP concentrations, however, $P2X_7$ receptor–induced osteoclast apoptosis may prevail over increased osteoclast formation that may occur due to ATP-induced RANKL expression or $P2X_7$ receptor–induced precursor cell fusion. Variation of the level of $P2X_7$ receptor expression by osteoclast precursors and mature osteoclasts may also affect the life span of the cell, although reports to date have suggested that the $P2X_7$ receptor is expressed throughout osteoclast development.[39]

Mice deficient in the $P2X_7$ receptor displayed significant reduction in total and cortical bone content and periosteal circumference in femurs, reduced periosteal bone formation, and increased trabecular bone resorption in tibias.[42] This phenotype is similar to that seen with disuse of the skeleton, indicating the involvement of $P2X_7$ receptors in the skeletal response to mechanical loading. Mechanically induced intercellular calcium signaling among osteoclasts has been shown to be inhibited by $P2X_7$ receptor blockade,[30] and the $P2X_7$ receptor is necessary for fluid shear stress–induced membrane pore formation and prostaglandin release in osteoblasts.[43]

Osteoclasts were present in $P2X_7$ receptor knockout mice *in vivo*, however, and monocytic osteoclast precursors from their blood were able to form multinucleated osteoclasts *in vitro*.[42,44] These data suggest that $P2X_7$ receptors are fundamental to normal bone development but indicate that this receptor is not essential for osteoclast fusion. It is suggested that an endogenous pore structure exists in osteoclast precursor cells that can be activated by the $P2X_7$ receptor, but in the absence of this receptor, alternative signals (for example, maitotoxin) can mediate pore formation and subsequent cell fusion.[44]

3.6 NUCLEOTIDE-INDUCED INTERLEUKIN PRODUCTION

Interleukin-6 is produced by many cells in the bone microenvironment, including osteoblasts and monocytes. It is known to stimulate osteoclast formation and resorption and has been implicated in the etiology of Paget's disease.[45] This condition is characterized by the presence of osteoclasts that are increased in size and with large numbers of nuclei and by unusually increased osteoclast formation and bone resorption followed by excessive new bone formation.

Stimulation of osteoblast P2Y receptors by ATP results in increased expression of IL-6, and this expression is dependent on mobilization of calcium from intracellular stores through inositol phosphate-3 receptor activation.[46] $P2X_7$ receptor expression by peripheral blood monocytes is reportedly upregulated by numerous interleukins, including IL-6.[47] In an area of bone where there is increased purinergic activity, either due to increased ATP release, prevention of its subsequent degradation, or increased receptor expression, this therefore represents a mechanism of increasing osteoclast resorption, as osteoblast P2Y-induced IL-6 release could stimulate osteoclast formation by $P2X_7$ receptor–mediated fusion of blood monocytes or enhance osteoclast resorptive activity by an as yet unidentified mechanism.

P2Y-induced IL-6 release from osteoblasts therefore represents the second mechanism by which ATP indirectly stimulates osteoclast activity via osteoblast gene expression, the other being via modulation of the expression ratio of RANKL/OPG.

3.7 BONE FORMATION

While much of the data involving P2 receptors in bone concerns the processes of osteoclast formation and resorption, numerous reports also describe a role for extracellular nucleotides in regulating bone formation. ATP stimulates the proliferation of MC3T3-E1 osteoblast-like cells,[48] and enhanced DNA synthesis has also been observed in osteoblast-like MG-63 cells following P2X receptor stimulation.[8] In assays to measure actual bone formation by osteoblasts, however, ATP and other nucleotide agonists were found to have an inhibitory effect.[49,50] These apparent discrepancies may reflect differences in P2 receptor subtype expression by the osteoblasts in the separate studies or differing concentrations of agonist application. It appears, however, that ATP can potentially simultaneously inhibit bone formation by osteoblasts and stimulate osteoclast formation and bone resorption, indicating a role for extracellular nucleotides signaling locally to induce catabolic changes in diseases of net bone loss.

3.8 SYNERGISTIC EFFECTS OF ATP AND PTH ON OSTEOBLASTS

In addition to the numerous effects that nucleotides are known to exert independently on bone cells, localized release of these molecules can sensitize osteoblasts to the activity of systemic factors. Parathyroid hormone plays a central role in plasma calcium homeostasis and is a principal regulator of bone remodeling. The action of PTH on skeletal cells is complex and can result in the stimulation of both resorption and new bone formation.[2,3] The PTH receptor is unusual in that it signals via two different second messenger pathways, coupling to both G_q and G_s, to elevate either intracellular calcium or cAMP, respectively.

It might be expected that PTH, by inducing release of calcium from intracellular stores, would reduce the subsequent response to nucleotide stimulation. Kaplan and colleagues, however, first reported that PTH can potentiate nucleotide-induced $[Ca^{2+}]_i$ release in osteoblasts, but in contrast had no effect on the cAMP response to PTH.[51] Potentiation of this signal was primarily dependent on intracellular store release and relied upon the activated PTH receptor coupling to G_s.[52,53] Elevated cAMP levels occurring following PTH stimulation of osteoblasts did not provoke the potentiation of nucleotide-induced $[Ca^{2+}]_i$ elevations by PTH, demonstrating that an as yet unidentified mechanism is responsible for this $[Ca^{2+}]_i$ potentiation observed upon dual activation of the PTH and P2 G protein–coupled receptors. It has been suggested that this potentiation may result from the ability of PTH to regulate Ca^{2+} mobilization by facilitating translocation of Ca^{2+} between discrete intracellular stores, thereby regulating the Ca^{2+} pool available to receptors linked to inositol 1,4,5-trisphosphate (IP_3) formation.[53] Alternatively, a direct G protein interaction has been implicated in inducing potentiated $[Ca^{2+}]_i$ release: in other systems, activated G_α or $G_{\beta\gamma}$ subunits can have synergistic or inhibitory interactions with other G proteins or their enzyme targets (Figure 3.4).[54]

Parathyroid hormone potentiation of nucleotide-induced $[Ca^{2+}]_i$ release by osteoblasts results in increased phosphorylation of the transcription factor CREB, which plays a key role in signaling-driven activation of the immediate early gene c-*fos*. This gene has been strongly implicated in driving many osteoblast functions, including proliferation and differentiation.[55,56] Synergistic induction of endogenous c-*fos* mRNA expression by osteoblasts occurs following their costimulation with PTH and ADP.[52,57] This potentiation of calcium and immediate early gene expression is likely to subsequently influence the induction of many genes of importance in the remodeling cycle;

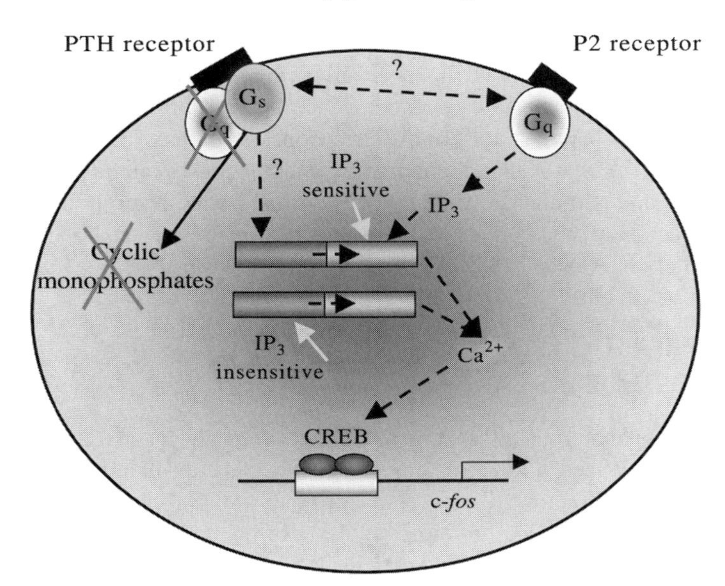

FIGURE 3.4 (See color insert following page 80) Possible mechanisms leading to potentiated $[Ca^{2+}]_i$ release and synergistic gene expression following nucleotide and PTH costimulation of osteoblasts. Co-activation of P2 and PTH G protein–coupled receptors leads to potentiated $[Ca^{2+}]_i$ release, which may be due to shuttling of calcium between intracellular stores or to a direct G protein subunit interaction.

indeed, how osteoblast expression of IL-6 following P2Y receptor stimulation is dependent on release of calcium from intracellular IP_3-sensitive stores has already been described,[46] and it follows that costimulation by PTH and nucleotides may serve to significantly increase osteoblast IL-6 release.

In contrast to the systemic action of PTH, remodeling events occur at small, localized sites in the skeleton, separated by distance and time. The sensitization of osteoblasts to surrounding systemic factors such as PTH, by localized lytic and nonlytic release of nucleotides, may allow the initiation of remodeling via osteoblast activation at these discrete loci. P2 receptor stimulation has been found to lower the "set point" of signal transduction pathways, allowing cells to be more receptive to other surrounding factors in the presence of extracellular nucleotides.[58] Release of ATP (and other nucleotides) therefore provides a means by which cells regulate their activation not only to nucleotides themselves, but also to agonists that act via hormone receptors.

3.9 CONCLUSIONS

Bone tissue is constantly being renewed, due to remodeling events throughout the skeleton, and the bone microenvironment is exposed to a host of local and systemic factors that are able to modulate this process. Purinergic signaling within the skeleton represents a convincing mechanism for allowing remodeling to occur at localized sites. P2 receptor subtypes have widespread expression in bone, and there are numerous mechanisms in place for ATP release. The subsequent metabolism of this extracellular nucleotide by catabolic and anabolic ecto-enzymes may act to selectively regulate the activation of the numerous P2 receptor subtypes via their differing agonist potency profiles.

The functional significance of P2 receptor expression in bone has recently become apparent, and the literature concerning the effects of extracellular nucleotides in bone is continuing to expand. Extracellular nucleotides may induce osteoblast gene expression and proliferation and play a critical role in the formation and stimulation of osteoclasts via regulation of the RANKL/OPG expression ratio, induction of IL production, and through $P2X_7$ receptor–induced pore formation. It appears that locally acting extracellular nucleotides are also able to interact with systemic factors, such as PTH, resulting in synergistic effects in osteoblasts that potentially provide cell-signaling responses of sufficient magnitude to induce cellular activity that may contribute to the remodeling cycle.

REFERENCES

1. Kumagai, H., Sacktor, B., and Filburn, C.R., Purinergic regulation of cytosolic calcium and phosphoinositide metabolism in rat osteoblast-like osteosarcoma cells, *J. Bone Miner. Res.*, 6, 697, 1991.
2. Kumagai, H. et al., Neurotransmitter regulation of cytosolic calcium in osteoblast-like bone cells, *Calcif. Tissue Int.*, 45, 251, 1989.

3. Schöfl, C. et al., Evidence for P_2-purinoceptors on human osteoblast-like cells, *J. Bone Miner. Res.*, 7, 485, 1992.
4. Reeve, J. et al., Anabolic effect of low doses of a fragment of human parathyroid hormone on the skeleton in postmenopausal osteoporosis, *Lancet*, 307, 1035, 1976.
5. Reimer, W.J. and Dixon, S.J., Extracellular nucleotides elevate $[Ca^{2+}]_i$ in rat osteoblastic cells by interaction with two receptor subtypes, *Am. J. Physiol.*, 263, C1040, 1992.
6. Bowler, W.B. et al., Identification and cloning of human P_{2U} purinoceptor present in osteoclastoma, bone and osteoblasts, *J. Bone Miner. Res.*, 10, 1137, 1995.
7. Maier, R. et al., Cloning of $P2Y_6$ cDNAs and identification of a pseudogene: comparison of P2Y receptor sub-type expression in bone and brain tissues, *Biochem. Biophys. Res. Commun.*, 237, 298, 1997.
8. Dixon, C.J. et al., Effects of extracellular nucleotides on single cell populations of human osteoblasts: contribution of cell heterogeneity to relative potencies, *Br. J. Pharmacol.*, 120, 777, 1997.
9. Nakamura, E. et al., ATP activates DNA synthesis by acting on P2X receptors in human osteoblast-like MG-63 cells, *Am. J. Physiol.*, 279, C510, 2000.
10. Hoebertz, A. et al., Expression of P2 receptors in bone and cultured bone cells, *Bone*, 27, 503, 2000.
11. Yu, H. and Ferrier, J., Mechanisms of ATP-induced Ca^{2+} signaling in osteoclasts, *Cell. Signal.*, 6, 905, 1994.
12. Wiebe, S.H., Sims, S.M., and Dixon, S.J., Calcium signaling via multiple P2 purinoceptor subtypes in rat osteoclasts, *Cell Physiol. Biochem.*, 9, 323, 1999.
13. Naemsch, L.N. et al., $P2X_4$ purinoceptors mediate an ATP-activated, non-selective cation current in rabbit osteoclasts, *J. Cell Sci.*, 112, 4425, 1999.
14. Weidema, A.F. et al., Extracellular nucleotides activate non-selective cation and Ca^{2+}-dependent K^+ channels in rat osteoclasts, *J. Physiol.*, 503, 303, 1997.
15. Gartland, A. et al., Expression of a $P2X_7$ receptor by a subpopulation of human osteoblasts, *J. Bone Miner. Res.*, 16, 846, 2001.
16. Buckley, K.A. et al., Adenosine triphosphate stimulates human osteoclast activity via upregulation of osteoblast-expressed receptor activator of nuclear factor-kB ligand, *Bone*, 31, 582, 2002.
17. Dixon, S.J. and Sims, S.M., P2 purinergic receptors on osteoblasts and osteoclasts: potential targets for drug development, *Drug Dev. Res.*, 49,187, 2000.
18. Buckley, K.A. et al., Release and interconversion of P2 receptor agonists by human osteoblast-like cells, *FASEB J.*, 17, 1401, 2003.
19. Lew, M.J. and White, T.D., Release of endogenous ATP during sympathetic nerve stimulation, *Br. J. Pharmacol.*, 92, 349, 1987.
20. Hillarp, N.A. and Thieme, G., Nucleotides in the catecholamine granules of the adrenal medulla, *Acta Physiol. Scand.*, 45, 328, 1959.
21. Todorov, L.D. et al., Evidence for the differential release of the cotransmitters ATP and noradrenaline from sympathetic nerves of the guinea-pig vas deferens, *J. Physiol.*, 496, 731, 1996.
22. Abraham, E.H. et al., The multidrug resistance (mdr1) gene product functions as an ATP channel, *Proc. Natl. Acad. Sci. U.S.A.*, 90, 312, 1993.
23. Bowler, W.B. et al., Real time measurement of ATP release from human osteoblasts, *J. Bone Miner. Res.*, 13, 524, 1998.
24. Roman, R.M. et al., Endogenous ATP release regulates Cl- secretion in cultured human and rat biliary epithelial cells, *Am. J. Physiol.*, 276, G1391, 1999.

25. Bowler, W.B. et al., Release of ATP by osteoblasts: modulation by fluid shear forces, *Bone*, 22, 3S, 1998.
26. Romanello, M. et al., Mechanically induced ATP release from human osteoblastic cells, *Biochem. Biophys. Res. Commun.*, 289, 1275, 2001.
27. Hayton, M.J. et al., Involvement of adenosine 5′–triphosphate in ultrasound-induced fracture repair, *Ultrasound Med. Biol.*, 31, 1131, 2005.
28. Osipchuk, Y. and Cahalan, M., Cell-to-cell spread of calcium signals mediated by ATP receptors in mast cells, *Nature*, 359, 241, 1992.
29. Jørgensen, N.R.et al., Human osteoblastic cells propagate intercellular calcium signals by two different mechanisms, *J. Bone Miner. Res.*, 15, 1024, 2000.
30. Jørgensen, N.R. et al., Intercellular calcium signaling occurs between human osteoblasts and osteoclasts and requires activation of osteoclast $P2X_7$ receptors, *J. Biol. Chem.*, 277, 7574, 2002.
31. Yasuda, H. et al., Osteoclast differentiation factor is a ligand for osteoprotegerin/osteoclastogenesis inhibitory factor and is identical to TRANCE/RANKL, *Proc. Natl. Acad. Sci. U.S.A.*, 95, 3597, 1998.
32. Myers, D.E. et al., Expression of functional RANK on mature rat and human osteoclasts, *FEBS Lett.*, 463, 295, 1999.
33. Tsuda, E. et al., Isolation of a novel cytokine from human fibroblasts that specifically inhibits osteoclastogenesis, *Biochem. Biophys. Res. Commun.*, 234, 137, 1997.
34. Horwood, N.J. et al., Osteotropic agents regulate the expression of osteoclast differentiation factor and osteoprotegerin in osteoblastic stromal cells, *Endocrinology*, 139, 4743, 1998.
35. Takai, H. et al., Transforming growth factor–β stimulates the production of osteoprotegerin/osteoclastogenesis inhibitory factor by bone marrow stromal cells, *J. Biol. Chem.*, 273, 27091, 1998.
36. Morrison, M.S. et al., ATP is a potent stimulator of the activation and formation of rodent osteoclasts, *J. Physiol.*, 511, 495, 1998.
37. Hoebertz, A. et al., Extracellular ADP is a powerful osteolytic agent: evidence for signaling through the $P2Y_1$ receptor on bone cells, *FASEB J.*, 15, 1139, 2001.
38. Kon, T. et al., Expression of osteoprotegerin, receptor activator of NF-κB ligand (osteoprotegerin ligand) and related proinflammatory cytokines during fracture healing, *J. Bone Miner. Res.*, 16, 1004, 2001.
39. Gartland, A. et al., Blockade of the pore-forming $P2X_7$ receptor inhibits formation of multinucleated human osteoclasts *in vitro*, *Calcif. Tissue Int.*, 73, 361, 2003.
40. Ferrari, D. et al., Extracellular ATP triggers IL-1 β release by activating the purinergic P2Z receptor of human macrophages, *J. Immunol.*, 159, 1451, 1997.
41. Ferrari, D. et al., ATP-mediated cytotoxicity in microglial cells, *Neuropharmacology*, 36, 1295, 1997.
42. Ke, H.Z. et al., Deletion of the $P2X_7$ nucleotide receptor reveals its regulatory roles in bone formation and resorption, *Mol. Endocrinol.*, 17, 1356, 2003.
43. Li, J. et al., Osteogenesis after mechanical loading requires the $P2X_7$ nucleotide receptor, *J. Bone Miner. Res.*, 19, S9, 2004.
44. Gartland, A. et al., Multinucleated osteoclast formation *in vivo* and *in vitro* by $P2X_7$ receptor–deficient mice, *Crit. Rev. Eukaryot. Gene Expr.*, 13, 243, 2003.
45. Reddy, S.V. et al., Cell biology of Paget's disease, *J. Bone Miner. Res.*, 14, 3, 1999.
46. Ihara, H. et al., ATP-stimulated interleukin-6 synthesis through P2Y receptors on human osteoblasts, *Biochem. Biophys. Res. Commun.*, 326, 329, 2005.
47. Zhang, X.J. et al., Effects of various inducers on the expression of $P2X_7$ receptor in human peripheral blood mononuclear cells, *Sheng Li Xue Bao*, 57, 193, 2005.

48. Shimegi, S., ATP and adenosine act as a mitogen for osteoblast-like cells (MC3T3-E1), *Calcif. Tissue Int.*, 58, 109, 1996.
49. Jones, S.J. et al., Purinergic transmitters inhibit bone formation by cultured osteoblasts, *Bone*, 21, 393, 1997.
50. Hoebertz, A. et al., ATP and UTP at low concentrations strongly inhibit bone formation by osteoblasts: a novel role for the $P2Y_2$ receptor in bone remodeling, *J. Cell Biochem.*, 86, 413, 2002.
51. Kaplan, A.D. et al., Extracellular nucleotides potentiate the cytosolic Ca^{2+}, but not cyclic adenosine 3′,5′–monophosphate response to parathyroid hormone in rat osteoblastic cells, *Endocrinology*, 136, 1674, 1995.
52. Buckley, K.A. et al., Parathyroid hormone potentiates nucleotide-induced $[Ca^{2+}]_i$ release in rat osteoblasts independently of Gq activation or cyclic monophosphate accumulation: a mechanism for localizing systemic responses in bone, *J. Biol. Chem.*, 276, 9565, 2001.
53. Short, A.D. and Taylor, C.T., Parathyroid hormone controls the size of the intracellular Ca^{2+} stores available to receptors linked to inositol trisphosphate formation, *J. Biol. Chem.*, 275, 1807, 2000.
54. Jiménez, A.I. et al., Potentiation of ATP calcium responses by A2B receptor stimulation and other signals coupled to Gs proteins in type-1 cerebellar astrocytes, *Glia*, 26, 119, 1999.
55. Lee, K. et al., Parathyroid hormone induces sequential c-*fos* expression in bone cells *in vivo–in situ* localization of its receptor and c-*fos* messenger ribonucleic-acids, *Endocrinology*, 134, 441, 1994.
56. Stein, G.S. et al., Transcriptional control of osteoblast growth and differentiation, *Physiol. Rev.*, 76, 593, 1996.
57. Bowler, W.B. et al., Signaling in human osteoblasts by extracellular nucleotides. Their weak induction of the c-fos proto-oncogene via Ca^{2+} mobilization is strongly potentiated by a parathyroid hormone/cAMP-dependent protein kinase pathway independently of mitogen-activated protein kinase, *J. Biol. Chem.*, 274, 14315, 1999.
58. Ostrom, R.S., Gregorian, C., and Insel, P.A., Cellular release of and response to ATP as key determinants of the set-point of signal transduction pathways, *J. Biol. Chem.*, 275, 11735, 2000.

4 Role of Purine and Pyrimidine Nucleotides and Nucleosides in Regulation of Cartilage Metabolism and Chondrocyte Function

Hilary P. Benton and Jessica M. Stokes

CONTENTS

4.1 INTRODUCTION

The developmental and physiological roles of bone and cartilage are so closely intertwined that it would be an omission to consider the roles of nucleotides in the regulation of bone formation and resorption without giving some consideration to their roles in cartilage formation and turnover. The focus in this chapter will be on cartilage at the articulating surfaces of long bones, but many of the principles discussed also apply to cartilage in locations such as the intervertebral discs.

Articular cartilage is a highly specialized tissue whose main function is to resist the biomechanical forces exerted during joint movement and to protect the underlying bone. The tissue is made up of a matrix consisting mostly of collagens and aggregating proteoglycan, which provide tensile strength and compressibility. Chondrocytes, the sole resident cells of cartilage, share their mesenchymal stem cell origin with osteoblasts, adipocytes, and muscle cells and bear the responsibility of secreting and maintaining a functional cartilage matrix, a balance that is disturbed in arthritic diseases. Chondrocytes exist as single cells and doublets embedded in an extensive matrix; consequently, they are not under the normal influences of direct cell–cell contact, important as a control mechanism for most cell types. Further unique features of their physiological regulation result from the fact that cartilage is usually both avascular and aneural.[1] For those reasons, chondrocytes experience unique circumstances with respect to nutrient and oxygen availability.[2] This contrasts with the majority of cells, which receive signals directly by close association with blood vessels and nerve endings. Mediators influencing chondrocytes are derived by diffusion from the overlying synovium and synovial fluid, from the underlying bone, or through secretion by the chondrocytes themselves, thus acting in an autocrine or paracrine fashion. There is significant evidence accumulating to demonstrate that chondrocytes possess complex mechanisms to regulate extracellular concentrations of ATP and associated nucleotides and nucleosides.[3] This knowledge, coupled with experimental data demonstrating that extracellular nucleotides and nucleosides influence chondrocyte metabolism, indicates that this class of molecules deserve recognition as extracellular signals influencing chondrocyte physiology and pathology.

4.2 P2 CLASS OF RECEPTORS

It has taken many decades for the P2 class of receptors that bind purine and pyrimidine nucleotides to be recognized as important players in biological and pathological processes.[4,5] Recognition of their roles in cartilage and arthritic diseases is still only in its infancy. Here we will discuss what is already known about these receptors on chondrocytes and draw attention to key areas where future investigation is warranted.

4.2.1 P2 Receptors Identified on Chondrocytes

Most of the studies documenting the presence of P2 receptors on chondrocytes were performed prior to the cloning of the majority of the individual P2X and P2Y subtypes now recognized and also prior to the availability of receptor-specific antibodies. Pharmacological and functional studies have shown chondrocyte responses

TABLE 4.1
Identification of P2 purine receptors on chondrocytes

Species	Model Used	Receptor Identified	Ref.
Bovine	Nasal and articular cartilage explants	P2Y	6, 7, 17
Bovine	Chondrocyte pellet cultures	P2Y	8
Ovine	Chondrocyte monolayer cultures	$P2Y_2$	9
Equine	Acute differentiated chondrocyte cultures	$P2Y_2$	10
Chick	Retinoic acid–treated embryonic chondrocytes	P2Y	11
Chick	Micromass cultures of limb mesenchymal cells	$P2Y_1$ in undifferentiated mesenchymal cells but receptor expression lost upon chondrogenic differentiation	12
Rat	Fetal and adult cartilage slices	P2Y	13
Rat	Growth plate chondrocytes	$P2X_2$	21
Human	Chondrocyte monolayer cultures	P2Y	14–16
Human	Isolated chondrons	$P2Y_2$	20
Human	Acute differentiated chondrocyte cultures	$P2Y_2$	18
Human	Acute differentiated chondrocyte cultures	$P2Y_2$ but not $P2Y_1$ gene expression assessed by RT-PCR	19

to ATP and related purine and pyrimidine nucleotides that are consistent with the presence of calcium-mobilizing P2Y receptors.[6–20] Experiments reported in these studies were performed using chondrocytes derived from a number of different species ranging from chick to human. Furthermore, observations were made using a variety of different tissue culture models, some in which the chondrocytes were fully differentiated and others in which they may have undergone varying degrees of dedifferentiation. These data are summarized in Table 4.1. Many of the studies reported identified the P2Y receptor type present as a $P2Y_2$ rather than a $P2Y_1$. This classification was based on pharmacological properties observed in functional studies, namely an equal activation by ATP and UTP with a significantly smaller or absent response to ADP.

Intracellular calcium concentration in response to P2Y receptor ligands was studied in differentiated sheep,[9] equine,[10] and human chondrocytes,[18,19] as well as in human chondrons cultured in three-dimensional agarose gel.[20] Increased intracellular calcium concentration measured using the fluorescent calcium-binding dye Fura-2 was observed in response to ATP and UTP in both the presence and absence of extracellular calcium, indicating that calcium was released from intracellular stores rather than entering from the extracellular environment. Heterologous desensitization experiments using ATP and UTP showed that ATP was able to completely abolish subsequent responses to UTP and vice versa, consistent with the conclusion that the two nucleotides were competing for the same receptor. The failure to obtain a response to the addition of the second nucleotide in these desensitization experiments was not due to depletion of intracellular calcium stores, because a subsequent calcium response could be obtained to thapsigargin or to unrelated ligands that act via G protein–coupled receptors such as bradykinin and lysophosphatidic acid. In addition, in single-cell imaging experiments, the calcium response to ATP and UTP showed a unique signature of oscillatory patterns when compared to the responses to bradykinin and lysophosphatidic acid.[18] Further evidence that signaling via P2 receptors represents a significant contribution to calcium-mediated signaling in chondrocytes comes from a study of calcium signal propagation in cocultures of a synovial cell line (HIG82) with isolated rabbit articular chondrocytes.[21] This work demonstrated that calcium waves can be propagated between chondrocytes and synovial fluids and pharmacological blocking experiments showed the propagation to be dependent on both gap junction communication and P2 receptor signaling. The P2X family of receptors is virtually unexplored in chondrocytes with the exception of the $P2X_2$ receptor shown to be present in rat growth plate chondrocytes.[22]

The highly specialized relationship between chondrocytes and their surrounding matrix and the inability of chondrocytes to remain differentiated in traditional monolayer culture systems often makes it difficult to assess the physiological importance of observations made *in vitro*. This problem is further confounded by the fact that it is impractical to isolate chondrocytes in sufficient numbers from small laboratory animals. Also of relevance is the fact that cartilage from those models has not been subjected to the same weight-bearing conditions as cartilage from humans and large animals. For this reason, investigators in the field have turned to larger species (bovine, porcine, equine, ovine) for chondrocyte isolation and they have developed a number of cell culture models that allow chondrocytes to maintain their normal three-dimensional structure or allow interaction with native and secreted matrix components. One such model that has been used to study ATP signaling via P2 receptors in chondrocytes involves the use of isolated chondrons[20] in a three-dimensional culture system. The term "chondron" was proposed to refer to the product of a partial cell isolation procedure from cartilage, which allows small groups of chondrocytes to be isolated with their immediate intercellular matrix intact.[23] A study using the chondron model confirmed human chondrocyte responsiveness to $P2Y_2$ receptor ligands with an equal magnitude and duration of intracellular calcium mobilization in response to ATP and UTP. The chondron model showed that cell surface purine receptors can undergo activation when the cells are embedded in their surrounding local matrix. In addition, ATP-induced calcium responses were elicited from explants of both human and porcine cartilage.

As described elsewhere in this volume, the P2Y receptor family has now expanded to at least eight members, but the majority of these have not been studied in chondrocytes. If we are to fully understand how purines regulate chondrocyte function, it will be important to elucidate the full pattern of P2 receptor expression by these cells and establish whether their expression is regulated in development, differentiation, and pathological conditions such as inflammation.

4.2.2 P2 Receptor Function in Cartilage Development

The cellular interactions and signaling pathways involved in cartilage development have been extensively studied to elucidate the complex development of the long bones of the adult skeleton via the process of endochondral ossification. Adhesion molecules, matrix molecules, and a plethora of growth factors contribute to the process.[24] In undifferentiated chick limb mesenchymal cells, the chick $P2Y_1$ ($cP2Y_1$) receptor is strongly expressed; this expression is lost as the cells differentiate.[12] In support of the hypothesis that purines may signal in an autocrine and paracrine manner, chick limb mesenchymal cells were shown to release ATP and to have an enzyme system in place to regulate ATP breakdown. Calcium was released in the immature chick limb bud cells in response to ATP, ADP, and 2-methylthio ATP (2-MeSATP), a synthetic analogue of ATP that activates $P2Y_1$ receptors. Responses to all three ligands were lost upon cellular differentiation. It is very well established that mobilization of intracellular calcium via receptor-mediated pathways plays a significant contribution to the process of cell growth and differentiation in many different tissue systems.[25] Given this established link and the observation that the $cP2Y_1$ expression loss was particularly predominant in cartilaginous regions, Meyer et al. speculated that the $P2Y_1$ receptor could be involved in the process of cartilage differentiation.[12] They tested this hypothesis using chick limb bud cells that had been maintained in culture for more than 24 h and had lost responsiveness to ATP, ADP, and 2-MeSATP. Transfection of these cells with a $cP2Y_1$ expression construct restored responsiveness to ATP, 2-MeSATP, and, to a lesser extent, ADP. Responsiveness in the transfected cells was maintained for periods of at least 3 days. The $cP2Y_1$–transfected cells showed a reduced amount of cartilage formation compared to cells transfected with a control construct lacking the $cP2Y_1$ cDNA. Cartilage formation was quantified by assessing the number of alcian blue staining nodules present in the cultures. Meyer et al. concluded that $cP2Y_1$ plays a role in maintaining the chick limb bud cells in an undifferentiated state.[12]

In contrast, using chick sternal chondrocytes from day 14 chick embryos,[11] immature caudal chondrocytes were unresponsive to ATP in untreated cultures. However, after retinoic acid treatment, calcium transients were elicited by ATP and a smaller but significant response was elicited by ADP. Retinoic acid causes maturation of chick cephalic chondrocytes and expression of hypertrophic chondrocyte markers.[26,27]

In a study designed to compare the reparative capacity of adult and fetal cartilage, superficial defects were created in the femoral knee cartilage of rats. In accord with a multitude of studies in different species, there were dramatic differences in the response to cartilage injury in fetal rats compared to adult rats, with a much greater capacity for repair by the immature tissue. One difference that was noted in the fetal tissue when compared to the adult model was a transient expression of *c-fos* immediately after injury in the fetal chondrocytes that surrounded the defect that was absent in the adult cartilage. Expression of *c-fos* was dependent on elevated intracellular calcium. This led to an investigation of the ability of fetal versus adult chondrocytes to mobilize calcium in response to calcium ionophore and ATP. Calcium transients were elicited by ATP (1 μM) in fetal cartilage slices and *c-fos* induction was observed in fetal cartilage slices, but a 1000-fold greater concentration of ATP (1 mM) was required to elicit calcium transients and *c-fos* induction in adult cartilage slices. The authors suggested that the higher cellularity of fetal cartilage may make paracrine communication more efficient than in the less cellular adult tissue. This is an interesting idea with respect to understanding how paracrine signals via purine receptors may be important in cartilage development and ultimately may be regulatory factors involved in generating the differences in tissue architecture that result in the inability of adult cartilage to undergo repair.

In tibial growth plate chondrocytes from sheep fetuses, ATP and UTP were shown to potentiate basic fibroblast growth factor–induced proliferation. Chondrocytes were obtained from fetuses between days 120 to 130 of gestation and used between passages two and ten. ATP has been implicated as a modulator of cell proliferation in many cell types, either alone or in combination with various growth factors.[28–33] In this study and in a study using chondrocytes isolated from day 19 chick embryo sternae,[34] chondrocyte proliferation was not influenced by addition of ATP alone. The potentiation of basic fibroblast growth factor–stimulated proliferation by ATP and UTP indicates that extracellular purine and pyrimidine nucleotides may influence chondrocyte proliferation by cross talk between P2 receptor signaling pathways and growth factor–stimulated signaling cascades.

4.2.3 P2 Receptor Function in Adult Chondrocytes

Given the well-established roles of P2 receptors in many organ systems,[35,36] relatively little is known regarding their role in adult cartilage. Due to the unique environment of chondrocytes within the cartilage matrix, a large body of information has been collected regarding the role of anabolic and catabolic growth factors[37–40] and cytokines[41–44] in regulating cartilage matrix homeostasis. Much less is known about small G protein–coupled receptors on chondrocytes and the effects of their activation. Small ligands, such as purines, may be very important in the fine-tuning of chondrocyte metabolism and modulation of cytokine and growth factor action. Hopefully, the next few years will see significant attention paid to developing this area of investigation. The information that has been collected regarding the functional effects P2 receptor activation in chondrocytes is summarized in Table 4.2.

4.2.3.1 P2 Receptors and Cartilage Matrix Turnover

Direct effects of ATP on cartilage matrix collagen and proteoglycan synthesis and resorption appear to be variable, depending on the cell model used and the duration of treatment. Brown and colleagues[6] showed that various proteinase inhibitors could block the activity of ATP in releasing proteoglycan from bovine nasal cartilage explants. This is the only report demonstrating the effects of ATP on cartilage matrix degrading enzymes. As described in section 4.2.1, P2 receptor ligands are known to mobilize intracellular calcium from endoplasmic reticular stores.[9,10,18–20] Experiments using the intracellular calcium chelator *bis*-(aminophenoxy)ethane-tetraacetic acid acetoxymethyl (BAPTA-AM) demonstrate suppression of the biomechanical induction of elevated aggrecan gene expression, indicating potential involvement of calcium signaling in mediating this response.[45,46] In bovine cartilage explants, buffering intracellular calcium suppressed the activation of aggrecan, type II collagen, link protein, c-Jun, and a number of metalloproteinase genes induced by mechanical compression.[46] These data imply that each of these genes may be sensitive to regulation by ligands that activate calcium-mobilizing receptors, such as the P2Y receptors, or calcium channels, such as the P2X receptors.

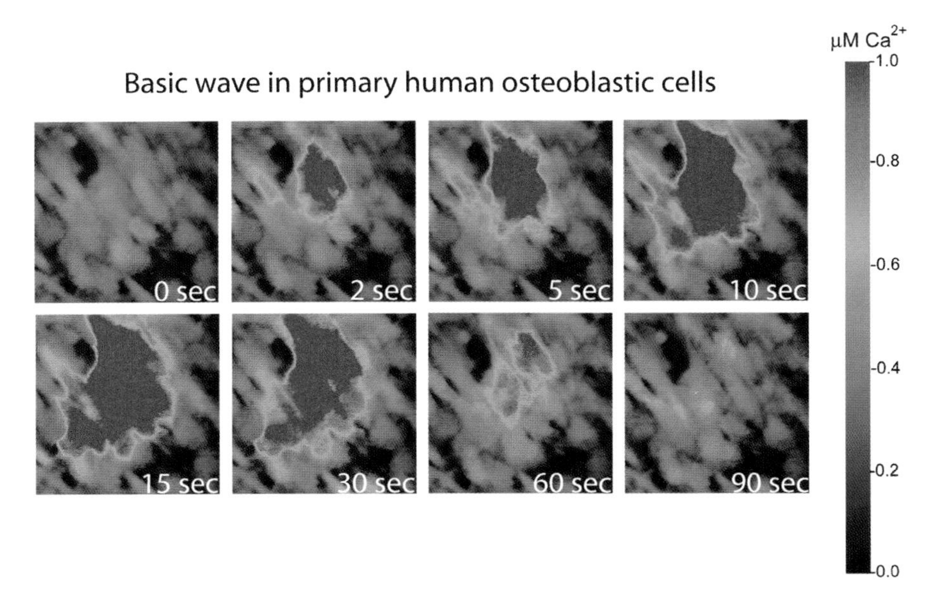

FIGURE 1.2 Intercellular calcium waves in human primary osteoblastic cells. A single cell is stimulated mechanically, causing an increase in the intracellular calcium concentration that subsequently propagates to adjacent cells. Numbers indicate seconds after mechanical stimulation. Scale bar indicates relation between pseudocolors and intracellular calcium concentrations.

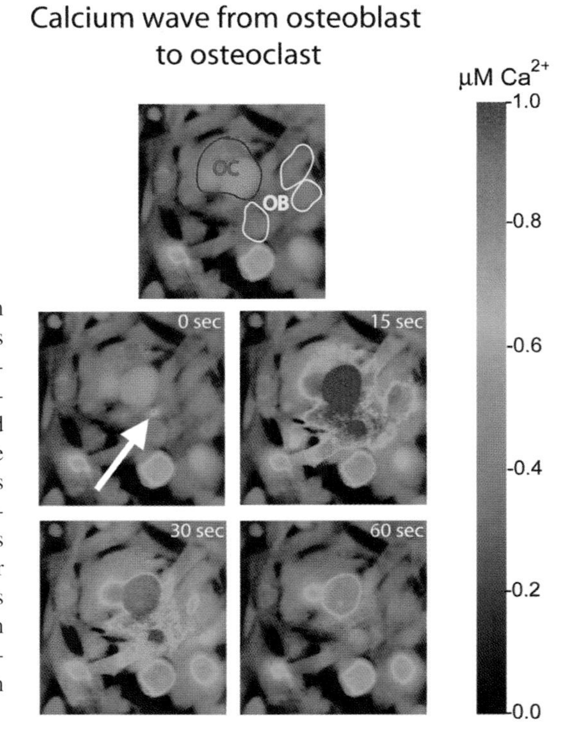

FIGURE 1.3 Intercellular calcium signaling occurs between osteoblasts and osteoclasts in coculture monolayers. Cocultures of human osteoblasts and osteoclasts were loaded with fura-2, single osteoblasts were mechanically stimulated with a glass pipette, and ratio imaging was performed. Numbers on the pictures indicate time in seconds after mechanical stimulation. Osteoblasts (OB) and osteoclasts (OC) are shown by arrows. Scale bar shows intracellular calcium concentrations in micromoles.

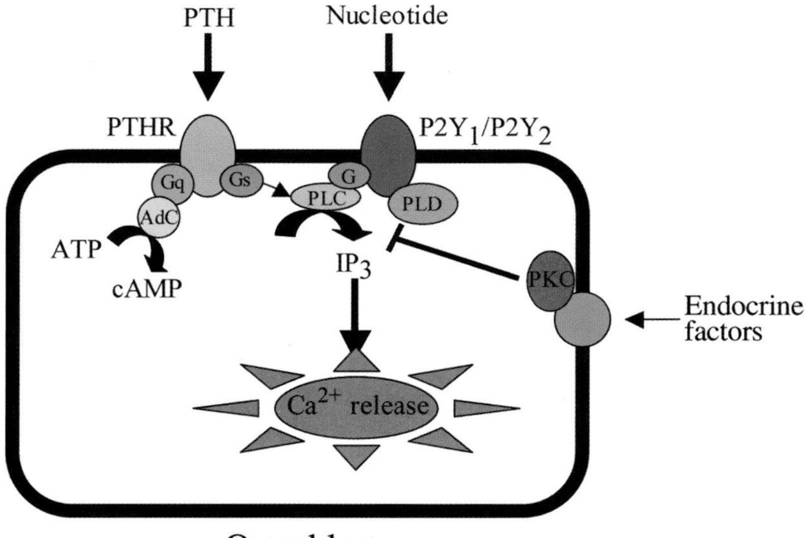

FIGURE 1.4 Shows the interacting pathways of parathyroid hormone (PTH) and purinergic $P2Y_1$ and $P2Y_2$ receptors. PTH acts via the PTH receptor (PTHR), which is a G protein–coupled receptor. Upon ligand binding, G_q and adenylate cyclase (AdC) are activated, inducing conversion of adenosine triphosphate (ATP) to cyclic adenosine monophosphate (cAMP). PTH binding also activates G_s and the phospholipase C (PLC) pathway, which is sensitized by nucleotide binding to P2Y receptors, enhancing PTH-induced inositol triphosphate (IP_3) generation and subsequent release of calcium from intracellular stores. Endocrine and paracrine factors like estradiol activates protein kinase C (PKC), which inhibits signal transduction from P2Y receptors and downstream. See text for further details.

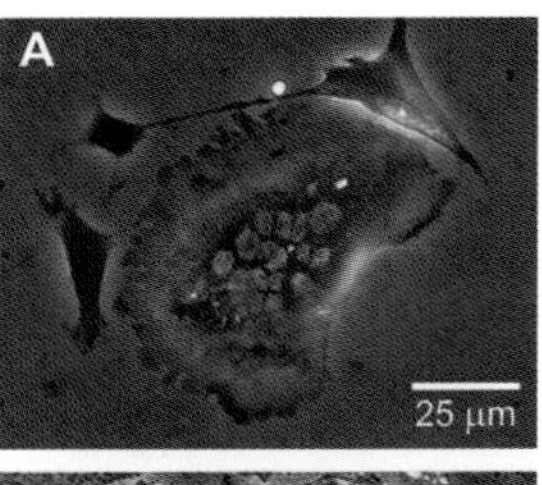

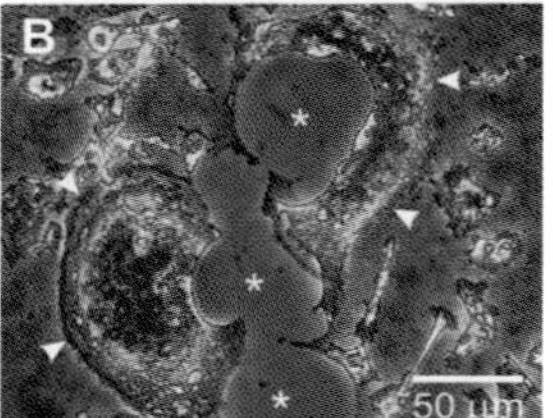

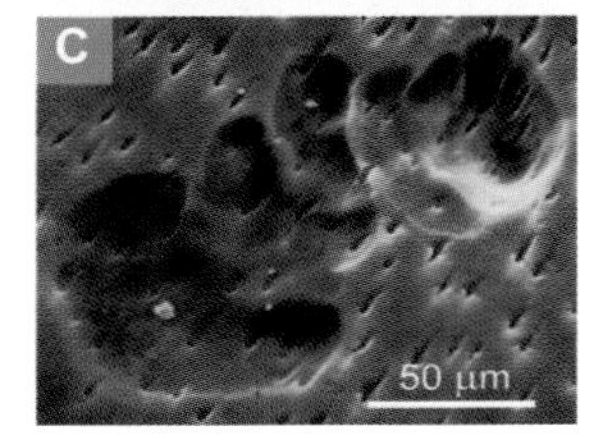

FIGURE 2.1 Morphology and resorptive activity of isolated osteoclasts. Osteoclasts were isolated from bone marrow of neonatal rats and rabbits. Shown are (A) A micrograph of a representative multinucleated rat osteoclast visualized by phase-contrast microscopy. (B) Isolated rabbit osteoclasts were plated on hydroxyapatite-coated discs. Two days following isolation, osteoclasts were washed, fixed, and stained for the osteoclast marker tartrate-resistant acid phosphatase (TRAP). TRAP-stained (maroon) rabbit osteoclasts (arrowheads) are associated with resorbed areas (asterisks). (C) A scanning electron micrograph of dentin slice reveals a resorption trail left behind by an osteoclast while resorbing matrix. Isolated rabbit osteoclasts were cultured for 48 h on dentin slices made from human teeth. Cells were removed, and slices were then examined using scanning electron microscopy.

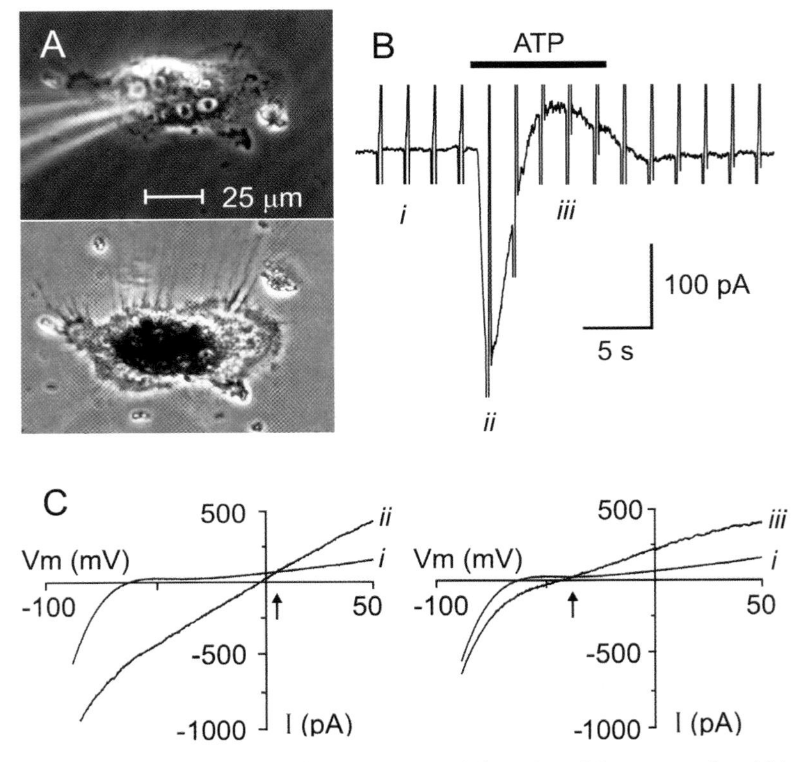

FIGURE 2.3 Extracellular ATP activates P2X and Ca^{2+}-activated K^+ currents in rabbit osteoclasts. (A) Phase-contrast photomicrograph of a multinucleated rabbit osteoclast (upper panel) used for electrophysiological studies in B and C. After the experiment, the cell was stained for TRAP, confirming its identity as an osteoclast (lower panel). (B) Whole-cell currents were recorded at –30 mV. ATP (100 μM) applied at the time indicated by the bar above the trace evoked an initial inward current (ii) that declined and was followed by a transient outward current (iii). To identify ATP-activated currents, voltage-ramp commands were used to obtain current–voltage (I-V) relationships. (C) Under control conditions, the I-V relationship showed an inwardly rectifying K^+ conductance (i). The initial ATP-activated current (ii) reversed near 0 mV (left panel). Rapid activation of this current is consistent with a ligand-gated P2X channel. An I-V relationship from the same cell shows the later part of the ATP-activated current (iii) reversing at more negative membrane potential (right panel), consistent with a K^+-selective current. Delayed activation of K^+ current suggests the involvement of a signaling cascade, consistent with activation of Ca^{2+}-dependent K^+ channels by P2Y receptor–mediated release of Ca^{2+} from intracellular stores. (Reprinted from Naemsch, L.N. et al., $P2X_4$ purinoceptors mediate an ATP-activated, non-selective cation current in rabbit osteoclasts, *J. Cell Sci.*, 112, 4425, 1999. With permission.)

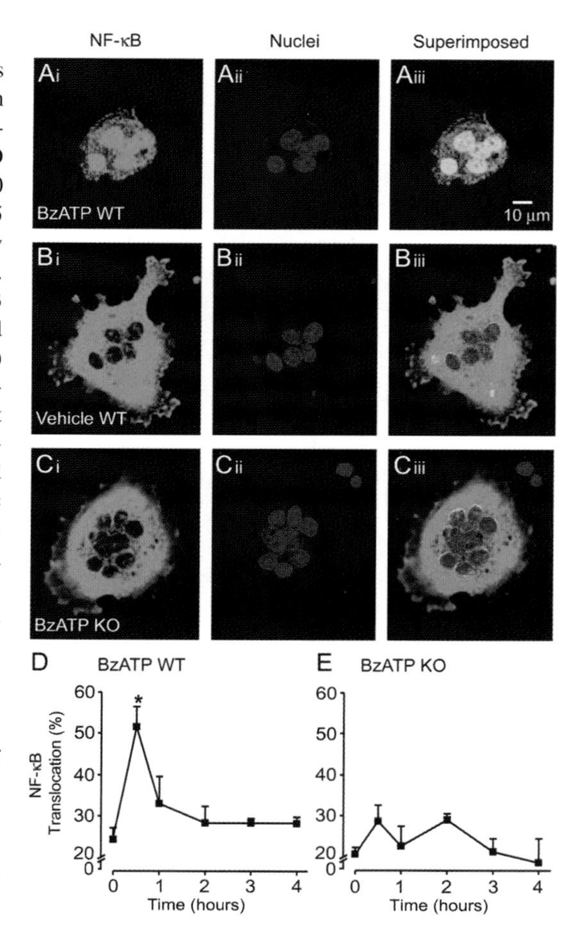

FIGURE 2.7 BzATP-induced nuclear translocation of NF-κB is mediated by the $P2X_7$ receptor in murine osteoclasts. Osteoclasts isolated from WT and $P2X_7$ receptor KO mice were treated with BzATP (300 μM) or vehicle for 0 to 4 h. The p65 subunit of NF-κB was visualized by immunofluorescence (green, left). Nuclei were stained with TOTO-3 (red, middle) and superimposed images are shown at right. A) BzATP-treated WT osteoclast exhibited nuclear localization of NF-κB at 30 min. B) and C) In contrast, vehicle-treated WT and BzATP-treated KO osteoclasts showed cytoplasmic localization of NF-κB. D) WT osteoclasts treated with BzATP exhibited a significant increase in nuclear translocation of NF-κB at 30 min, compared to time 0 (* $p < .05$). E) In contrast, KO osteoclasts did not show a significant change in NF-κB translocation at any time point following BzATP treatment. Data are the percentage of osteoclasts with nuclear localization of NF-κB. (Reproduced from Korcok, J. et al., Extracellular nucleotides act through $P2X_7$ receptors to activate NF-κB in osteoclasts, *J. Bone Miner. Res.*, 19, 642, 2004. With permission of the American Society for Bone and Mineral Research.)

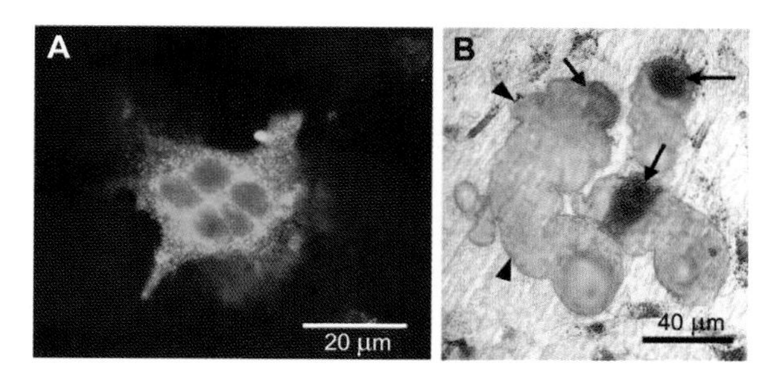

FIGURE 2.12 Activation of $P2Y_1$ receptor on rat osteoclasts stimulates bone resorption. (A) Immunofluorescence shows localization of $P2Y_1$ receptor on an isolated rat osteoclast. (B) A representative photomicrograph of a 26-h culture of rat long bone cells on dentin discs shows ADP-stimulated, TRAP-stained multinucleated osteoclasts (arrows) with corresponding resorption pits (arrowheads), visualized by reflective light microscopy. (Modified from Hoebertz, A. et al., Extracellular ADP is a powerful osteolytic agent: evidence for signaling through the $P2Y_1$ receptor on bone cells, *FASEB J.*, 15, 1139, 2001. With permission.)

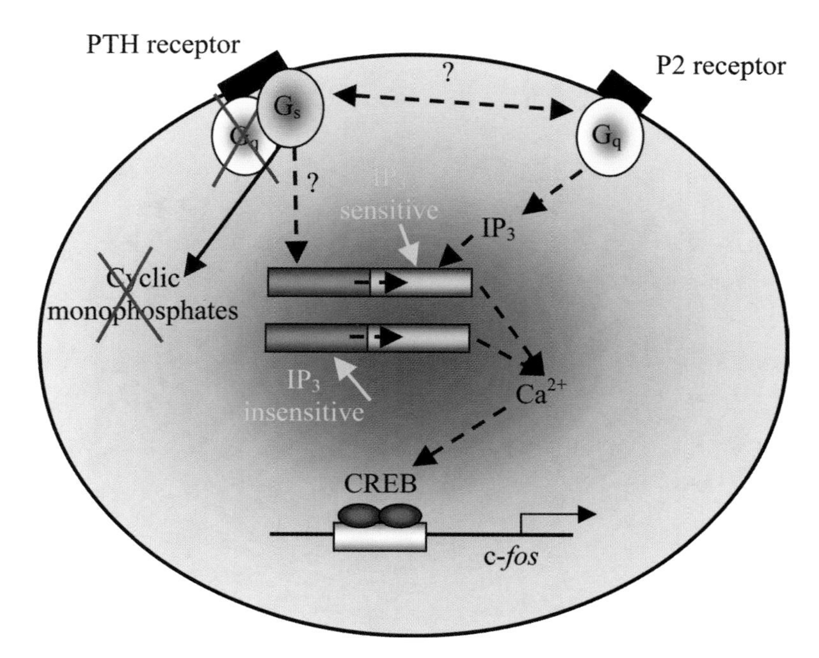

FIGURE 3.4 Possible mechanisms leading to potentiated $[Ca^{2+}]_i$ release and synergistic gene expression following nucleotide and PTH costimulation of osteoblasts. Co-activation of P2 and PTH G protein–coupled receptors leads to potentiated $[Ca^{2+}]_i$ release, which may be due to shuttling of calcium between intracellular stores or to a direct G protein subunit interaction.

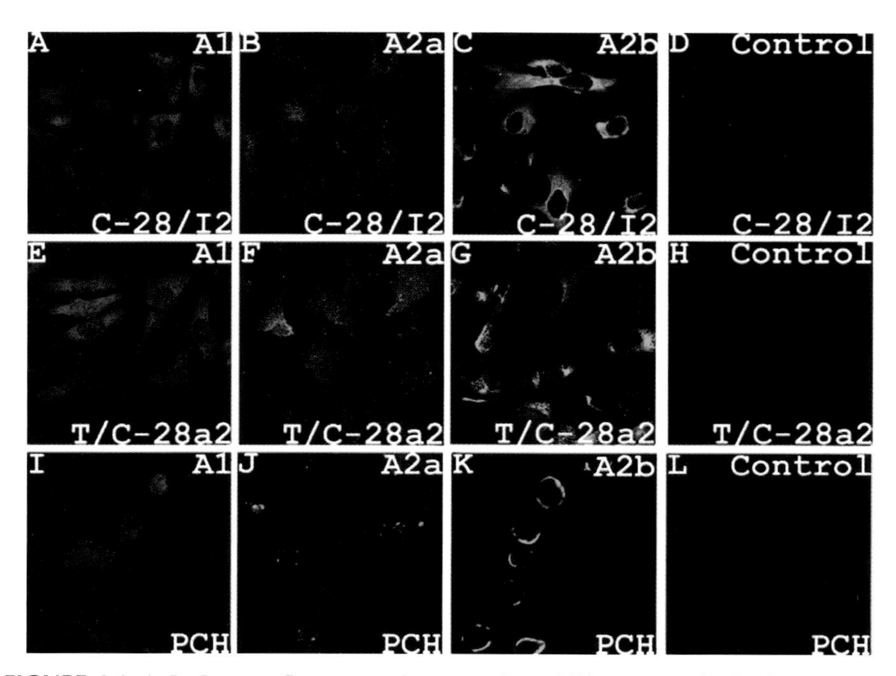

FIGURE 4.1 A–L: Immunofluorescent demonstration of P1 receptors in the human costochondral cell lines C28-I2 and T/C-28-a2 and in primary differentiated human chondrocytes (PCH). Note that the differentiated primary cells are still in a rounded state.

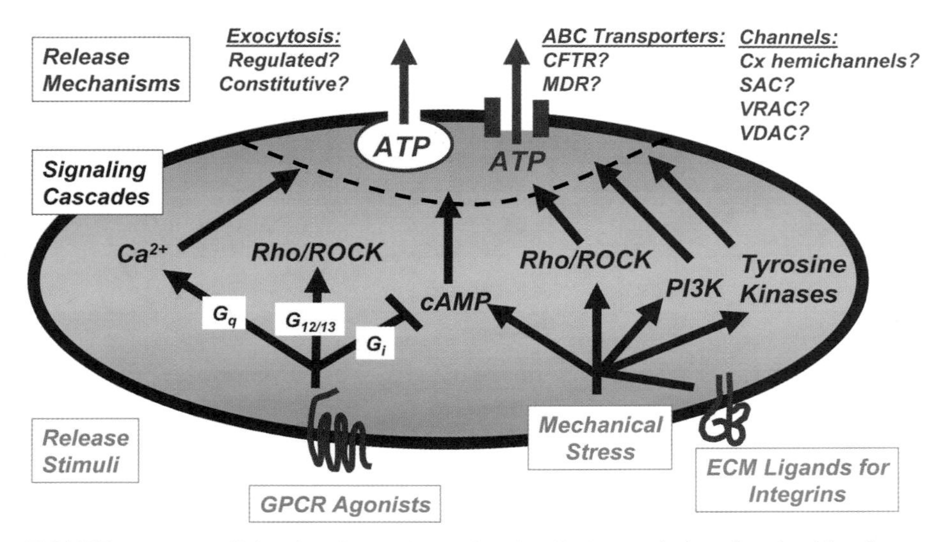

FIGURE 5.7 Intracellular signaling pathways involved in the regulation of nucleotide release. Mechanisms for release of ATP (and other nucleotides) can include constitutive or regulated exocytosis or nucleotides concentrated within intracellular granules or vesicles. Alternatively, cytosolic ATP (or other cytosolic nucleotides) can flow down its concentration gradient across the plasma membrane during the transient activation of various nucleotide-permeable channels or transporters. These ATP-conductive transport mechanisms may include ABC family transporters such as the cystic fibrosis transmembrane regulator (CFTR) anion channel or various multidrug-resistance (MDR) transporters. Other ATP-conductive pathways that have been studied include connexin (Cx) hemichannels, stretch-activated anion channels (SAC), volume-regulated anion channels (VRAC), and plasma membrane variants of the mitochondrial voltage-dependent anion channels (VDAC). Extrinsic stimuli for increased ATP release may include: (1) neurotransmitters, hormones, or local mediators that target diverse G protein–coupled receptors (GPCR); (2) different types of mechanical stress (see Figure 5.3); and (3) perturbed interaction (possibly via mechanical stress) of cell surface integrins with components of the extracellular matrix (ECM). Most of these extrinsic stimuli are coupled to exocytotic or conductive ATP release via intracellular signaling pathways based on transient or sustained increases in cytosolic [Ca^{2+}]. However, a growing body of data suggests that the central Ca^{2+}-based regulatory pathway acts synergistically with other signaling pathways that—depending on tissue or cell type—may include rho family GTPases, rho-operated protein kinases (ROCK), phosphatidylinositol-3-kinases (PI3K), and protein tyrosine kinases.

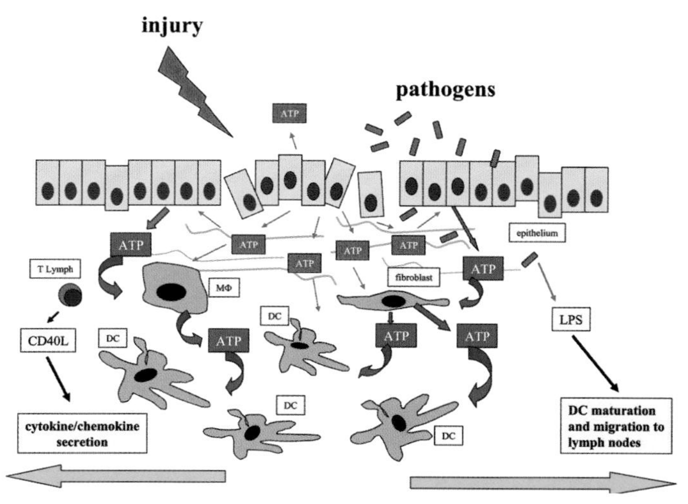

FIGURE 6.1 Cell damage caused by physical trauma or pathogens causes an initial wave of ATP release (red) from injured cells. This first wave of ATP release diffuses through the pericellular space and activates adjacent cells (e.g., epithelial cells, macrophages, fibroblasts). P2 receptors expressed on the plasma membrane of the bystander cells will be activated and trigger the release of several bioactive agents, among which ATP itself (ATP-induced ATP release, dark blue). Accumulation of ATP into the extracellular milieu will prime Langerhans/dendritic cells (DC) and modify their responses to bacterial endotoxin (LPS) or T lymphocyte–derived CD40 ligand (CD40L). The combined action of these agents drives DC maturation and the secretion of cytokines and chemokines. (Reprinted from Di Virgilio, F., *Purinergic Signalling*, in press. With permission of Springer Science and Business Media.)

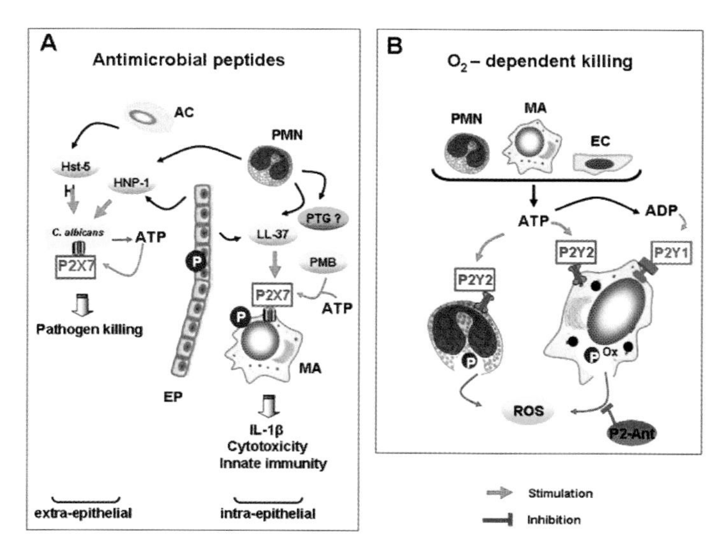

FIGURE 6.2 Schematic representation of the potential implication of P2 receptors in extracellular killing of pathogens. (A) P2 receptors and antimicrobial peptides. (B) P2 receptors and O_2-dependent killing. AC = salivary acinar cells; P = pathogen; EP = epithelium; PMN = neutrophil; MA = monocyte/macrophage; EC = endothelial cells; Ox = NADPH oxidase; P2-Ant = P2 antagonist; Hst-5 = Histatin-5; HNP-1 = alpha-defensin 1; PTG = protegrin; LL-37 = human cathelicidin peptide; ROS = reactive oxygen species.

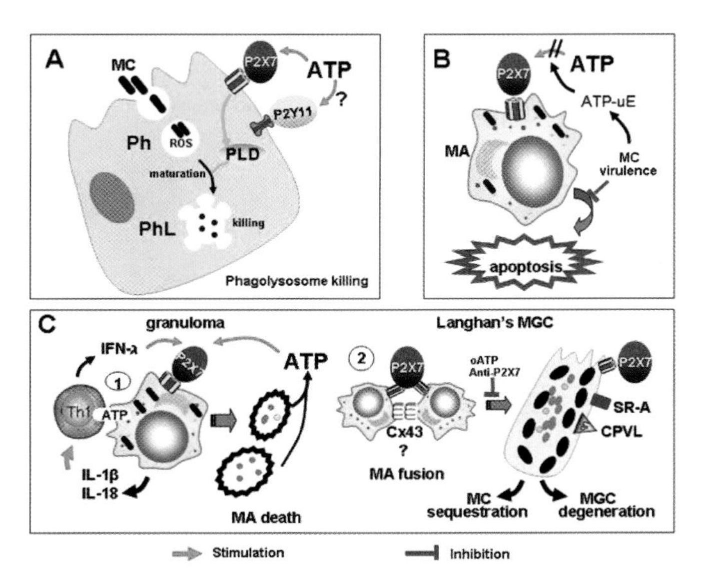

FIGURE 6.3 Schematic diagram illustrating the potential role of P2 receptors in intracellular killing of pathogens. (A) P2 receptors and endocytic phagolysosome killing. MC = mycobacteria; Ph = phagosome; PhL - phagolysosome. (B) $P2X_7$-mediated apoptosis. ATP-uE = ATP-utilizing enzymes. (C) Infective granulomas and multinucleated giant cells. (1) ATP-mediated signaling in macrophage–lymphocyte coordinated activation of mycobactericidal activity. (2) $P2X_7$ receptor and MGC formation. Cx-43 = connexin-43; SR-A = type-A scavenger receptor; CPVL = carboxypeptidase-related vitellogenin-like molecule; MA = macrophage; oATP = oxidized ATP; Th1 = T-helper 1; PLD = phospholipase D.

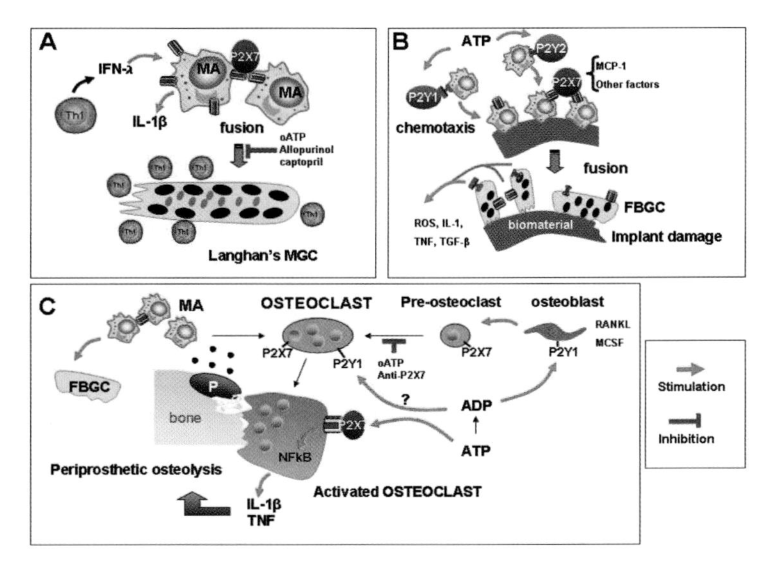

FIGURE 6.4 General scheme illustrating the potential involvement of P2 receptors in the elimination of noninfectious agents. (A) Sarcoidosis and Langhan's MGC. (B) Foreign body reaction to implanted biomaterials and FBGC. (C) Periprosthetic bone destruction and osteoclasts. P = prosthesis; MCP-1 = monocyte chemoattractant protein–1; FBGC = foreign body giant cell.

TABLE 4.2
Documented functions of P2 receptors in adult chondrocytes

Species	Model Used	Function	Ref.
Human	Chondrocyte monolayer cultures	Prostaglandin release and potentiation of IL-1– and TNFα–induced prostaglandin release and stimulation of cartilage resorption	14–16
Human	Acute differentiated chondrocyte cultures	Intracellular calcium mobilization, prostaglandin release, and potentiation of IL-1–induced prostaglandin release	19
Bovine	Nasal cartilage explants	Proteoglycan release and aggrecanase release	6, 7, 17
Bovine	Articular cartilage explants	Increased proteoglycan synthesis	6
Bovine	Chondrocyte pellet cultures	Stimulation of proteoglycan and collagen accumulation	8
Equine	Acute differentiated chondrocyte cultures	Intracellular calcium mobilization and prostaglandin release	10
Rabbit	Chondrocyte monolayer cultures	Prostaglandin and arachidonic acid release. Activation of cytoplasmic phospholipase A2 via ERK1/2 and p38 MAPK signaling	63

The regulation of extracellular matrix genes and associated matrix proteins, degrading enzymes, and receptors by P2 receptors remains almost entirely unexplored, with the exception of a clear demonstration of a role for $P2Y_1$ and $P2Y_{12}$ receptors in platelets. Stimulation of these receptors has been shown to be involved in activation of the platelet collagen receptor integrin $\alpha_2\beta_1$, a key component in the process of platelet activation.[47–53] The $P2X_1$ receptor calcium channel has also been implicated in this response.[54]

4.2.3.2 P2 Receptors and Eicosanoid Synthesis and Release

One observation that has been verified in a number of chondrocyte systems, including chondrocytes from human cartilage, is that ATP and UTP are capable of stimulating prostaglandin release. ATP and UTP also potentiate the prostaglandin release that is stimulated by inflammatory cytokines such as interleukin 1 (IL-1) and tumor necrosis factor α (TNFα) (see Table 4.2). ATP has been shown to regulate eicosanoid production in other systems, including the demonstrated stimulation of cyclo-oxygenase 2 (COX-2) expression and prostaglandin E_2 (PGE_2) release in astrocytes[55] and stimulation of prostaglandin release by rat inner medullary collecting duct preparations.[56] In MC3T3-E1 osteoblast-like cells, prostaglandin release has been shown to be stimulated by the ATP that is released in response to fluid shear.[57] This latter study is complementary to the studies showing biomechanical regulation of ATP release in chondrocytes. This emphasizes the importance of considering mechanisms controlling ligand availability when assessing how P2 receptors influence physiological and pathological processes.

A healthy cartilage matrix is dependent on an extensively regulated balance between anabolic and catabolic pathways. The effects of prostaglandins on this pathway are complex.[58,59] Both anabolic and catabolic effects of prostaglandins on cartilage have been documented and are shown to depend on the other matrix-influencing factors within the microenvironment. In experiments using mice deficient in the prostaglandin EP4 receptor, induction of collagen antibody–induced arthritis was significantly suppressed when compared to wild-type controls.[60] In general, the balance tends toward a pro-inflammatory, catabolic action of prostaglandins in the articular joint.[61,62] These observations led to the extensive use of COX-2 inhibitors in osteoarthritis. The adverse side effects that have arisen using this class of therapeutic agents bring into focus the need to continue the search for novel approaches to modulate prostaglandin pathways.

The mechanism by which ATP stimulates prostaglandin release has been explored in some detail.[19,63] The ATP-stimulated release of PGE_2 from rabbit chondrocytes in monolayer culture was shown to occur rapidly, reaching a peak by one hour after initiation of ATP treatment. Stimulation of chondrocytes by ATP resulted in the activation of an extracellular-regulated protein kinase (ERK) known as ERK1/2, and the mitogen-activated protein kinase (MAPK) known as p38 MAPK, as well as the mobilization of intracellular calcium. These cascades led to activation of cytosolic phospholipase A_2 ($cPLA_2$) but not activation of the secreted form of phospholipase A_2. This is consistent with a number of published reports that have shown G protein–coupled calcium-mobilizing receptors to be involved in the activation of $cPLA_2$, leading to the rapid production of prostaglandins.[64–66] The demonstration of the involvement of the p38 MAPK in the action of ATP on cartilage is of particular interest. This kinase is proving to be a major stimulant of cytokine-induced inflammation and it is a current target for new therapeutic approaches for the treatment of arthritis and other inflammatory conditions.[67–69]

The synergy seen between ATP-induced prostaglandin release and inflammatory cytokines[15,16,19] should also be noted, due to potential implications for ATP as a paracrine and autocrine regulator in cartilage. In differentiated human articular chondrocytes, agents that raise cytosolic calcium such as thapsigargin, cyclopiazonic acid, and the calcium ionophore A23187 did not show synergistic effects with IL-1 in stimulating prostaglandin release.[19] However, synergistic effects upon release of prostaglandin were seen in response to co-addition of IL-1 and phorbol myristate acetate, an activator of protein kinase C.

In summary, the role of ATP in stimulating prostaglandin release via activation of $P2Y_2$ receptors is the most studied function of ATP in chondrocytes and provides strong evidence that purine receptors can stimulate inflammatory pathways in cartilage.

4.2.3.3 P2 Receptors and Cytokine Synthesis and Release

The work described in section 4.2.3.2 used chondrocytes to demonstrate synergy between exogenous ATP and inflammatory cytokines in stimulating eicosanoid release, but data have not been collected to determine whether ATP and other P2 receptor ligands can directly stimulate cytokine release from chondrocytes. Evidence from other cell systems has indicated that P2 receptor stimulation is capable of influencing cellular cytokine release. ATP stimulates IL-6 production in osteoblasts,[70] thyrocytes,[71] and macrophages[72] and also stimulates IL-8 release from eosinophils.[73] Secretion of IL-2 and interferon γ (IFNγ) from T cells appears to require extracellular ATP.[74,75] In contrast, exogenous ATP was found to inhibit pro-inflammatory cytokines in whole blood and in dendritic cells, potentially via a $P2Y_{11}$ receptor.[76,77] ATP regulation of several cytokines, including IL-4, IL-10, and IL-1, has been proposed to occur in an autocrine manner in different cell and tissue systems.[78–80] Probably the best developed story with respect to cytokines and P2 receptors is the involvement of the $P2X_7$ receptor in the release of multiple cytokines including IL-1 and the IL-1 receptor antagonist (IL-1ra).[81–85] In many of these cases there seems to be an absolute requirement for the $P2X_7$ receptor. Given the importance of IL-1 autocrine regulation within the cartilage matrix, this may be a particularly significant receptor that warrants future investigation in chondrocytes.

4.3 P1 CLASS OF RECEPTORS

In comparison to the very large families of P2Y and P2X receptors, the P1 family of receptors, which primarily recognizes adenosine, seems simple with only four identified receptors. However, the coexistence of the multiple receptor types on the same cells, the presence of alternatively spliced forms of the receptors, and factors that alter the relative expression of each receptor type make analysis of functional pathways following P1 receptor stimulation far from simple. The four cloned subtypes in the P1 receptor family are termed A_1, A_{2A}, A_{2B}, and A_3.[86–88] All four receptor subtypes are G protein–coupled members of the seven-transmembrane superfamily of receptors and share 50 to 70% homology in their membrane-spanning domains.[86–88] Specific adenosine receptor agonists and antagonists have been identified and their therapeutic potential has been recognized over the last decade.[87,89] Clinically important applications have been identified in many medical disciplines including cardiovascular medicine,[90–92] asthma,[93] inflammatory diseases,[94,95] and septic shock.[96] All the adenosine receptors are linked to cyclic adenosine monophosphate (cAMP)-dependent intracellular pathways.[87,88] A_{2A} and A_{2B} receptor activation results in an increase in cAMP, and A_1 and A_3 receptor activation results in a decrease in cAMP. There is also evidence that inosine is a potent agonist for the A_3 receptor.[97–99]

4.3.1 P1 Receptors Identified on Chondrocytes

Reverse transcriptase polymerase chain reaction (RT-PCR) identified A_{2A} and A_{2B} receptor transcripts using cDNA obtained from freshly isolated differentiated human chondrocytes and from passaged cultured human chondrocytes obtained from patients undergoing joint replacement surgery.[19] In the same series of experiments, the A_1 and A_3 receptor transcripts were not found.[19] Using the more sensitive approach of Taqman quantitative real-time PCR, we have recently confirmed expression of the A_{2A} and A_{2B} genes by human chondrocytes and also consistently found

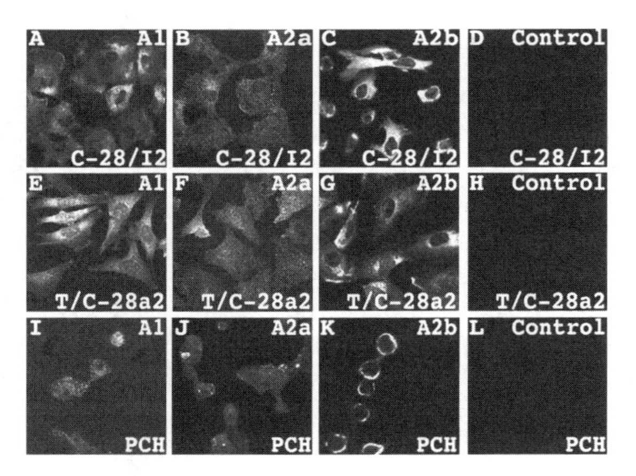

FIGURE 4.1 (See color insert following page 80) A–L: Immunofluorescent demonstration of P1 receptors in the human costochondral cell lines C28-I2 and T/C-28-a2 and in primary differentiated human chondrocytes (PCH). Note that the differentiated primary cells are still in a rounded state.

a low level of transcription of the A_1 receptor gene. No expression of the A_3 receptor gene was seen. This same pattern of gene expression was seen in osteoarthritic chondrocytes isolated from tissue removed during joint replacement surgery as well as in the T/C-28a2 and C-28/I2 human costochondral chondrocyte cell lines. Using receptor-specific antibodies, at the protein level (AbCam, Cambridge, U.K.) we identified A_1, A_{2A} and A_{2B} receptors on the surface of human osteoarthritic chondrocytes and T/C-28a2 and C-28/I2 cells (Figure 4.1). There is a disparity in the correlation between the A_1 receptor expression seen at the protein level compared to the receptor gene transcript level. There are several potential explanations for this disparity that we are currently investigating including posttranscriptional regulation or expression of an alternatively spliced form of the receptor. The human A_1 receptor transcript can be alternatively spliced to regulate cell surface receptor expression in other cell systems.[100] The demonstration of adenosine receptor expression by the immortalized human juvenile costochondral chondrocyte lines T/C-28a2 and C-28/I2 suggests that these cells may offer a reproducible, reliable model for mechanistic studies of adenosine receptor function in chondrocytes.

These are SV40 immortalized cells that have been developed specifically for the purpose of studying cartilage-specific gene regulation.[101] SV40 immortalization stabilizes proliferative capacity, not phenotype. However, if proliferation is slowed in these cell lines by incubation in serum-free conditions with a serum substitute for one to three days, cartilage matrix genes are expressed. Alternatively, cartilage matrix-specific gene expression may be studied by transfecting the cells with cDNA or the promoter of interest. These cell lines have been used extensively in published experiments for the study of regulation of cartilage matrix gene expression,[102–106] matrix metalloproteinase expression,[107–109] and cyclo-oxygenase expression.[110,111] Adenosine receptor expression has been demonstrated in the T/C-28a2 and C-28/I2 chondrocyte cell lines; these lines have not yet been used to investigate adenosine function in chondrocytes.

4.3.2 P1 Receptor Function in Chondrocytes

Extracellular adenosine has been shown to initiate anti-inflammatory responses in a wide range of cells,[94,95] a role that may be particularly important in preventing articular cartilage damage. The attenuation of inflammatory activity by adenosine has been particularly well studied in immune cells.[112–115] Much of this activity is attributed to modulation of cytokine release[116–120] but actions on the synthesis of arachidonic acid metabolites,[121–124] superoxide generation,[125] and adhesion proteins have been reported, which may or may not be secondary to an altered cytokine profile. In chondrocytes direct actions of adenosine on cytokine release has not been demonstrated but adenosine and P1 receptor agonists have been shown to attenuate other mediators known to be key players in cartilage biology; namely prostaglandins and nitric oxide. In isolated equine articular chondrocytes adenosine alone had no significant effect on nitric oxide release but both adenosine and the adenosine receptor agonist 5′–N-ethylcarboxyaminoadenosine (NECA) inhibited bacterial lipopolysaccharide (LPS) and IL-1–mediated nitric oxide release[126] via inhibition of inducible nitric oxide synthase activity. ATP but not UTP showed a small but significant inhibition of LPS-stimulated nitric oxide release after a 24-h incubation. This was most likely due to rephosphorylation forming extracellular adenosine. Using specific receptor agonists, the attenuation of chondrocyte nitric oxide release was found to occur via the A_{2A} receptor and to be mimicked by other agents that increase intracellular cAMP accumulation.[127]

In the extracellular environment adenosine is rapidly degraded to form inosine. Incubation of chondrocytes with an adenosine deaminase inhibitor, 9-(2-hydroxy-3-nonyl)adenine hydrochloride (EHNA), indicated that protection of adenosine against degradation resulted in a decrease in LPS-stimulated nitric oxide release in both the presence and absence of exogenously added adenosine. This led to the hypothesis that endogenously produced adenosine may regulate chondrocyte activity in an autocrine manner. Indeed, chondrocytes were shown to accumulate physiologically relevant concentrations of adenosine by inhibition of extracellular breakdown using EHNA and by inhibition of adenosine phosphorylation to form ATP by use of the adenosine kinase inhibitor 5′–iodotubericidin (ITU).[128,129] Evidence that adenosine may play a role in cartilage matrix homeostasis comes from experiments showing that enzymatic removal of exogenous adenosine increases glycosaminoglycan release from cartilage explant cultures, and this effect is mimicked by the A_{2A} receptor antagonist ZM241385.[129] Culture of cartilage explants with the adenosine kinase inhibitor ITU results in an increase in extracellular adenosine and inhibits both IL-1– and LPS-induced glycosaminoglycan release.[130] Another process that is fundamental to tissue repair and remodeling is apoptosis or programmed cell death. The role of adenosine has not been studied in modulating apoptosis in chondrocytes, but there are a number of studies showing that ligands that activate cAMP can regulate apoptosis in other cell systems.[131–133] This suggests that regulation of apoptotic pathways may be a potential function of P1 receptor activation in cartilage that is worthy of future experimental investigation.

4.4 ATP AND ADENOSINE AVAILABILITY IN THE ARTICULAR JOINT

Ultimately, the question of whether or not purine nucleotides and nucleosides exert a significant influence on chondrocyte activity via purine receptors is dependent on how their concentrations are regulated at or near the cell surface. There is extensive evidence that extracellular nucleotide concentration is regulated at multiple levels within cartilage, and this has been the topic of an excellent recent review.[3] There is also a significant literature indicating that purine nucleotide metabolic pathways may be disrupted in arthritic conditions.[3] Mechanical load is important in regulating extracellular ATP concentration surrounding chondrocytes,[20] and accumulation of ATP, ADP, AMP, and adenosine has been shown to occur in the media of cultured chondrocytes from several species.[20,34,128] Chondrocytes are embedded in their matrix in a rounded conformation that is required for maintenance of their differentiated function; Figure 4.2 shows an electron micrograph of a chondrocyte cultured within

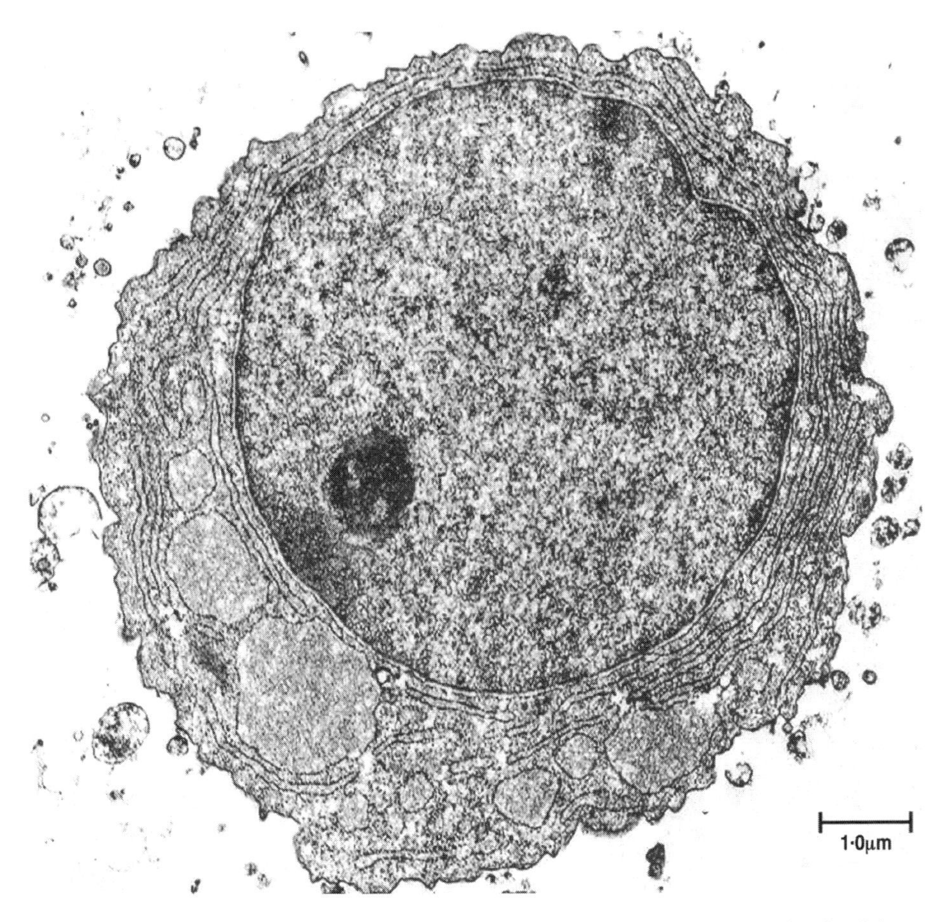

FIGURE 4.2 Electron micrograph of a primary porcine chondrocyte maintained in a three-dimensional conformation by culture within an agarose gel.

an agarose gel to maintain this three-dimensional conformation. Note the extensive endoplasmic reticulum associated with the matrix-producing function of the cell and the accumulation of a number of extracellular matrix vesicles near the cell surface. These matrix vesicles have been shown to be important in maintaining the balance between extracellular ATP and extracellular inorganic phosphate and inorganic pyrophosphate.

In many arthritic conditions, calcium phosphate–containing crystals accumulate in cartilage. The inorganic phosphate and pyrophosphates necessary for generation of these crystals is obtained by direct release of pyrophosphate from the chondrocytes as well as by dephosphorylation of ATP at the cell surface and at the membrane surface of the matrix vesicles.[3] The adenosine generated as a result of ATP dephosphorylation may be rapidly converted to inosine by the action of adenosine deaminases[134] or may be taken back up into the cell to restore intracellular pools via specific nucleoside transporters.[135] Serum and synovial fluid concentrations of adenosine deaminase have been shown to correlate with disease severity in rheumatoid arthritis,[136–138] osteoarthritis,[139] and other conditions associated with joint pathology such as systemic lupus erythematosus.[140] Two adenosine deaminase isoenzymes, ADA1 and ADA2, have been measured in synovial fluids and peripheral blood from osteoarthritic and rheumatoid patients; ADA1 is the higher affinity enzyme. Concentrations of both isoenzymes were found to be higher in the synovial fluids from rheumatoid arthritis compared to osteoarthritis patients and ADA1 was also found to be produced in large amounts by synovial fibroblasts isolated from rheumatoid arthritis patients.[138] Adenosine that is taken back up into the cell undergoes phosphorylation by intracellular adenosine kinases to form intracellular ATP. When intracellular phosphorylation of adenosine to form ATP is inhibited, high concentrations of adenosine build up inside the cell, resulting in transport to the outside and an increase in local extracellular adenosine. The recognition of the therapeutic potential of adenosine kinase inhibitors[141] has resulted in the development of a number of very specific adenosine kinase inhibitors, some of which are orally active.[142] A nonnucleoside adenosine kinase inhibitor, ABT-702, attenuates inflammation in rat adjuvant arthritis,[143] consistent with a potential role for adenosine in alleviating inflammation-induced tissue damage within articular joint tissues, including cartilage.

4.5 SIGNIFICANCE OF P1 AND P2 RECEPTORS IN ARTHRITIC DISEASES

So far in this chapter, the role of ATP and related nucleotides and nucleosides has been considered solely with respect to their direct actions on cartilage. However, we need to understand how purines influence physiological homeostasis within the articular joint and the imbalance that occurs during the progression of articular disease. In order to understand physiological regulation, all the various tissues resident within the joint must be considered, as well as influences from cells infiltrating the synovial fluid in the joint space. Figure 4.3 shows a schematic diagram of an articular joint for the purpose of highlighting all the different potential cellular influences on cartilage integrity. Articular cartilage may be influenced not only by the chondrocytes themselves

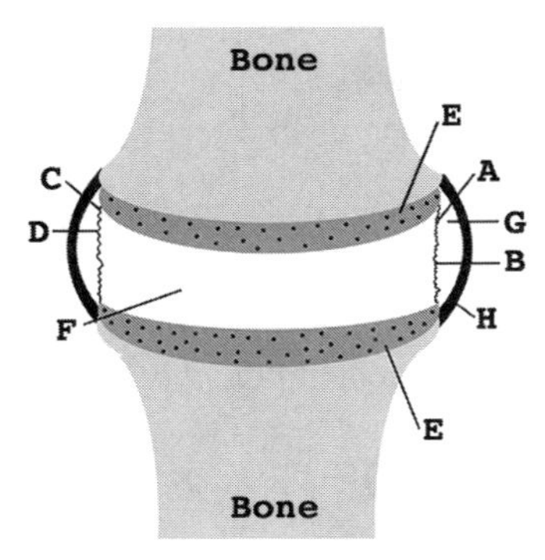

FIGURE 4.3 Schematic diagram of an articular joint to show local cell populations with capacity to influence cartilage matrix. A = synovial membrane; B = resident synovial cells within the synovial membrane; C = adipocytes; D = endothelial cells lining blood vessels within the synovial membrane; E = articular cartilage; F = blood-borne cells present in synovial fluid; neutrophils, monocytes, and lymphocytes; G = nerves in the joint capsule and synovial membrane; H = joint capsule. The diagram is not to scale. For illustrative purposes, the usually closely opposing articular surfaces are drawn apart.

but also by factors secreted from the underlying bone or the synovial membrane present in the joint space. Endothelial cells lining blood vessels within the synovial membrane, joint capsule, and adipose tissue all represent potential sources of growth factors and cytokines. The evidence for a role for adenosine in endothelial cell function during inflammatory processes has recently been reviewed.[144] ATP is itself released from sympathetic nerve endings,[145] and tissue damage to any areas within the joint space could potentially result in cell death and release of intracellular nucleotides and nucleosides as a result of cell disruption. Synovial fluid is an ultrafiltrate of blood in which white blood cells, including neutrophils, monocytes, and lymphocytes, are found to be present in varying quantities, dependent on the state of health of the joint. All these cellular components represent potential sources of purine nucleotides and nucleosides as well as potential targets for purine receptor–mediated actions.

4.5.1 ATP and Adenosine in Synovial Fluids

The presence of ATP has been demonstrated in synovial fluids derived from the articular joints of patients with arthritic conditions, including rheumatoid arthritis and osteoarthritis. Concentrations measured are in the range of 10 to 100 μM in normal and osteoarthritic joints, with higher values recorded in patients with calcium pyrophosphate crystal deposition.[146,147] Methods have been developed to accurately quantify adenosine in synovial fluids,[148,149] and concentrations measured (8.72 ± 3.26 μg/ml)[148] are consistent with concentrations needed for cell surface receptor activation.

4.5.2 P1 and P2 Receptors in Synovial Tissue

The synovial membrane that surrounds and encloses diarthrodial joints proliferates in both osteoarthritis and rheumatoid arthritis, but in rheumatoid arthritis this process is much more extensive, with a proliferation of the resident cells, formation of extensive granulation tissue, and development of invasive properties allowing attachment to the articular surface. The resulting tissue is referred to as "pannus" and contributes significantly to the process of cartilage damage. Adherent human rheumatoid synovial cells have been shown to express calcium-mobilizing receptors with properties consistent with the $P2Y_2$ receptor and to release prostaglandin in response to both ATP and UTP.[150] Adenosine responsive receptors have also been identified on rheumatoid synovial cells and adenosine treatment of these cells was shown to inhibit collagenase gene expression.[151]

4.5.3 P1 and P2 Receptors in Animal Models of Arthritic Disease

Evidence for involvement of P1 and P2 receptors in arthritic diseases has been accumulated using rodent models of acute joint inflammation. Support for the hypothesis that the $P2X_7$ receptor is involved in inflammation-induced pain comes from studies treating arthritic rats with oxidized ATP, a $P2X_7$ antagonist. In these studies, blockade of the $P2X_7$ receptor was shown to significantly decrease nociceptive responses.[152,153] The use of specific adenosine receptor agonists in rodent models of collagen-induced septic arthritis and adjuvant-induced arthritis supports a role for adenosine in attenuating inflammation in the joint; however, multiple receptor subtypes appear to be involved with protective responses as demonstrated by stimulation of A_3 receptors[154,155] and A_{2A} receptors.[156] In addition, targeting A_1 receptors in the spinal cord attenuates inflammation in rat adjuvant arthritis,[157] providing support for a role of neurogenic inflammation in the joint. In addition to direct stimulation of purine receptors, extracellular adenosine concentrations have been manipulated in rat adjuvant arthritis using an adenosine kinase inhibitor, resulting in a reduced inflammatory response,[143] as discussed earlier in this chapter. The ability of the A_{2A} receptor to downregulate inflammation *in vivo* and potentially protect against tissue damage[158] has led to the proposal that activity at this receptor may be a good target to manipulate the course of inflammation to optimize the initial beneficial protective effects of an inflammatory response but then attenuate the response before the long-term deleterious effects begin to take hold.[94] Our understanding of the temporal patterning of the systemic cellular response to inflammation, lymphocyte trafficking, and localized tissue-specific inflammatory processes is becoming more and more sophisticated. The concept of developing multistep therapeutic approaches is attractive and appears attainable, although more research is necessary.

4.5.4 Adenosine and Mechanisms of Action of Methotrexate

The mechanism of action of methotrexate represents the most extensive area of investigation regarding a role for adenosine in arthritis. Methotrexate is still widely used for treatment of rheumatoid arthritis due to its ability to reduce inflammation. As a result of interference in purine metabolic pathways, methotrexate treatment leads to increased extracellular adenosine concentrations. It was proposed over a decade ago that the increased availability of adenosine for interaction with cell surface P2 receptors accounts for many of the anti-inflammatory actions of adenosine.[159] One study has shown that adenosine receptors are required for the anti-inflammatory actions of adenosine.[160] In rat adjuvant arthritis, the anti-inflammatory effects of adenosine can be reversed by the nonselective adenosine receptor antagonists theophylline and caffeine.[161] These observations provide support for pursuing purine receptors as targets for development of novel therapeutic approaches to alleviate symptoms of inflammatory joint diseases, and clinical studies are underway to begin to explore these possibilities.[162]

4.6 SUMMARY

Accumulating data provide strong support for the hypothesis that purine receptors play an active role in modulating pathological processes in the joint. Moreover, the ability of chondrocytes to release physiologically relevant amounts of ATP and adenosine indicates that purine receptor activation can contribute to normal physiological control in chondrocytes and in cartilage development via autocrine and paracrine mechanisms. Purine nucleotides and nucleosides and their associated cell surface receptors have not received as much attention in cartilage and associated articular joint tissues as they have received in other organ systems. The time has come to acknowledge the presence of this complex class of receptors on the chondrocyte surface and focus on elucidating how they interact with established mediators of cartilage homeostasis.

REFERENCES

1. Stockwell, R., *Biology of Cartilage Cells*, Cambridge University Press, Cambridge, 1979.
2. Lee, R.B. and Urban, J.P., Functional replacement of oxygen by other oxidants in articular cartilage, *Arthritis Rheum.*, 46, 3190, 2002.
3. Picher, M., Graff, R.D., and Lee, G.M., Extracellular nucleotide metabolism and signaling in the pathophysiology of articular cartilage, *Arthritis Rheum.*, 48, 2722, 2003.
4. Burnstock, G., Introduction: ATP and its metabolites as potent extracellular agonists, in *Current Topics in Membranes*, Vol 54., Schwiebert, E.M., Ed., Academic Press, San Diego, 2003, p. 1.
5. Burnstock, G., Purinergic P2 receptors as targets for novel analgesics, *Pharmacol. Therap.*, 110, 433, 2006.
6. Brown, C.J. et al., Proteoglycan breakdown from bovine nasal cartilage is increased, and from articular cartilage is decreased, by extracellular ATP, *Biochim. Biophys. Acta.*, 1362, 208, 1997.
7. Leong, W.S., Russell, R.G., and Caswell, A.M., Extracellular ATP stimulates resorption of bovine nasal cartilage, *Biochem. Soc. Trans.*, 18, 951, 1990.
8. Croucher, L.J. et al., Extracellular ATP and UTP stimulate cartilage proteoglycan and collagen accumulation in bovine articular chondrocyte pellet cultures, *Biochim. Biophys. Acta.*, 1502, 297, 2000.
9. Kaplan, A.D. et al., Extracellular nucleotides act through P2U purinoceptors to elevate $[Ca^{2+}]_i$ and enhance basic fibroblast growth factor–induced proliferation in sheep chondrocytes, *Endocrinology*, 137, 4757, 1996.

10. Koolpe, M. and Benton, H.P., Calcium-mobilizing purine receptors on the surface of mammalian articular chondrocytes, *J. Orthop. Res.*, 15, 204, 1997.
11. Hung, C.T. et al., Extracellular ATP modulates $[Ca^{2+}]_i$ in retinoic acid–treated embryonic chondrocytes, *Am. J. Physiol.*, 272, C1611, 1997.
12. Meyer, M.P. et al., The extracellular ATP receptor, $cP2Y_1$, inhibits cartilage formation in micromass cultures of chick limb mesenchyme, *Dev. Dyn.*, 222, 494, 2001.
13. Kumahashi, N. et al., Involvement of ATP, increase of intracellular calcium and the early expression of c-fos in the repair of rat fetal articular cartilage, *Cell Tissue Res.*, 317, 117, 2004.
14. Caswell, A.M., Leong, W.S., and Russell, R.G., Evidence for the presence of P2-purinoceptors at the surface of human articular chondrocytes in monolayer culture, *Biochim. Biophys. Acta,* 1074, 151, 1991.
15. Caswell, A.M., Leong, W.S., and Russell, R.G., Interleukin-1β enhances the response of human articular chondrocytes to extracellular ATP, *Biochim. Biophys. Acta*, 1137, 52, 1992.
16. Leong, W.S., Russell, R.G., and Caswell, A.M., Induction of enhanced responsiveness of human articular chondrocytes to extracellular ATP by tumour necrosis factor-α, *Clin. Sci. (Lond)*, 85, 569, 1993.
17. Leong, W.S., Russell, R.G., and Caswell, A.M., Stimulation of cartilage resorption by extracellular ATP acting at P2-purinoceptors, *Biochim. Biophys. Acta*, 1201, 298, 1994.
18. Koolpe, M., Rodrigo, J.J., and Benton, H.P., Adenosine 5′–triphosphate, uridine 5′–triphosphate, bradykinin, and lysophosphatidic acid induce different patterns of calcium responses by human articular chondrocytes, *J. Orthop. Res.*, 16, 217, 1998.
19. Koolpe, M., Pearson, D., and Benton, H.P., Expression of both P1 and P2 purine receptor genes by human articular chondrocytes and profile of ligand-mediated prostaglandin E2 release, *Arthritis Rheum.*, 42, 258, 1999.
20. Graff, R.D. et al., ATP release by mechanically loaded porcine chondrons in pellet culture, *Arthritis Rheum.*, 43, 1571, 2000.
21. D'Andrea, P., Calabrese, A., and Grandolfo, M., Intercellular calcium signalling between chondrocytes and synovial cells in co-culture, *Biochem. J.*, 329, 681, 1998.
22. Hoebertz, A. et al., Expression of P2 receptors in bone and cultured bone cells, *Bone*, 27, 503, 2000.
23. Lee, G.M. et al., Isolated chondrons: a viable alternative for studies of chondrocyte metabolism *in vitro*, *Osteoarthritis Cartilage*, 5, 261, 1997.
24. DeLise, A.M., Fischer, L., and Tuan, R.S., Cellular interactions and signaling in cartilage development, *Osteoarthritis Cartilage*, 8, 309, 2000.
25. Berridge, M.J., Lipp, P., and Bootman, M.D., The versatility and universality of calcium signalling, *Nat. Rev. Mol. Cell. Biol.*, 1, 11, 2000.
26. Iwamoto, M., et al., Retinoic acid induces rapid mineralization and expression of mineralization-related genes in chondrocytes, *Exp. Cell Res.*, 207,413, 1993.
27. Iwamoto, M., et al., Retinoic acid is a major regulator of chondrocyte maturation and matrix mineralization, *Microsc. Res. Tech.*, 28, 483, 1994.
28. Huang, N., Wang, D.J., and Heppel, L.A., Extracellular ATP is a mitogen for 3T3, 3T6, and A431 cells and acts synergistically with other growth factors, *Proc. Natl. Acad. Sci. USA*, 86, 7904, 1989.
29. Huwiler, A. and Pfeilschifter, J., Stimulation by extracellular ATP and UTP of the mitogen-activated protein kinase cascade and proliferation of rat renal mesangial cells, *Br. J. Pharmacol.*, 113, 1455, 1994.

30. Agresti, C. et al., ATP regulates oligodendrocyte progenitor migration, proliferation, and differentiation: involvement of metabotropic P2 receptors, *Brain Res. Rev.*, 48, 157, 2005.
31. Jacques-Silva, M.C. et al., ERK, PKC and PI3K/Akt pathways mediate extracellular ATP and adenosine-induced proliferation of U138-MG human glioma cell line, *Oncology*, 67, 450, 2004.
32. Di Virgilio, F. et al., Leukocyte P2 receptors: a novel target for anti-inflammatory and anti-tumor therapy, *Curr. Drug Targets Cardiovasc. Haematol. Disord.*, 5, 85, 2005.
33. Coutinho-Silva, R. et al., P2X and P2Y purinergic receptors on human intestinal epithelial carcinoma cells: effects of extracellular nucleotides on apoptosis and cell proliferation, *Am. J. Physiol.*, 288, G1024, 2005.
34. Hatori, M. et al., Adenine nucleotide metabolism by chondrocytes *in vitro*: role of ATP in chondrocyte maturation and matrix mineralization, *J. Cell. Physiol.*, 165, 468, 1995.
35. Ralevic, V. and Burnstock, G., Receptors for purines and pyrimidines, *Pharmacol. Rev.*, 50, 413, 1998.
36. Burnstock, G. and Knight, G.E., Cellular distribution and functions of P2 receptor subtypes in different systems, *Int. Rev. Cytol.*, 240, 31, 2004.
37. Laron, Z., IGF-1 and insulin as growth hormones, *Novartis Found. Symp.*, 262, 56; discussion 77, 265, 2004.
38. Trippel, S.B., Growth factor inhibition: potential role in the etiopathogenesis of osteoarthritis, *Clin. Orthop. Relat. Res.*, 427, S47, 2004.
39. Grimaud, E., Heymann, D., and Redini, F., Recent advances in TGF-β effects on chondrocyte metabolism. Potential therapeutic roles of TGF-β in cartilage disorders. *Cytokine Growth Factor Rev.*, 13, 241, 2002.
40. van der Kraan, P.M. et al., Interaction of chondrocytes, extracellular matrix and growth factors: relevance for articular cartilage tissue engineering, *Osteoarthritis Cartilage*, 10, 631, 2002.
41. Goldring, S.R. and Goldring, M.B., The role of cytokines in cartilage matrix degeneration in osteoarthritis, *Clin. Orthop. Relat. Res.*, 427, S27, 2004.
42. Steinmeyer, J., Cytokines in osteoarthritis-current status on the pharmacological intervention, *Front. Biosci.*, 9, 575, 2004.
43. Cawston, T.E. et al.,Cytokine synergy, collagenases and cartilage collagen breakdown, *Biochem. Soc. Symp.*, 70, 125, 2003.
44. Fernandes, J.C., Martel-Pelletier, J., and Pelletier, J.P., The role of cytokines in osteoarthritis pathophysiology, *Biorheology*, 39, 237, 2002.
45. Valhmu, W.B. and Raia, F.J., myo-Inositol 1,4,5-trisphosphate and Ca^{2+}/calmodulin-dependent factors mediate transduction of compression-induced signals in bovine articular chondrocytes, *Biochem. J.*, 361, 689, 2002.
46. Fitzgerald, J.B. et al., Mechanical compression of cartilage explants induces multiple time-dependent gene expression patterns and involves intracellular calcium and cyclic AMP, *J. Biol. Chem.*, 279, 19502, 2004.
47. Jung, S.M. and Moroi, M., Platelet collagen receptor integrin $\alpha_2\beta_1$ activation involves differential participation of ADP-receptor subtypes $P2Y_1$ and $P2Y_{12}$ but not intracellular calcium change, *Eur. J. Biochem.*, 268, 3513, 2001.
48. Remijn, J.A. et al., Role of ADP receptor $P2Y_{12}$ in platelet adhesion and thrombus formation in flowing blood, *Arterioscler. Thromb. Vasc. Biol.*, 22, 686, 2002.

49. Oury, C. et al., $P2X_1$-mediated activation of extracellular signal–regulated kinase 2 contributes to platelet secretion and aggregation induced by collagen, *Blood*, 100, 2499, 2002.
50. Oury, C. et al., Overexpression of the platelet $P2X_1$ ion channel in transgenic mice generates a novel prothrombotic phenotype, *Blood,* 101, 3969, 2003.
51. Baurand, A. and Gachet, C., The $P2Y_1$ receptor as a target for new antithrombotic drugs: a review of the $P2Y_1$ antagonist MRS-2179, *Cardiovasc. Drug Rev.,* 21, 67, 2003.
52. Andre, P. et al., $P2Y_{12}$ regulates platelet adhesion/activation, thrombus growth, and thrombus stability in injured arteries, *J. Clin. Invest.*, 112, 398, 2003.
53. Takano, K. et al., Collagen-induced generation of platelet-derived microparticles in whole blood is dependent on ADP released from red blood cells and calcium ions, *Platelets*, 15, 223, 2004.
54. Oury, C. et al., The ATP-gated $P2X_1$ ion channel acts as a positive regulator of platelet responses to collagen, *Thromb. Haemost.*, 86, 1264, 2001.
55. Xu, J. et al., Prostaglandin E2 production in astrocytes: regulation by cytokines, extracellular ATP, and oxidative agents, *Prostaglandins Leukot. Essent. Fatty Acids*, 69, 437, 2003.
56. Welch, B.D. et al., $P2Y_2$ receptor–stimulated release of prostaglandin E_2 by rat inner medullary collecting duct preparations, *Am. J. Physiol.*, 285, F711, 2001.
57. Genetos, D.C. et al., Fluid shear–induced ATP secretion mediates prostaglandin release in MC3T3-E1 osteoblasts, *J. Bone Miner. Res.*, 20, 41, 2005.
58. Goldring, M.B. and Berenbaum, F., The regulation of chondrocyte function by proinflammatory mediators: prostaglandins and nitric oxide, *Clin. Orthop. Relat. Res.*, 427, S37, 2004.
59. Martel-Pelletier, J., Pelletier, J.P., and Fahmi, H., Cyclooxygenase-2 and prostaglandins in articular tissues, *Semin. Arthritis Rheum.*, 33, 155, 2003.
60. McCoy, J.M., Wicks, J.R., and Audoly, L.P., The role of prostaglandin E_2 receptors in the pathogenesis of rheumatoid arthritis, *J. Clin. Invest.*, 110, 651, 2002.
61. Laufer, S., Role of eicosanoids in structural degradation in osteoarthritis, *Curr. Opin. Rheumatol.*, 15, 623, 2003.
62. Amin, A.R. et al., COX-2, NO, and cartilage damage and repair, *Curr. Rheumatol. Rep.*, 2, 447, 2000.
63. Berenbaum, F. et al., Concomitant recruitment of ERK1/2 and p38 MAPK signalling pathway is required for activation of cytoplasmic phospholipase A2 via ATP in articular chondrocytes, *J. Biol. Chem.*, 278, 13680, 2003.
64. Sundqvist, G. and Lerner, U.H., Bradykinin and thrombin synergistically potentiate interleukin 1 and tumour necrosis factor induced prostanoid biosynthesis in human dental pulp fibroblasts, *Cytokine*, 8, 168, 1996.
65. Bathon, J.M. et al., Mechanisms of prostanoid synthesis in human synovial cells: cytokine-peptide synergism, *Inflammation*, 20, 537, 1996.
66. Kramer, R.M. et al., Thrombin-induced phosphorylation and activation of Ca^{2+}-sensitive cytosolic phospholipase A2 in human platelets, *J. Biol. Chem.*, 268, 26796, 1993.
67. Zarubin, T. and Han, J., Activation and signaling of the p38 MAP kinase pathway, *Cell Res.*, 15, 11, 2005.
68. Wada, Y. et al., Novel p38 MAP kinase inhibitor R-130823 suppresses IL-6, IL-8 and MMP-13 production in spheroid culture of human synovial sarcoma cell line SW 982, *Immunol. Lett.*, 101, 50, 2005.

69. Saklatvala, J., The p38 MAP kinase pathway as a therapeutic target in inflammatory disease, *Curr. Opin. Pharmacol.*, 4, 372, 2004.
70. Ihara, H. et al., ATP-stimulated interleukin-6 synthesis through P2Y receptors on human osteoblasts, *Biochem. Biophys. Res. Commun.*, 326, 329, 2005.
71. Caraccio, N. et al., Extracellular ATP modulates interleukin-6 production by human thyrocytes through functional purinergic P2 receptors, *Endocrinology*, 146, 3172, 2005.
72. Hanley, P.J. et al., Extracellular ATP induces oscillations of intracellular Ca^{2+} and membrane potential and promotes transcription of IL-6 in macrophages, *Proc. Natl. Acad. Sci. USA*, 101, 9479, 2004.
73. Idzko, M. et al., Stimulation of P2 purinergic receptors induces the release of eosinophil cationic protein and interleukin-8 from human eosinophils, *Br. J. Pharmacol.*, 138, 1244, 2003.
74. Langston, H.P. et al., Secretion of IL-2 and IFN-γ, but not IL-4, by antigen-specific T cells requires extracellular ATP, *J. Immunol.*, 170, 2962, 2003.
75. Loomis, W.H. et al., Hypertonic stress increases T cell interleukin-2 expression through a mechanism that involves ATP release, P2 receptor, and p38 MAPK activation, *J. Biol. Chem.*, 278, 4590, 2003.
76. Swennen, E.L., Bast, A., and Dagnelie, P.C., Immunoregulatory effects of adenosine 5′–triphosphate on cytokine release from stimulated whole blood, *Eur. J. Immunol.*, 35, 852, 2005.
77. Marteau, F. et al., Involvement of multiple P2Y receptors and signaling pathways in the action of adenine nucleotides diphosphates on human monocyte-derived dendritic cells, *J. Leukoc. Biol.*, 76, 796, 2004.
78. Tsuzaki, M. et al., ATP modulates load-inducible IL-1β, COX 2, and MMP-3 gene expression in human tendon cells, *J. Cell. Biochem.*, 89, 556, 2003.
79. Palaga, T., Kataoka, T., and Nagai, K., Extracellular ATP inhibits apoptosis and maintains cell viability by inducing autocrine production of interleukin-4 in a myeloid progenitor cell line, *Int. Immunopharmacol.*, 4, 953, 2004.
80. Seo, D.R., Kim, K.Y., and Lee, Y. B., Interleukin-10 expression in lipopolysaccharide-activated microglia is mediated by extracellular ATP in an autocrine fashion, *Neuroreport*, 15, 1157, 2004.
81. Bulanova, E. et al., Extracellular ATP induces cytokine expression and apoptosis through $P2X_7$ receptor in murine mast cells, *J. Immunol.*, 174, 3880, 2005.
82. Sluyter, R., Shemon, A.N., and Wiley, J.S., Glu496 to Ala polymorphism in the $P2X_7$ receptor impairs ATP-induced IL-1β release from human monocytes, *J. Immunol.*, 172, 3399, 2004.
83. Derks, R. and Beaman, K., Regeneration and tolerance factor modulates the effect of adenosine triphosphate–induced interleukin 1β secretion in human macrophages, *Hum. Immunol.*, 65, 676, 2004.
84. Wilson, H.L. et al., Secretion of intracellular IL-1 receptor antagonist (type 1) is dependent on $P2X_7$ receptor activation, *J. Immunol.*, 173, 1202, 2004.
85. Chakfe, Y. et al., ADP and AMP induce interleukin-1β release from microglial cells through activation of ATP-primed $P2X_7$ receptor channels, *J. Neurosci.*, 22, 3061, 2002.
86. Collis, M.G. and Hourani, S.M., Adenosine receptor subtypes, *Trends Pharmacol. Sci.*, 14, 360, 1993.
87. Olah, M.E. and Stiles, G.L., Adenosine receptor subtypes: characterization and therapeutic regulation, *Annu. Rev. Pharmacol. Toxicol.*, 35, 581, 1995.

88. Linden, J., Molecular approach to adenosine receptors: receptor-mediated mechanisms of tissue protection, *Annu. Rev. Pharmacol. Toxicol.*, 41, 775, 2001.
89. Pintor, J., Miras-Portugal, T., and Fredholm, B.B., Research on purines and their receptors comes of age, *Trends Pharmacol. Sci.*, 21, 453, 2000.
90. Le, D.E. et al., A1-receptor blockade: a novel approach for assessing myocardial viability in chronic ischemic cardiomyopathy, *J. Am. Soc. Echocardiogr.*, 16, 764, 2003.
91. Linden, J., Autoregulation of cyclic AMP in vascular smooth muscle: a role for adenosine receptors, *Mol. Pharmacol.*, 62, 969, 2002.
92. Riou, L.M. et al., Influence of propranolol, enalaprilat, verapamil, and caffeine on adenosine A_{2A}-receptor–mediated coronary vasodilation, *J. Am. Coll. Cardiol.*, 40, 1687, 2002.
93. Pahl, A. and Szelenyi, I., Asthma therapy in the new millennium, *Inflamm. Res.*, 51, 273, 2002.
94. Gomez, G. and Sitkovsky, M.V., Targeting G protein–coupled A2a adenosine receptors to engineer inflammation *in vivo*, *Int. J. Biochem. Cell Biol.*, 35, 410, 2003.
95. Sitkovsky, M.V., Use of the A_{2A} adenosine receptor as a physiological immunosuppressor and to engineer inflammation *in vivo*, *Biochem. Pharmacol.*, 65, 493, 2003.
96. Volpini, R. et al., Medicinal chemistry and pharmacology of A2B adenosine receptors, *Curr. Top. Med. Chem.*, 3, 427, 2003.
97. Hasko, G. et al., Inosine inhibits inflammatory cytokine production by a posttranscriptional mechanism and protects against endotoxin-induced shock, *J. Immunol.*, 164, 1013, 2000.
98. Hasko, G., Sitkovsky, M.V., and Szabo, C., Immunomodulatory and neuroprotective effects of inosine, *Trends Pharmacol. Sci.*, 25, 152, 2004.
99. la Sala, A. et al., Alerting and tuning the immune response by extracellular nucleotides, *J. Leukoc. Biol.*, 73, 339, 2003.
100. Ren, H. and Stiles, G.L., Characterization of the human A1 adenosine receptor gene. Evidence for alternative splicing, *J. Biol. Chem.*, 269, 3104, 1994.
101. Goldring, M.B., Culture of immortalized chondrocytes and their use as models of chondrocyte function, *Methods Mol. Med.*, 100, 37, 2004.
102. Kokenyesi, R. et al., Proteoglycan production by immortalized human chondrocyte cell lines cultured under conditions that promote expression of the differentiated phenotype, *Arch. Biochem. Biophys.*, 383, 79, 2000.
103. Robbins, J.R. et al., Immortalized human adult articular chondrocytes maintain cartilage-specific phenotype and responses to interleukin-1β, *Arthritis Rheum.*, 43, 2189, 2000.
104. Osaki, M. et al., The TATA-containing core promoter of the type II collagen gene (COL2A1) is the target of interferon-γ∠mediated inhibition in human chondrocytes: requirement for Stat1α, Jak1 and Jak2, *Biochem. J.*, 369, 103, 2003.
105. Tan, L. et al., Egr-1 mediates transcriptional repression of COL2A1 promoter activity by interleukin-1β, *J. Biol. Chem.*, 278, 17688, 2003.
106. Sen, M. et al., WISP3-dependent regulation of type II collagen and aggrecan production in chondrocytes, *Arthritis Rheum.*, 50, 488, 2004.
107. Yokota, H., Goldring, M.B., and Sun, H.B., CITED2-mediated regulation of MMP-1 and MMP-13 in human chondrocytes under flow shear, *J. Biol. Chem.*, 278, 47275, 2003.
108. Pufe, T. et al., Vascular endothelial growth factor (VEGF) induces matrix metalloproteinase expression in immortalized chondrocytes, *J. Pathol.*, 202, 367, 2004.

109. Koshy, P.J. et al., The modulation of matrix metalloproteinase and ADAM gene expression in human chondrocytes by interleukin-1 and oncostatin M: a time-course study using real-time quantitative reverse transcription–polymerase chain reaction, *Arthritis Rheum.*, 46, 961, 2002.
110. Thomas, B. et al., Differentiation regulates interleukin-1β∠induced cyclo-oxygenase-2 in human articular chondrocytes: role of p38 mitogen–activated protein kinase, *Biochem. J.*, 362, 367, 2002.
111. Abulencia, J.P. et al., Shear-induced cyclooxygenase-2 via a JNK2/c-Jun–dependent pathway regulates prostaglandin receptor expression in chondrocytic cells, *J. Biol. Chem.*, 278, 28388, 2003.
112. Barankiewicz, J. et al., Selective adenosine release from human B but not T lymphoid cell line, *J. Biol. Chem.*, 265, 15738, 1990.
113. Cronstein, B.N. et al., Neutrophil adherence to endothelium is enhanced via adenosine A1 receptors and inhibited via adenosine A2 receptors, *J. Immunol.*, 148, 2201, 1992.
114. Merrill, J.T. et al., Adenosine A1 receptor promotion of multinucleated giant cell formation by human monocytes: a mechanism for methotrexate-induced nodulosis in rheumatoid arthritis, *Arthritis Rheum.*, 40, 1308, 1997.
115. Zalavary, S. and Bengtsson, T., Modulation of the chemotactic peptide- and immunoglobulin G–triggered respiratory burst in human neutrophils by exogenous and endogenous adenosine, *Eur. J. Pharmacol.*, 354, 215, 1998.
116. Zhong, H. et al., A_{2B} adenosine receptors increase cytokine release by bronchial smooth muscle cells, *Am. J. Respir. Cell Mol. Biol.*, 30, 118, 2004.
117. Soop, A. et al., Adenosine treatment attenuates cytokine interleukin-6 responses to endotoxin challenge in healthy volunteers, *Shock*, 19, 503, 2003.
118. Le Moine, O. et al., Adenosine enhances IL-10 secretion by human monocytes, *J. Immunol.*, 156, 4408, 1996.
119. Bouma, M.G. et al., Differential regulatory effects of adenosine on cytokine release by activated human monocytes, *J. Immunol.*, 153, 4159, 1994.
120. Bshesh, K. et al., The A2A receptor mediates an endogenous regulatory pathway of cytokine expression in THP-1 cells, *J. Leukoc. Biol.*, 72, 1027, 2002.
121. Krump, E. and Borgeat, P., Adenosine, An endogenous inhibitor of arachidonic acid release and leukotriene biosynthesis in human neutrophils, *Adv. Exp. Med. Biol.*, 447, 107, 1999.
122. Flamand, N. et al., Adenosine, a potent natural suppressor of arachidonic acid release and leukotriene biosynthesis in human neutrophils, *Am. J. Respir. Crit. Care Med.*, 161, S88, 2000.
123. Surette, M.E. et al., Activation of leukotriene synthesis in human neutrophils by exogenous arachidonic acid: inhibition by adenosine A_{2a} receptor agonists and crucial role of autocrine activation by leukotriene B_4, *Mol. Pharmacol.*, 56, 1055, 1999.
124. Pouliot, M. et al., Adenosine up-regulates cyclooxygenase-2 in human granulocytes: impact on the balance of eicosanoid generation, *J. Immunol.*, 169, 5279, 2002.
125. Cronstein, B.N. et al., Adenosine: a physiological modulator of superoxide anion generation by human neutrophils, *J. Exp. Med.*, 158, 1160, 1983.
126. Benton, H.P., MacDonald, M.H., and Tesch, A.M., Effects of adenosine on bacterial lipopolysaccharide- and interleukin 1–induced nitric oxide release from equine articular chondrocytes, *Am. J. Vet. Res.*, 63, 204, 2002.
127. Tesch, A.M. et al., Chondrocytes respond to adenosine via A_2 receptors and activity is potentiated by an adenosine deaminase inhibitor and a phosphodiesterase inhibitor, *Osteoarthritis Cartilage*, 10, 34, 2002.

128. Tesch, A.M. et al., Effects of an adenosine kinase inhibitor and an adenosine deaminase inhibitor on accumulation of extracellular adenosine by equine articular chondrocytes, *Am. J. Vet. Res.*, 63, 1512. 2002.
129. Tesch, A.M. et al., Endogenously produced adenosine regulates articular cartilage matrix homeostasis: enzymatic depletion of adenosine stimulates matrix degradation, *Osteoarthritis Cartilage*, 12, 349, 2004.
130. Petrov, R. et al., Inhibition of adenosine kinase attenuates interleukin-1– and lipopolysaccharide-induced alterations in articular cartilage metabolism, *Osteoarthritis Cartilage*, 13, 250, 2005.
131. Seite, P. et al., Ectopic expression of Bcl-2 switches over nuclear signalling for cAMP-induced apoptosis to granulocytic differentiation, *Cell Death Differ.*, 7, 1081, 2000.
132. Turner, P.R. et al., Apoptosis mediated by activation of the G protein–coupled receptor for parathyroid hormone (PTH)/PTH-related protein (PTHrP), *Mol. Endocrinol.*, 14, 241, 2000.
133. von Knethen, A. and Brune, B., Attenuation of macrophage apoptosis by the cAMP-signaling system, *Mol. Cell. Biochem.*, 212, 35, 2000.
134. Cristalli, G. et al., Adenosine deaminase: functional implications and different classes of inhibitors, *Med. Res. Rev.*, 21, 105, 2001.
135. Kong, W., Engel, K., and Wang, J., Mammalian nucleoside transporters, *Curr. Drug Metab.*, 5, 63, 2004.
136. Iwaki-Egawa, S., Watanabe,Y., and Matsuno, H., Correlations between matrix metalloproteinase-9 and adenosine deaminase isozymes in synovial fluid from patients with rheumatoid arthritis, *J. Rheumatol.*, 28, 485, 2001.
137. Sari, R.A. et al., Correlation of serum levels of adenosine deaminase activity and its isoenzymes with disease activity in rheumatoid arthritis, *Clin. Exp. Rheumatol.*, 21, 87, 2003.
138. Nakamachi, Y. et al., Specific increase in enzymatic activity of adenosine deaminase 1 in rheumatoid synovial fibroblasts, *Arthritis Rheum.*, 48, 668, 2003.
139. Yuksel, H. and Akoglu, T.F., Serum and synovial fluid adenosine deaminase activity in patients with rheumatoid arthritis, osteoarthritis, and reactive arthritis, *Ann. Rheum. Dis.*, 47, 492, 1988.
140. Hitoglou, S. et al., Adenosine deaminase activity and its isoenzyme pattern in patients with juvenile rheumatoid arthritis and systemic lupus erythematosus, *Clin. Rheumatol.*, 20, 411, 2001.
141. Kowaluk, E.A. and Jarvis, M.F., Therapeutic potential of adenosine kinase inhibitors, *Expert Opin. Investig. Drugs*, 9, 551, 2000.
142. Lee, C.H. et al., Discovery of 4-amino-5-(3-bromophenyl)-7-(6-morpholino-pyridin-3-yl)pyrido[2,3-d]pyrimidine, an orally active, non-nucleoside adenosine kinase inhibitor, *J. Med. Chem.*, 44, 2133, 2001.
143. Boyle, D.L. et al., Anti-inflammatory effects of ABT-702, a novel non-nucleoside adenosine kinase inhibitor, in rat adjuvant arthritis, *J. Pharmacol. Exp. Ther.*, 296, 495, 2001.
144. Sands, W.A. and Palmer, T.M., Adenosine receptors and the control of endothelial cell function in inflammatory disease, *Immunol. Lett.*, 101, 1, 2005.
145. Westfall, D.P., Todorov, L.D., and Mihaylova-Todorova, S.T., ATP as a co-transmitter in sympathetic nerves and its inactivation by releasable enzymes, *J. Pharmacol. Exp. Ther.*, 303, 439, 2002.
146. Fam, A.G., Calcium pyrophosphate crystal deposition disease and other crystal deposition diseases, *Curr. Opin. Rheumatol.*, 4, 574, 1992.

147. Ryan, L.M., Rachow, J.W., and McCarty, D.J., Synovial fluid ATP: a potential substrate for the production of inorganic pyrophosphate, *J. Rheumatol.*, 18, 716, 1991.
148. Huang, L.-F. et al., Simple and rapid determination of adenosine in human synovial fluid with high performance liquid chromatography–mass spectrometry, *J. Pharm. Biomed. Anal.*, 36, 877, 2004.
149. Sottofattori, E., Anzaldi, M., and Ottonello, L., HPLC determination of adenosine in human synovial fluid, *J. Pharm. Biomed. Anal.*, 24, 1143, 2004.
150. Loredo, G.A. and Benton, H.P., ATP and UTP activate calcium-mobilizing P2U-like receptors and act synergistically with interleukin-1 to stimulate prostaglandin E_2 release from human rheumatoid synovial cells, *Arthritis Rheum.*, 41, 246, 1998.
151. Boyle, D.L., Sajjadi, F.G., and Firestein, G.S., Inhibition of synoviocyte collagenase gene expression by adenosine receptor stimulation, *Arthritis Rheum.*, 39, 923, 1996.
152. Dell'Antonio, G. et al., Relief of inflammatory pain in rats by local use of the selective $P2X_7$ ATP receptor inhibitor, oxidized ATP, *Arthritis Rheum.*, 46, 3378, 2002.
153. Dell'Antonio, G. et al., Antinociceptive effect of a new $P_{2Z}/P2X_7$ antagonist, oxidized ATP, in arthritic rats, *Neurosci. Lett.*, 327, 87, 2002.
154. Szabo, C. et al., Suppression of macrophage inflammatory protein (MIP)-1α production and collagen-induced arthritis by adenosine receptor agonists, *Br. J. Pharmacol.*, 125, 379, 1998.
155. Baharav, E. et al., Antiinflammatory effect of A3 adenosine receptor agonists in murine autoimmune arthritis models, *J. Rheumatol.*, 32, 469, 2005.
156. Cohen, S.B. et al., Reducing joint destruction due to septic arthrosis using an adenosine2A receptor agonist, *J. Orthop. Res.*, 22, 427, 2004.
157. Boyle, D.L. et al., Spinal adenosine receptor activation inhibits inflammation and joint destruction in rat adjuvant-induced arthritis, *Arthritis Rheum.*, 46, 3076, 2002.
158. Ohta, A. and Sitkovsky, M., Role of G-protein–coupled adenosine receptors in downregulation of inflammation and protection from tissue damage, *Nature*, 414, 916, 2001.
159. Cronstein, B.N., Low-dose methotrexate: a mainstay in the treatment of rheumatoid arthritis, *Pharmacol. Rev.*, 57, 163, 2005.
160. Montesinos, M.C. et al., Adenosine A2A or A3 receptors are required for inhibition of inflammation by methotrexate and its analog MX-68, *Arthritis Rheum.*, 48, 240, 2003.
161. Montesinos, M.C. et al., Reversal of the antiinflammatory effects of methotrexate by the nonselective adenosine receptor antagonists theophylline and caffeine: evidence that the antiinflammatory effects of methotrexate are mediated via multiple adenosine receptors in rat adjuvant arthritis, *Arthritis Rheum.*, 43, 656, 2000.
162. Forrest, C.M. et al., Modulation of cytokine release by purine receptors in patients with rheumatoid arthritis, *Clin. Exp. Rheumatol.*, 23, 89, 2005.

5 ATP Release Mechanisms

George R. Dubyak

CONTENTS

5.1 RELEASED NUCLEOTIDES AS EXTRACELLULAR SIGNALING MOLECULES: GENERAL CONSIDERATIONS

In addition to their roles in intracellular energy metabolism and nucleic acid synthesis, ATP and other nucleotides play important functions as extracellular signaling molecules (Figure 5.1). It is now more than 30 years since Burnstock[1] first proposed that extracellular adenine nucleotides and nucleosides can be used for signal transduction at nerve endings in diverse tissues. Implicit in this notion of purine-based

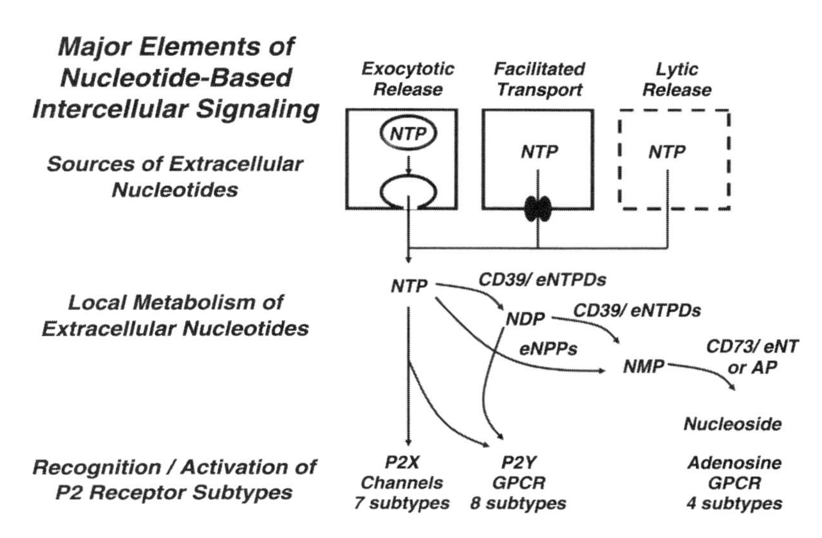

FIGURE 5.1 Major elements of intercellular signaling by extracellular nucleotides. Nucleotide triphosphates (NTP), predominantly ATP, may be released to extracellular compartments by lytic, channel-mediated, or exocytotic pathways. The released NTPs can be initially hydrolyzed to nucleotide diphosphates (NDPs) by CD39 family nucleoside triphosphate diphosphohydrolases (eNTPDs) or they can be directly degraded to nucleotide monophosphates (NMPs) by ecto-nucleotide pyrophosphatase/phosphodiesterases (eNPP). Extracellular NMPs can be dephosphorylated to the corresponding nucleoside via the action of the CD73 ecto-5′nucleotidase (eNT) or extracellular alkaline phosphatases (AP). Released nucleotides activate signaling via the ionotropic P2X family receptors or the P2Y family G protein–coupled receptors (GPCR)

neurotransmission was a requirement for ATP (or other nucleotides) to be released and then degraded in a highly localized fashion at sites of cell-to-cell communication. Given the early emphasis on the role of purines in neuronal signaling, initial studies were focused on nucleotide/nucleoside release at neuron-to-neuron synapses, neuron-to-tissue varicosities, or the immediate vicinity of neuroendocrine cells (e.g., adrenal chromaffin cells).[2,3] At approximately the same time, studies from the hematological literature were showing that platelets also release large amounts of ATP and ADP during activation of hemostasis and degranulation of dense granules.[4–6] These early investigations demonstrated that neurons, neuroendocrine cells, and platelets could all release ATP via classical mechanisms involving exocytotic release of nucleotides copackaged with biogenic amines or other neurotransmitters within specialized secretory vesicles or granules (Figure 5.1 and Figure 5.2).

As schematically summarized in Figure 5.1, vertebrate genomes now include 15 distinct ATP/nucleotide receptor genes,[7–11] 4 adenosine receptor genes,[12,13] and at least 9 different ecto-nucleotidase genes.[14–18] Almost every mammalian cell type expresses one or more subtypes of nucleotide or nucleoside receptor along with one or more of the ecto-nucleotidases used for scavenging extracellular nucleotides. The majority of these cell types lack direct physical proximity to neurons, neuroendocrine cells, or degranulating platelets. Thus, the identification of alternative

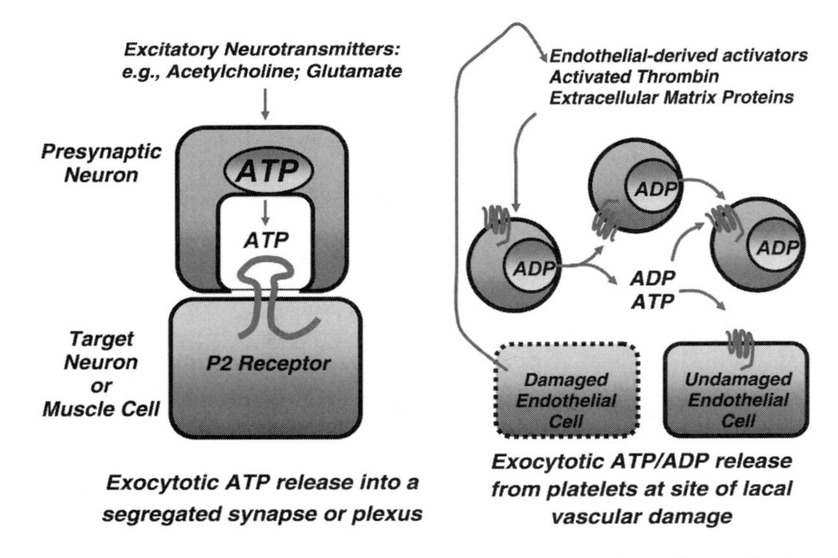

FIGURE 5.2 Exocytotic ATP release from neurons, neuroendocrine cells, or platelets. Left panel: In response to membrane depolarization, excitatory cells, such as neurons and neuroendocrine cells, release secretory vesicles or granules containing compartmentalized ATP stores at nerve–nerve or nerve–muscle synapses or into a nerve-ending plexus within a peripheral tissue. Right panel: In response to vasoactive mediators released at sites of vascular damage, platelets release dense-core granules containing ATP and ADP that is copackaged with serotonin.

sources of extracellular nucleotides has become a significant area of investigation. This chapter will review salient examples of the mechanisms known or proposed to mediate nucleotide release from cell types other than neurons, neuroendocrine cells, or platelets.[9]

A recurring observation from many recent studies is that mechanically stressed cells either *in vitro* or within different tissues can locally release ATP via mechanisms that do not involve readily measured exocytosis of nucleotide-containing vesicles or granules.[19–28] Mechanical stresses shown to produce nonlytic nucleotide release include direct membrane deformation (e.g., by a micropipette positioned to touch but not penetrate a cell), shear forces generated by fluid flow, or cell volume changes triggered by osmotic swelling or shrinking. Given that activation of most P2Y and P2X receptor subtypes results in transient elevation of cytosolic Ca^{2+}, the ATP/nucleotide release elicited by a cell or cells directly subjected to mechanical stress has often been associated with the phenomenon of "Ca^{2+} wave propagation" observed in multiple models of cell-to-cell communication.[29] This has suggested that regenerative processes of "nucleotide-induced nucleotide release" may be a widespread response to local tissue stress due to the paracrine activation of P2 receptors in adjacent or nearby cells (Figure 5.3).

All cells contain millimolar levels of cytosolic ATP in contrast to the submicromolar levels of ATP measured in most extracellular compartments. Since most cells also maintain negative membrane potentials, there is a very high electrochemical

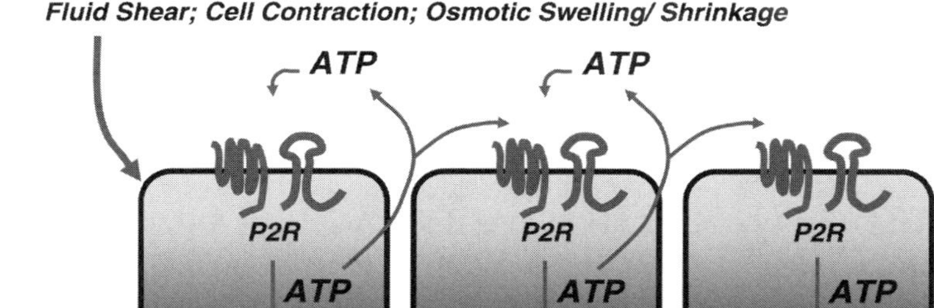

FIGURE 5.3 ATP released from nonexcitatory cells in response to mechanical stimulation can trigger autocrine and paracrine signaling cascades. ATP can be released from many types of cells subjected to diverse mechanical stimuli. ATP released from the single cell or several cells that were the direct targets of the mechanical stimulus can activate P2 receptors (P2R) in nearby cells. Activation of most P2 receptor subtypes triggers increases in cytosolic [Ca^{2+}] due either to mobilization of intracellular Ca^{2+} stores by various P2Y receptors or to Ca^{2+} influx elicited by P2X receptors. ATP release pathways (exocytotic or channel-mediated) in most cells are activated by increased cytosolic [Ca^{2+}]. Thus, ATP locally released from one or several mechanically stressed cells within a discrete tissue locus can trigger spreading "waves" of ATP-induced ATP release.

gradient for ATP efflux across the plasma membrane. Consistent with the critical bioenergetic role of intracellular ATP, efflux of cytosolic ATP is generally very low due to the limited permeability of the plasma membrane to nucleotides. However, a growing number of studies of stimulated or stressed cells have suggested that plasma membrane permeability to ATP can be transiently increased due to the presumed activation/gating of transporters or channels that can accommodate the size and charge of nucleotides.[25,30–33] Given the electrochemical driving force for ATP efflux, even brief activation of such facilitated transport systems could rapidly increase the rate of ATP release to extracellular compartments (Figure 5.1). Thus, the identification and characterization of membrane transport proteins that can act as potential conduits for the regulated efflux of cytosolic nucleotides has become an active area of investigation.

Another obvious but poorly characterized pool of extracellular ATP is that which is released from irreversibly lysed cells in tissues subjected to strong mechanical trauma (Figure 5.1). Other lytic or necrotic sources of extracellular nucleotide may occur during massive infection of host tissue with microbial pathogens. In this case, both the invading microorganisms and host cells are possible sources.[34] Bacteria will be rapidly killed by innate immune mechanisms during the initial phases of infection, while host cells may be increasingly killed at later stages by bacterially derived toxins or, as bystanders, by host-derived lytic molecules directed against the microbes.

Regardless of the mechanism by which ATP or other nucleotides might be released from cells in different tissues or physiological contexts, their half-lives within extracellular compartments will generally be short due to rapid catabolism by ecto-nucleotidases. Moreover, it is increasingly evident that analysis and interpretation of nucleotide release mechanisms requires an understanding of the biochemical properties of these ecto-enzymes, as well as their localization within subdomains of the surface plasma membrane. Thus, this chapter will also provide brief descriptions of the major ecto-nucleotidase families.

5.2 RELEASED NUCLEOTIDES AS EXTRACELLULAR SIGNALING MOLECULES: WHICH NUCLEOTIDES AND IN WHAT CONCENTRATIONS?

Table 5.1 illustrates the selectivities of the 15 molecularly defined P2 receptor subtypes for physiological nucleotide agonists. While all seven ionotropic P2X receptor subtypes utilize ATP as their principal agonistic ligand, most of the G protein–coupled P2Y subtypes exhibit a preferred, and sometimes absolute, selectivity for nucleotides other than ATP *per se*.[9,10] These additional P2Y agonists include UTP, UDP, ADP, and UDP glucose. Thus, it is reasonable to assume that, in addition to ATP (the most abundant physiological nucleotide), these latter nucleotides also transiently accumulate in the extracellular compartments which comprise the local environments of cells expressing $P2Y_1$, $P2Y_4$, $P2Y_6$, $P2Y_{12}$, $P2Y_{13}$, or $P2Y_{14}$ receptors. It is important to stress that the extracellular accumulation of nucleotides such as UTP, UDP, and ADP may not necessarily involve a requirement for their selective release from intracellular pools. Rather, these nucleotides may accumulate as a secondary consequence of the extracellular metabolism (degradation or synthesis) of nucleotide precursors that are directly released in various physiological or pathological conditions.[35–39] Thus, ADP or UDP may either be released directly or be generated via the extracellular hydrolysis of directly released ATP or UTP. It is also

TABLE 5.1
Agonist Selectivities and Affinities of the P2 Receptor Subtypes

Receptor Subtype	Signaling Mechanism	Selectivity for Physiological Nucleotide Agonists	EC_{50} for Physiological Nucleotide Agonist
$P2X_{1-7}$	Cation channel: depolarization and/or Ca^{2+} influx	ATP >> ADP, UTP, UDP	ATP: 1–10 μM ($P2X_{1-6}$); ATP: 100 μM ($P2X_7$)
$P2Y_1$	Gq → PI-PLC	ADP > ATP >> UDP, UTP	ADP: 0.1 μM
$P2Y_2$	Gq → PI-PLC	ATP = UTP >> ADP, UDP	ATP, UTP: 1–3 μM
$P2Y_4$	Gq → PI-PLC	UTP >> ATP, UDP, ADP (human); UTP = ATP >> UDP, ADP (rodent)	UTP: 1–10 μM
$P2Y_6$	Gq → PI-PLC	UDP > UTP > ADP >> ATP	UDP: 0.1–1 μM
$P2Y_{11}$	Gq → PI-PLC and Gs → AC	ADP > ATP >> UDP, UTP	ATP, ADP: 10 μM
$P2Y_{12}$	Gi → AC/others	ADP > ATP >> UDP, UTP	ADP: 0.1–1 μM
$P2Y_{13}$	Gi → AC/others	ADP > ATP >> UDP, UTP	ADP: 1–10 μM
$P2Y_{14}$	Gi → AC/others	UDP-glucose >> UTP, ATP, UDP, ADP	UDP-glucose: 0.1–1 μM

possible that extracellular UTP may be secondarily generated via an ecto-nucleoside diphosphokinase (NDPK)-mediated transphosphorylation of directly released UDP by ATP that is coreleased from the same cell or released coincidently from adjacent cells.[40–43] These multiple mechanisms for extracellular accumulation of particular P2 nucleotide agonists underscore why the characterization of potential nucleotide release mechanisms in a particular tissue or cell model usually involves a corresponding analysis of extracellular nucleotide metabolism in that cell system.

Given that intracellular ATP levels are 10 to 50 times higher than the other purine or pyrimidine nucleoside triphosphate species (GTP, UTP, CTP, TTP), it is not surprising that evolutionary selection has resulted in adenine nucleotide–selective binding pockets for 11 of the 15 known P2 receptor subtypes. However, it is interesting to consider why the remaining four P2 receptor subtypes have evolved pyrimidine nucleotide–binding sites that selectively recognize uridine- versus cytidine- or thymidine-containing species.[44] Given the important role of UDP sugars in protein glycosylation within the endoplasmic reticulum (ER) and Golgi, it is possible that residual uridine nucleotides contained within the Golgi-derived constitutive secretory vesicles are coreleased with the major protein cargo of those vesicles. As discussed in a later section, Lazarowski and colleagues have demonstrated that UDP glucose is constitutively released from several cell types examined thus far.[45] This continuous delivery of uridine nucleotides to extracellular compartments as an ancillary "wastage cost" to the critical and ubiquitous function of specific glycoprotein/protein secretion may have favored the evolution of G protein–coupled receptors that are selectively activated by uridine nucleotides.

Table 5.1 also indicates that the EC_{50} values for nucleotide agonists at all P2Y and P2X receptor subtypes (with the exception of $P2X_7$) are in the 0.1 to 10 μM range. As described below, the basal concentration of ATP (and presumably other nucleotides) in generic extracellular compartments is maintained in the 1 to 10 nM via the actions of the various ecto-nucleotidases. Thus, for the activation of P2Y and P2X receptors, the concentrations of ATP, ADP, UTP, UDP, and UDP glucose must transiently rise by 10- to 100-fold to micromolar levels within particular intercellular or interstitial compartments during physiological signal transduction.

5.3 ANALYSIS OF NUCLEOTIDE RELEASE/ACCUMULATION IN EXTRACELLULAR COMPARTMENTS

5.3.1 *In Situ* Tissues: Temporal and Spatial Considerations

It is important to consider how much nucleotide mass must be transferred from intracellular compartments in order for the extracellular concentration of a nucleotide to vary from a basal level in the nanomolar range to a stimulated level in the micromolar range. Figure 5.4 illustrates this, with a generic case of ATP being released into the interstitial compartment of a solid tissue volume containing 10^6 cells. Given an average cellular diameter of ~10 μm, this tissue volume will contain approximately 1 μL of intracellular compartment. Assuming a homogenous intracellular concentration of 5 mM ATP, there will be 5000 pmol of ATP mass within this 1-μL tissue

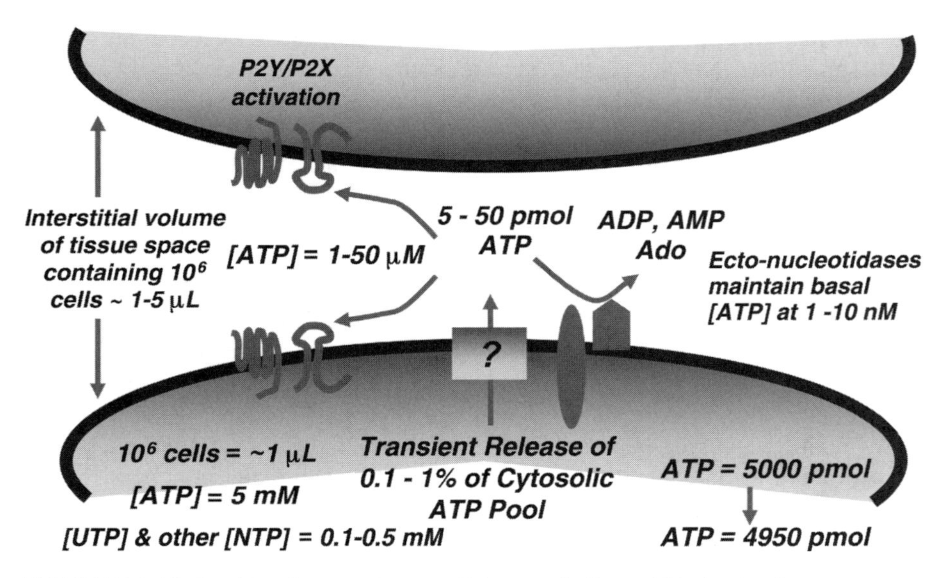

FIGURE 5.4 Nucleotide release and extracellular metabolism within *in situ* tissue compartments: Temporal and spatial considerations. A "compartmental analysis" view of ATP release into the interstitial compartments of a typical *in situ* tissue. The question mark indicates that the actual pathway for ATP release from most nonexcitatory cells remains incompletely characterized. See text for discussion and details.

compartment. Depending on the packing density of the cells, the extracellular (i.e., interstitial) volume of this tissue sample will be in the 1 to 5 μL range. Assuming that the basal extracellular ATP concentration is 10 nM, this extracellular volume will contain 10 to 50 fmol (0.01 to 0.05 pmol) of ATP mass under resting conditions. To achieve a stimulatory ATP concentration of 5 to 10 μM (i.e., the EC_{50} for some P2 receptors) would require an increase of 5 to 50 pmol of ATP mass within this 1- to 5-μL extracellular compartment. Thus, the transfer or release of only 0.01 to 0.1% of the 5000 pmol of intracellular ATP mass will be sufficient to raise the extracellular ATP concentration within this tissue to the level required for P2 receptor activation. Under dynamic physiological conditions, this released ATP will also be rapidly metabolized by cell surface or locally secreted ecto-nucleotidases; this serves to both terminate P2 receptor signaling and to convert the extracellular ATP to adenosine, which can be readily transported back into the cells for serial rephosphorylation and regeneration of the intracellular ATP pool.

Thus, the physiologically relevant accumulation of extracellular ATP within any given tissue compartment *in vivo* will reflect time-dependent and nonlinear changes in the rates of ATP release, hydrolysis by ecto-nucleotidases, and synthesis by ecto-nucleoside diphosphokinases. However, with very few exceptions (described in a later section), the direct measurement of ATP or other nucleotides within *in situ* extracellular compartments has been largely precluded by the technical limitations inherent in assaying low levels of metabolically labile, small organic molecules within very restricted interstitial spaces. Thus, most *in situ* studies of nucleotide release have employed indirect approaches based on the perturbation of physiological cell-to-cell communication within semi-intact preparations (e.g., isolated, perfused organs or perifused tissue slices) by exogenously added P2 receptor antagonists or soluble nucleotidases. The relative paucity of P2 receptor subtype–selective antagonists has been a limiting factor for the first type of approach. In contrast, the use of extrinsic nucleotidases, such as potato apyrase, which can rapidly scavenge endogenously released ATP, ADP, UTP, or UDP, has been widely and successfully used to provide support for P2 receptor–based paracrine signaling pathways in a broad range of isolated tissues.[22,26,27,29,46–61] However, because apyrase will scavenge the nucleoside triphosphate (NTP) and nucleoside diphosphate (NDP) ligands for all P2 receptor subtypes (except $P2Y_{14}$), this approach provides no direct information regarding the species or quantities of nucleotides that are released into, or generated within, the relevant extracellular compartments.

5.3.2 *In Vitro* Model Systems: Temporal and Spatial Considerations

Due to the technical limitations of directly assaying nucleotides within the spatially restricted extracellular compartments of intact tissues or *in situ* organs, most studies of nucleotide release have been performed using isolated cells or tissues maintained under defined *in vitro* culture conditions. As illustrated in Figure 5.5, this approach usually involves incubation of the isolated cells/tissues within relatively large volumes of extracellular media and then sampling or assaying this extracellular compartment for nucleotides released from the cells under basal conditions or following

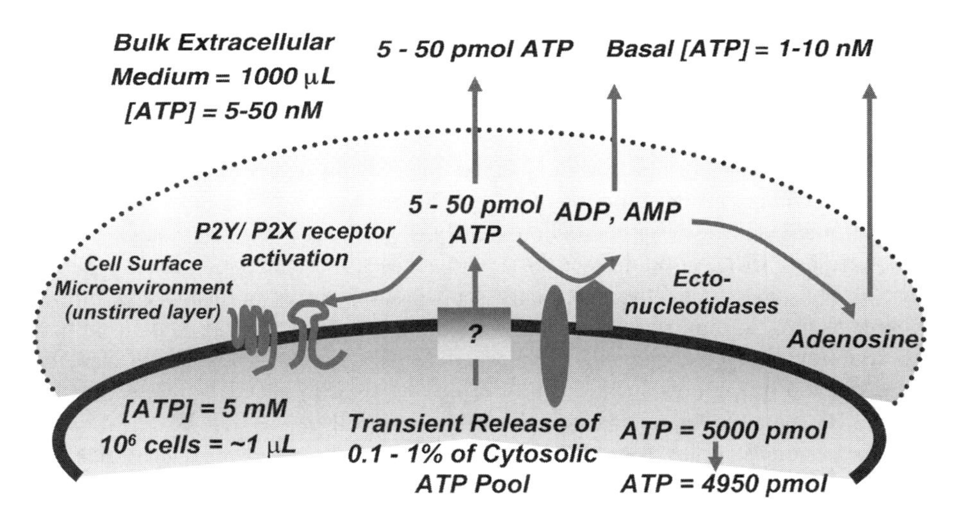

FIGURE 5.5 Nucleotide release and extracellular metabolism under experimental conditions with tissue-cultured cells or isolated tissue slices: Temporal and spatial considerations. A "compartmental analysis" view of ATP release into the bulk extracellular compartment versus the cell surface microenvironment under the standard experimental conditions used for most *in vitro* analyses of nucleotide release. The question mark indicates that the actual pathway for ATP release from most nonexcitatory cells remains incompletely characterized. See text for discussion and details.

stimulation. As noted above, the ability of ATP or other nucleotides to act as intercellular signaling molecules depends on the kinetics and magnitude of accumulation within interstitial tissue spaces and intercellular junctions. It is unclear whether the dynamics and magnitudes that characterize extracellular nucleotide accumulation within such *in vivo* compartments are effectively preserved in the experimentally tractable *in vitro* models. Several factors complicate or limit studies aimed at the direct quantitative evaluation of physiologically relevant nucleotide accumulation/release in such *in vitro* assay systems. Most methods for determining nucleotide concentrations involve analysis of samples drawn from the bulk extracellular solution media bathing cells. The ratio of extracellular volume to intracellular volume is usually very large (~1000:1) in these *in vitro* systems when compared to the ~1:1 to 10:1 ratio that characterizes an intact tissue. Thus, there will be a massive dilution of any released nucleotides into the large bulk extracellular volume. For example, the release of 5 to 50 pmol ATP from 10^6 adherent cells on a tissue culture surface (e.g., a 35-mm dish) will raise the extracellular ATP to only 5 to 50 nM when diluted into a 1-mL bulk extracellular volume of bathing medium. Consequently, it is important to consider the disparity between observed changes in concentrations of released nucleotides within such bulk extracellular volumes and the concentrations required for paracrine or autocrine activation of P2 receptors.

Another important factor concerns the rates of diffusion or convective transport that will characterize equilibration of released nucleotides into a bulk extracellular volume. Without convective mixing, the rate of nucleotide diffusion from the cell surface microenvironment into the bulk assay volume may be limited by biophysical factors, such as the presence of extensive cell surface membrane invaginations, or diffusional barriers, such as secreted mucus layers or generic unstirred layers. In such cases, released nucleotides may remain effectively "trapped" at the surface of the releasing cells and thereby equilibrate with the bulk extracellular solution relatively slowly. This may be offset in *in vitro* experiments using suspended cells that are stirred, or with adherent cell monolayers that are physically agitated, to produce a more efficient convective mixing of released nucleotides with the bulk extracellular compartment. However, as noted previously, mechanical stresses, such as those produced by stirring or fluid shear force, will act as strong nucleotide release stimuli in most cell types. Thus, measurements of nucleotide release in response to various nonmechanical stimuli (e.g., neurotransmitters, hormones, or local mediators) may be confounded by the unavoidable mechanical stimulation produced by standard experimental procedures such as the brief mixing used to ensure homogenous dispersion of the added neurotransmitter/hormone stimuli or the withdrawal of extracellular media samples for subsequent analysis of the released nucleotides. As described in a later section, these complicating factors have spurred the development of methods that permit the real-time measurement of nucleotides within, or close to, the microenvironment of the immediate cell surface.

Further discrepancies between cell surface and bulk phase nucleotide concentrations will be produced by rapid metabolism (degradation or synthesis) of released nucleotides by the cell surface ecto-nucleotidases and ecto-nucleoside kinases. If the rate at which a released nucleotide is metabolized by these ecto-enzymes is rapid relative to the rate at which the nucleotide equilibrates with the bulk medium, then the latter compartment will predominantly accumulate nucleotide metabolites, rather than the released nucleotide *per se*. If the released nucleotide is rapidly metabolized to its component nucleoside (e.g., ATP → adenosine), it is also possible that the generated nucleoside may be rapidly re-accumulated by the releasing cells via the widely expressed concentrative nucleoside transporters. In extreme cases, only minor amounts of the released nucleotide or its component nucleoside may accumulate in the bulk extracellular medium despite a significant rate of nucleotide release from the stimulated cells. For these reasons, studies aimed at the direct measurement of nucleotide/nucleoside accumulation in the bulk extracellular compartment can be significantly facilitated by the use of pharmacological agents that inhibit the ecto-nucleotidases and/or the concentrative nucleoside transporters. The appropriate application of this type of pharmacological approach requires an understanding of which ecto-nucleotidase subtypes and nucleoside transporters are expressed in the cells or tissue being investigated. Thus, the following overview of these ecto-nucleotidases and nucleoside transporters may provide useful background information for the subsequent discussion of experimental approaches for assaying nucleotide release and extracellular accumulation.

5.4 ROLES OF EXTRACELLULAR NUCLEOTIDE METABOLISM IN THE REGULATION OF NUCLEOTIDE LEVELS IN EXTRACELLULAR COMPARTMENTS

Four major families of ecto-nucleotidases have been identified to date:[15] (1) the ecto-nucleoside 5′ triphosphate diphosphohydrolases (eNTPDases/CD39 family of ecto-apyrases);[17] (2) the ecto-nucleotide pyrophosphatases/phosphodiesterases (eNPPs);[18] (3) the glycosylphosphatidylinositol (GPI)-anchored ecto-5′-nucleotidase (also known as CD73);[14] and (4) the GPI-anchored alkaline phosphatases (APs).[62] The eNTPDase, eNPP, and AP families include multiple subtypes encoded by separate genes and all plasma membrane ecto-enzymes are oriented with their active sites facing the extracellular compartment.

5.4.1 CD39 Family Ecto-Nucleotidases

The original CD39 gene product was identified as a cell surface marker for certain subsets of differentiated B and T lymphocytes.[63] CD39 structure is characterized by intracellular amino- and carboxy-termini, two transmembrane segments, and a large extracellular loop. Subsequent studies demonstrated that CD39 acts as a cell surface apyrase (ATP, ADP → AMP + phosphate) that can hydrolyze all naturally occurring NTPs and NDPs with similar kinetic constants (Km ~ 20 to 60 μM).[64–66] Thus, this enzyme will serially convert released ATP to ADP and then to AMP. Given its nonselectivity for the nucleotide bases, CD39 also acts as an effective scavenger for UTP and UDP, the preferred physiological ligands for the $P2Y_4$ and $P2Y_6$ receptors. Despite its initial identification as a lymphocyte marker antigen, CD39 is expressed in many tissues and at particularly high levels in vascular endothelial cells.[67–69] Within the vasculature, CD39 plays an important role in scavenging extracellular ADP, a potent activator of platelet aggregation and secretion. Consistent with this latter function, CD39 knockout mice exhibit marked alterations in normal hemostasis and prothrombotic susceptibility.[68] Biochemical analyses of cells expressing native or recombinant CD39 have demonstrated that plasma membrane CD39 exists as a tetrameric complex, and disruption of this oligomerization by detergent extraction represses its catalytic activity.[70] However, when only the extracellular loop of CD39 is fused to the signal peptide of CD4, the resulting chimera is expressed as a constitutively secreted monomer with high ecto-ATPase activity.

In addition to CD39 *per se*, the eNTPDase family of ecto-nucleotidases includes five other structurally related enzymes.[17,71] These include the CD39-like or CD39L1-4 gene products that contain a homologous "apyrase conserved region," or ACR. Like CD39/NTPDase1, CD39L1/NTPDase2[72–74] and CD39L3/NTPDase3[71,75] have been verified as cell surface enzymes that degrade extracellular NTP to NDP, and NDP to nucleoside monophosphate (NMP), with varying substrate specificities, releasing inorganic phosphate (Pi). CD39L1/NTPDase2 and CD39L3/NTPDase3 also share the two-transmembrane domain topology of CD39/NTPDase1. NTPDase4 is another related protein with an ACR and two predicted transmembrane domain that functions primarily as a Golgi-localized UDPase.[76]

The other family members, CD39L2/NTPDase6 and CD39L4/NTPDase5, are predicted to contain only one N-terminal membrane-spanning region.[71,77–80] Moreover, there is a cleavage site near this N-terminal domain that results in processing and release of secreted form of these proteins. Secretion of human CD39L4/NTPDase5 as an enzymatically active nucleotidase has been demonstrated using CD39L4-transfected COS cells.[78,81] Likewise, both human and rat orthologues of CD39L2/NTPDase6 can be expressed in membrane-associated forms and soluble forms that are secreted to the extracellular medium.[79,80,82] When heterologously expressed as an epitope-tagged construct in CHO cells or PC12 cells, most of the membrane-associated rat CD39L2/NTPDase6 is associated with intracellular membranes (ER and Golgi) rather the plasma membrane.[82] Thus, certain CD39 family ecto-nucleotidases may be constitutively secreted into interstitial compartments to provide additional nucleotide-scavenging capacity at sites other than the immediate cell surface. Although the various CD39L proteins function as ecto-nucleotidases that recognize all naturally occurring purine and pyrimidine nucleotides, their particular substrate selectivities distinguish them from the original CD39 prototype.

CD39L1/NTPDase2 has 30-fold higher activity with triphosphate substrates versus diphosphates,[15,16,83,84] and CD39L4/NTPDase5[81] and CD39L2/NTPDase6[80,82] have 20- to 100-fold higher activities with diphosphate substrates over triphosphates. Thus, some CD39 isoforms function predominantly as NTPases and others as NDPases.

5.4.2 eNPP Family Ecto-Nucleotidases

Studies of extracellular nucleotide metabolism in some cells perfused with ATPase-resistant synthetic ATP analogs indicated the presence of an ATP pyrophosphatase activity that catalyses the reaction: ATP $\rightarrow$ AMP + pyrophosphate.[16] Members of the ecto-nucleotide pyrophosphatase/phosphodiesterase gene family hydrolyze internal phosphodiester bonds of various adenine nucleotides (including ATP and cyclic AMP) to directly yield AMP.[18] Thus, to varying degrees, these NPP enzymes are capable of hydrolyzing ATP to AMP and PPi, as well as ADP to AMP and Pi, 3′,5′-cAMP to AMP, and NAD^+ to AMP. This family includes eNPP1 (formerly termed PC-1), eNPP2 (formerly termed PD-1α or autotaxin), and eNPP3 (formerly termed PD-1β). eNPP1 and eNPP3 are expressed as type-2 membrane proteins with a cytosolic NH_2 locus and a single transmembrane domain. eNPP1/PC-1 and eNPP2/autotaxin are widely expressed in many tissues, while eNPP3/PD-1β appears more restricted.[85]

eNPP1/PC-1 plays particularly important roles in bone and other mineralizing tissues due its ability to hydrolyze released ATP into AMP and pyrophosphate. The resulting pyrophosphate is further hydrolyzed to phosphate by tissue nonspecific alkaline phosphatase (TNAP).[86] Because pyrophosphate acts as a potent inhibitor of mineralization, modulation of the relative activities or expression of eNPP1 and TNAP within the bone matrix regulates net mineralization. Recent studies have also indicated that eNPP1 expressed in peripheral tissues other than bones may serve as a general physiological inhibitor of calcification, likely by generating local pools of extracellular pyrophosphate.[87–90] Significantly, mutations that produce human NPP1 deficiency have been linked to a syndrome of spontaneous infantile arterial and periarticular calcification.[89] Another important function of eNPP1/PC-1 is its ability to act as an inhibitor of the insulin receptor activation steps (e.g., autophosphorylation and tyrosine kinase induction) normally triggered by insulin binding.[91,92] Thus, overexpression of eNPP1/PC-1 in insulin-target tissues has been associated with insulin resistance and type-2 diabetes in humans[93] and animal models.[94] This may involve a direct interaction of eNPP1/PC-1 with the insulin receptor and appears to be independent of the enzyme's nucleotide pyrophosphatase activity.

eNPP2 or autotaxin functions primarily as a secreted ecto-lysophospholipase D for generating extracellular lysophospholipids, rather than as an ecto-nucleotidase.[95,96] NPP3 was initially described as $gp130^{RB13\text{-}6}$ since it is the same as a plasma membrane–associated glycoprotein of 130 kDa that is recognized by the monoclonal antibody RB13-6.[97,98] This protein is also identical to B10, a protein on the apical surface of rat hepatocytes.[85] In nonactivated human basophils, NPP3 is retained in an intracellular membrane site but is rapidly upregulated to the cell surface when these cells are stimulated by anti-IgE.[99,100] This upregulation of NPP3, now also designated as CD203c, is being used clinically as a possible blood marker of asthmatic activation. Like eNPP1, eNPP3 can degrade ATP to AMP. Four other members (NPP4 to NPP7) are less characterized but are predicted to be type-1 membrane proteins with extracellular catalytic domains. Choline phosphate esters appear to be the major substrates for NPP6 and NPP7, which act as lyso-phospholipase C enzymes.[18]

Like some CD39L family ecto-nucleotidases, both soluble and membrane-associated forms of eNPP1 and eNPP3 have been described or proposed.[101,102] eNPP2 appears to be predominantly secreted as a soluble ecto-enzyme.[103] Stefan et al.[85] used Northern analysis to screen all of the major rat organs for the three known eNPP gene products. While mRNAs for all three genes were detected in rat heart, the signals for eNPP1/PC-1 and eNPP3/PD1β were particularly strong. These mRNAs also show striking changes in hepatic expression during rat development and following liver injury. This raises the possibility that expression of these genes in other tissues will exhibit regulation by developmental and tissue injury signals.

5.4.3 CD73 Ecto-Nucleotidase and Ecto-Alkaline Phosphatases

The ecto-5′nucleotidase activity associated with CD73 leukocyte marker antigen has been extensively characterized.[14,104] CD73 is a prototypic GPI-anchored plasma membrane protein and 5′-nucleotidase activity has often been used as a plasma membrane marker.[105,106] This enzyme hydrolyzes extracellular AMP (Km ~ 1 to 50 μM) rather than ATP or ADP (which act as inhibitors), and thus acts in concert with the CD39 or eNPP ecto-nucleotidases to generate adenosine from adenine nucleotides released to extracellular compartments. Finally, extracellular AMP can also be hydrolyzed to adenosine by any of four ecto-alkaline phosphatase isoforms.[62,86] Like CD73, alkaline phosphatases are GPI-anchored ectoenzymes and hence are known to be colocalized with detergent-resistant and glycolipid-rich membrane subdomains such as lipid rafts.[15,16] The fate of extracellular adenosine additionally depends on the presence of extracellular adenosine deaminases that can further degrade the molecule to inosine, as well as nucleoside transporters that allow the accumulated adenosine (and inosine) to be taken up by cells for reincorporation into the intracellular nucleotide pools.

5.4.4 Ecto-NDPK and Ecto-Adenylate Kinases

The identification of ATP/NTP synthetic pathways, characterized by ecto-NDPK or ecto-adenylate kinase (ecto-AK) activities, have further complicated the interpretation of extracellular ATP/NTP accumulation in different tissues.[35,37,38,40] Ecto-NDPKs facilitate the transphosphorylation of extracellular ADP/NDP by other nucleotide triphosphates, while ecto-AKs can reversibly convert two extracellular ADP equivalents into ATP and AMP. Ecto-NDPK and/or ecto-AK activity may be important components of nucleotide-entrapment cycles that prolong the elevation of ATP in the extracellular compartments, and significant ecto-NDPK activity has been reported in 1321N1 human astrocytes, C6 rat glioma, and endothelial cells.[35–39,43]

5.5 Methods for Studying Release and Accumulation of Nucleotides in Extracellular Compartments

5.5.1 General Methodological Considerations

Experimental approaches for characterizing nucleotide release and accumulation in extracellular compartments can be broadly divided into two categories: 1) "offline" protocols that involve serial sampling of an extracellular compartment at selected time points followed by subsequent chemical or biochemical analyses of the nucleotides in the collected samples; and 2) "online" protocols that involve direct and time-resolved detection of nucleotide content within a given extracellular compartment as it is continuously conditioned by the cells being studied. Each approach presents several advantages and disadvantages.

5.5.1.1 Offline Protocols: Advantage and Disadvantages

Major advantages of the offline protocols are: 1) The serial extracellular samples can be assayed for multiple nucleotides and their metabolites. For example, the same sample might be analyzed for directly released ATP as well as the ADP, AMP, and adenosine resulting from its rapid metabolism by ecto-nucleotidases. Alternatively, a given sample could be assayed for both adenine and uridine nucleotides. 2) Because the extracellular media samples are no longer in contact with living cells, they can be processed as needed (e.g., by strong acid extraction, organic extraction, or chemical derivatization) to optimize the chemical or biochemical detection of nucleotides. The major disadvantages of offline protocols concern the actual collection of the extracellular samples and, as discussed previously, the degree to which nucleotide content in the collected samples reflects transient and highly localized changes in nucleotide content at cell surfaces or interstitial compartments of tissues.

5.5.1.2 Online Protocols: Advantage and Disadvantages

Major advantages of the online protocols are: 1) They are well suited for following the highly dynamic changes in extracellular nucleotide concentration produced by cells or tissues under minimally invasive experimental conditions. 2) They can provide unique spatial information by being adapted to monitor nucleotide levels at the immediate or near immediate surface of intact cells or tissues. However, several disadvantages offset these strengths: 1) At present, all available online methods are limited to the direct assay of extracellular ATP only; these include firefly luciferase–based bioluminescence protocols, amperometry based on ATP-sensitive enzyme electrodes, cellular biosensors based on ATP-selective ionotropic P2X receptors, and atomic force microscopy (AFM) sensors based on cantilevers coated with ATP-sensitive myosin heads. 2) As evident by this latter list, the currently available online protocols generally require quite specialized instrumentation and expertise. 3) The online sensors of extracellular nucleotide concentration need to operate under physical (e.g., temperature, pH, ionic strength) and biochemical (presence of other extracellular metabolites and macromolecules) conditions required for normal cell function but which may reduce efficacy of the nucleotide sensor.

5.5.2 Offline Protocols for Measuring Extracellular Nucleotides

Two major issues need to be considered in the evaluation of the offline methods: 1) how and when the extracellular samples are collected; and 2) how these samples are assayed for nucleotide content.

5.5.2.1 Methods for Collecting Extracellular Medium Samples

The most straightforward procedure is the removal of aliquots from a static extracellular medium compartment. The major variable here is the volume of sampled aliquot(s) relative to the overall volume of extracellular medium bathing the cells or tissue. In simple endpoint experiments, the entire extracellular volume can be removed, while in time course experiments, small aliquots can be serially removed from the bulk medium. The minimum volume of such aliquots will largely depend on the sensitivity of the assay system used for measuring released nucleotides and the absolute magnitude of extracellular nucleotide release or accumulation in the particular model system. A more involved media collection method involves the classic approach used by endocrinologists of continuously perfusing/supefusing intact tissues or cells with fresh extracellular medium and periodically assaying aliquots of the perfusate/superfusate for nucleotide content.[107] Critical variables for this approach are the perfusion rate and the relative mechanical shear that may be produced by perfusion rate. Guyot and Hanrahan have described a particularly innovative variant of this method that involved growing 10^8 airway epithelial cells as confluent monolayers within the inner lumens (100 cm^2 total surface area) of the permeable hollow-fiber capillaries in a cell culture "bioreactor" system.[108] This permitted support of basic metabolism via basolateral exchange with the large reservoir (500 mL) of tissue culture medium that bathes the capillaries while allowing simultaneous perfusion and sampling (for offline assay of nucleotides) of the apical extracellular compartment (1.4 mL total volume) at different luminal flow rates. These bioreactor cultures can be maintained for many months and the authors used them to study how changes in flow rate and/or perfusate osmolarity trigger rapid increases in nonlytic ATP release.

For the analysis of nucleotide levels within the interstitial compartments of solid tissues, several investigators have successfully adapted microdialysis-type methods that are routinely used for analyzing released neurotransmitters within the central nervous system.[109] A microdialysis probe usually consists of a thin (~100 to 300 μm diameter) concentric capillary probe (i.e., a narrower central rigid tube within a wider outer rigid tube). At the probe tip, the outer tube is replaced by a short length of semipermeable dialysis membrane. Following insertion of the probe into a solid tissue, the inner tube is perfused with saline, which is delivered to the probe tip and then returned via the outer tube to a fraction collector. Depending on the rate of perfusion (usually a few microliters per minute) and the permeability of the dialysis membrane tip, small molecules present in the interstitial compartment will equilibratively diffuse into the perfusing fluid. This collected fluid is then assayed for the endogenous extracellular molecules of interest. Such microdialysis approaches have been used to assay interstitial nucleotide levels (mainly ATP and ATP metabolites) in the brains,[110–112] kidneys,[113] and hearts[114–121] of living animals or isolated perfused organs.

5.5.2.2 Offline Methods for Analyzing Nucleotide Content in Extracellular Medium Samples

A wide range of analytical methods have been developed for measuring the generally low concentrations (submicromolar) of nucleotides and nucleosides present in collected samples of bulk extracellular medium or microdialysates. These can be broadly divided into chemically versus biochemically based procedures.

5.5.2.2.1 HPLC Analysis of Nucleotides

High-performance liquid chromatography, based on ion-exchange or reversed-phase resins and absorbance detection in the 250 to 260 nm wavelength range, has been widely used for the separation and mass quantitation of nucleotides and nucleosides (reviewed in Zakaria and Brown[122]) and can be readily adapted for analysis of extracellular media. The major advantage of HPLC-based methods is that, in principle, all extracellular nucleotide and nucleoside species of possible interest (ATP, ADP, AMP, adenosine, UTP, UDP, and UDP sugars) can be assayed within a single sample. However, given UV extinction coefficients in the 10 to 15 mM^{-1} cm^{-1} range and HPLC flow cell path lengths in the submillimeter range, the effective detection limit for nucleotide mass is in the 0.1 to 1 nmol range. This is at the upper range of the quantities (0.1 to 100 pmol) likely to be present in reasonable injected volumes (100 μL) of extracellular samples containing nanomolar to micromolar concentrations of various nucleotides. For this reason, some studies of nucleotide release and accumulation have employed metabolic labeling of intracellular nucleotide pools with ^3H-adenine or ^3H-uridine to increase detection sensitivity.[123] Indeed, the initial description of adenine nucleotide release from vascular cells by Pearson and Gordon utilized this approach.[124] However, the interpretation of such metabolic labeling approaches can be complicated by uncertainties regarding the kinetics and extent of isotopic equilibration with all relevant pools of intracellular nucleotide. For example, while cytosolic ATP pools will be rapidly labeled to steady-state by accumulated ^3H-adenine, ATP sequestered within granules or vesicles might require much longer periods of metabolic labeling.

Another very useful approach for increasing the detection sensitivity of HPLC-resolved adenine nucleotides involves their reaction with chloroacetaldehyde (prior to HPLC separation) to generate highly fluorescent etheno-derivatives.[125–133] This post-collection derivatization reaction is simple, reproducible, and quantitative. The etheno-derivatized forms of ATP, ADP, AMP, and adenosine in the 1 to 10 pmol range can be easily detected within a single extracellular sample by HPLC fluorescence detectors (270 nm $\lambda_{ex} \rightarrow$ 410 nm λ_{em}). For example, Lazarowski et al.[133] have recently adapted this approach to measure nanomolar levels of adenine nucleotides and adenosine in 20-μL samples of extracellular media collected from the air–liquid surface microenvironment of cultured airway epithelial monolayers.

5.5.2.2.2 Enzymatic Assay of Adenine Nucleotides by Luciferase-Based Methods

Firefly (*Photinus*) luciferase has been the most widely used reagent for the measurement of extracellular ATP release and accumulation in model systems ranging from single cells to intact tissues (as discussed in later sections).[134–137] In the presence of molecular oxygen, this enzyme catalyzes the following reaction between luciferin and MgATP:

$$\text{D-luciferin} + \text{ATP} \xrightarrow[\text{Mg}^{2+},\,\text{O}_2]{\text{Firefly luciferase}} \text{oxy-luciferin} + \text{AMP} + \text{PPi} + h\nu\ (562\text{ nm})$$

The reaction is both exquisitely sensitive and highly specific for ATP and yields 562-nm light with a high quantum yield using standard luminometers, microplate luminometers, or other photon-counting instruments such as liquid scintillation counters.[138,139] This reaction is so widely used to detect ATP because the luminescent intensity varies linearly over a range of picomolar (fmol/ml) to micromolar (nmol/ml) ATP concentrations.[135,137] Given its sensitivity, luciferase can be used to reliably measure the ATP content of small extracellular samples, such as the microliter volume samples collected using *in situ* microdialysis probes. Also, other nucleotides that may be released into extracellular compartments have negligible effects on the ATP-dependent light production at physiologically relevant concentrations.[140–143] Despite this very high selectivity for ATP, luciferase-based methods can be readily adapted for the indirect assay of extracellular ADP and AMP through the use of coupled enzymatic reactions that quantitatively convert AMP and/or ADP to ATP.[144–148] This "rephosphorylation" approach involves treating parallel aliquots of an extracellular sample with adenylate kinase and pyruvate kinase in the presence of phosphoenolpyruvate and a low (nanomolar) concentration of ATP to serially convert all AMP to ADP and then the ADP to ATP; this is then assayed by the luciferase reaction. A second sample can be treated with only pyruvate kinase plus phosphoenolpyruvate to convert only the extracellular ADP to ATP followed by the luciferase reaction. A third sample is directly assayed for ATP by the luciferase reactions. The bioluminescence signal from the first sample yields the total AMP + ADP + ATP content; the second sample yields the total ADP + ATP content, and the third yields the ATP content.

5.5.2.2.3 Enzymatic Assay of Uridine Nucleotides

A significant limitation of the luciferase method is that, at present, it cannot be used for the assay of the extracellular uridine nucleotides, which act as key agonistic ligands for the $P2Y_2$, $P2Y_4$, and $P2Y_6$ receptors. To address this issue, Lazarowski and Harden[41] developed an elegant enzyme-linked assay with high selectivity and sensitivity (picomolar to nanomolar) for UTP. The assay is based on the ability of UDP glucose pyrophosphorylase to quantitatively convert UTP to UDP glucose in the presence of glucose-1-phosphate (producing pyrophosphate as an additional product). To pull the reaction in the direction of UDP glucose synthesis, pyrophosphatase is also included to prevent accumulation of the pyrophosphate, which would otherwise limit the forward reaction. In practice, this assay also involves the use of ^{14}C-labeled glucose-1-phosphate; HPLC separation of the UTP, UDP glucose, glucose-1-P, and pyrophosphate reactants; and radiometric analysis of the HPLC-fractioned samples. This enzyme can also be adapted for the analysis of extracellular UDP levels by first treating collected samples with NDPK and exogenous ATP to convert ambient UDP to UTP prior to the assay of UTP. Differences in the UTP content of parallel samples processed with or without the NDPK step provide an index of UDP context. Finally, Lazarowski et al.[45] have capitalized on the ability of UDP-glucose pyrophosphorylase to catalyze the reverse reaction (pyrophosphate + UDP glucose $\rightarrow$ UTP + glucose-1-P) to develop a quantitative assay for UDP glucose within extracellular media samples.

This involves incubation of the samples with the enzyme and an excess of ^{32}P-pyrophosphate to drive the quantitative conversion of UDP glucose to ^{32}P-UTP.

5.5.3 Online Protocols for Measuring Extracellular Nucleotides

5.5.3.1 Luciferase-Based Protocols—Bulk Phase Measurements

Because the luciferase reaction can efficiently proceed under conditions of physiological ionic strength and pH, it can be used to directly and continuously monitor the ATP concentration in extracellular medium bathing cells or tissues. Indeed, some of the earliest studies of agonist-induced ATP release were (and continue to be) performed in suspensions of intact platelets, supplemented with extracellular luciferase/luciferin, in cuvettes that can be fitted into dual-purpose spectrophotometer/luminometers. Such optical systems have been routinely used to assess the kinetics and magnitude of key platelet functions such as shape change, aggregation, and secretion. Application of similar online luciferase-based methods was used by White and colleagues to monitor neurotransmitter-induced release of ATP from purified synaptosomes isolated from various peripheral nerve plexi.[149–151] The Rojas group used another adaptation of this approach to assay the time course and magnitude of exocytotic ATP release from small populations of adrenal chromaffin cells.[152–154] More recently, Schwiebert and colleagues pioneered the development of online luciferase methods suitable for monitoring ATP release from adherent monolayers of various cell types maintained in 35-mm dishes that can be directly placed into the sample chamber of a Turner TD20 luminometer.[155] This approach has been readily adapted for cells grown on either conventional plastic tissue culture surfaces or on permeable filters to permit the analysis of polarized epithelia or endothelia with maintained separation of the apical and basolateral extracellular compartments.[155,156]

Several caveats and limitations should be noted regarding online luciferase assays. First, luciferase undergoes a rapid ($t_{1/2}$ ~ 3 min) but reversible unfolding at temperatures >30°C and this results in greatly diminished activity.[157] Thus, the assay of ATP release/accumulation from the intact cells is best performed at room temperature to ensure stable luminescence readout. However, reduced temperatures may alter the rate and extent of ATP release or extracellular metabolism in cells being studied. Second, the extrinsic enzyme activity of luciferase can be sensitive to various pharmacological agents used to inhibit putative ATP release mechanisms. For example, gadolinium (Gd^{3+}) has been employed to test the role of mechanosensitive channels as possible regulators of, or conduits for, the ATP released in response to mechanical or hypotonic stress. However, Gd^{3+} also exerts an allosteric effect on luciferase that results in a right shift in affinity for ATP.[158,159] Thus, a Gd^{3+}-induced decrease in the luminescence triggered by a particular stimulus may reflect a reduction in luciferase activity rather than reduced ATP release. Another example concerns the effect of altered ionic strength on luciferase activity. In this case, a reduction in ionic strength, such as occurs during application of a hypotonic stress stimulus, increases the ATP affinity of luciferase.[160] Thus, an increase in luminescence in response to the hypotonic stimulus may reflect increased activity of luciferase by the ambient ATP present in bathing medium rather than (or in addition to) increased delivery of ATP to the bathing medium. All of these potential complications can be easily assessed by establishing parallel calibrations for the ATP-dependent luciferase activity in the presence and absence of each compound or condition used in a particular set of ATP release experiments. Finally, the exquisite ATP sensitivity of luciferase raises the possibility that an increased luminescence observed in response to addition of various hormones, growth factors, or cytokines may be due to low-level ATP contamination in the hormone, etc., rather than an increased rate of ATP release. This may be a particularly salient caveat when testing recombinant proteins expressed in bacteria or bacterially derived immune response modifiers (e.g., lipopolysaccharides or peptidoglycan) as potential regulators of nucleotide release.[161]

5.5.3.2 Luciferase-Based Protocols—Pericellular Measurements

As discussed previously, the presence of physical barriers and high ecto-nucleotidase activity at the cell surface may limit the delivery of released ATP to the bulk extracellular fluid phase. Thus, stimulus-induced changes in luciferase activity within this extracellular compartment may considerably underestimate the magnitude and dynamics of ATP release/accumulation at the cell surface. This has prompted the development of luciferase-based protocols for monitoring ATP within cell surface or immediate pericellular compartments. One approach has been to use modified forms of luciferase that can be directly or indirectly associated with the cell surface (Figure 5.6). Our group has used a chimeric protein (proA-luc) that consists of the immunoglobulin (Ig)-binding domain of *Staphylococcus aureus* protein A as an N-terminal fusion with luciferase.[162] Because protein A has very high affinity for Ig, the proA-luc will strongly bind to any immobilized fraction of Ig, such as antibodies directed against protein epitopes on the surface of intact cells. Thus, intact cells (adherent or in suspension) can be first treated with specific antibodies against natively expressed cell surface proteins or ectopically overexpressed proteins to provide binding sites for the proA-luc. This effectively concentrates and confines the proA-luc ATP sensor to the cell surface microenvironment. Using this method with platelets or astrocytes, we have reported that cell surface ATP concentrations rapidly increase to micromolar levels in response to the activation of certain G protein–coupled receptors.[162,163] One limitation of this approach is that it requires suitable and specific antibodies against high copy number surface proteins. Identifying such proteins and readily available antibodies may be difficult for some types of cells or tissues. To address this problem, Godecke et al.[164] further modified this approach by using nonspecific Ig protein chemically cross-linked to wheat germ agglutinin (WGA-Ig) as the Ig docking sites for proA-luc. This takes advantage of the ability of WGA to avidly bind to the generic N-acetylglucosamine residues present on most glycosylated proteins of the cell surface. Thus, WGA-Ig comprises a single reagent that can be used to localize high amounts of proA-luc on the surfaces

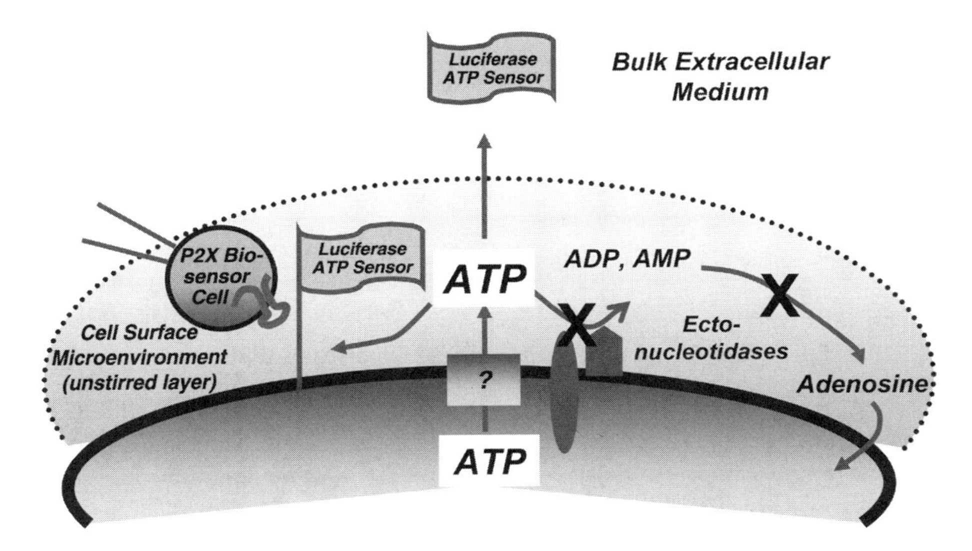

FIGURE 5.6 Experimental approaches for assaying released ATP within the pericellular or cell surface microenvironment. Two basic approaches have been developed for measuring ATP levels within localized, pericellular compartments. One method involves the use of P2X receptor–expressing single cells as ATP "biosensors" that can be both patch-clamped and physically micromanipulated into the pericellular space surrounding the cell being assayed for induced ATP release. ATP-evoked inward currents in the biosensor cell will provide a real-time index of the ATP released from the cell under investigation. An alternative approach involves the targeting or anchoring of genetically engineered versions of firefly luciferase onto the cell surface. In the presence of added extracellular luciferin, the bioluminescence of this surface-restricted luciferase will provide a real-time index of the ATP released from the cells before it is significantly metabolized or diluted into the bulk extracellular medium. See text for discussion and details.

of most cell types. For example, the authors employed this WGA-Ig to localize proA-luc to the surfaces of freshly isolated rat ventricular myocytes or cultured monolayers of human umbilical vein endothelial cells.[164] They showed that healthy myocytes do not release ATP during the normal contraction–relaxation cycle but that hypercontracted myocytes (an indicator of cell damage) are surrounded by a halo of increased luminescence that suggested an elevated cell surface level of ATP. In contrast to the absence of ATP release during the normal cycle of mechanical contraction in myocytes, application of brief mechanical shear force to the endothelial cells triggered a large and transient release of ATP.

A different approach for localizing luciferase to the cell has recently been reported by Pellegatti et al.[165] These authors generated an expression plasmid that encodes a protein chimera consisting of N-terminal luciferase fused to the GPI-anchor domain of the folate receptor. Upon transfection into mammalian cells, the GPI-anchored luciferase traffics to the plasma membrane where it can act as a sensor of cell surface ATP levels.

Another way of using luciferase to probe the pericellular or cell surface compartments for ATP is to use highly sensitive, cooled charge-coupled device (CCD) microscope cameras to spatially image localized luminescence signals from extracellular (bulk phase) luciferase/luciferin in the immediate vicinity of single cells. This approach has been successfully used to image ATP signals in the immediate pericellular spaces of single rat ventricular myocytes,[164] single astrocytes,[166,167] molluscan neurons,[168] and retinal Müller cells.[169] These studies have provided unique information regarding spatial and temporal aspects of ATP release that can be correlated with other functions of imaged single cells such as general permeability/integrity of the plasma membrane, shape changes, and release of other paracrine signaling molecules.

5.5.3.3 Luciferase-Based Protocols—*In Situ* Superfused Tissues

The vast majority of luciferase-based experiments have characterized ATP release from various types of isolated cells maintained in tissue culture conditions. However, a recent study by Nedergaard and colleagues has extended the use of luciferase for the analysis of localized ATP release in an *in situ* extracellular tissue compartment.[170] These elegant experiments involved exposed, but intact and *in situ* spinal cords (of anaesthetized rats) that were superfused with artificial cerebrospinal fluid (ACSF) containing luciferase/luciferin. Significantly, the luciferase-containing ACSF was delivered via thin tubing inserted through the otherwise intact dura that envelopes the spinal cord. Luminescence within discrete superficial areas (~5 × 8 mm) of the superfused spinal cord was imaged with a cooled CCD camera; the imaged regions included the midline artery for demarcation. The cord region on one side of the midline artery was subjected to a single, very brief mechanical trauma induced by a weight-drop device. Significantly, increased luminescence, indicative of elevated extracellular ATP, was maintained for several hours after the weight drop in a nominally undamaged zone (~1.4 × 0.2 mm) in the immediate region surrounding the point of mechanical impact. Luminescence was low at the immediate point of impart and in the contralateral (relative to the midline artery) region of the superfused cord. If a rat was subjected to acute cardiac arrest during the posttrauma period of imaging, the local zone of increased luminescence rapidly decreased to baseline levels consistent with a loss of intracellular ATP pools within those cells that were releasing ATP into the peritraumatic interstitial compartment. Besides being a technical *tour de force*, this study suggested that paracrine factors initially derived from acutely damaged cells trigger long-lasting periods of increased ATP release from nearby cells. As discussed in a later section, this result may be consistent with a model of regenerative ATP-induced ATP release in the peritraumatic zone.

5.5.3.4 Luciferase-Based Protocols—Luciferin Fluorescence

Sorensen and Novak have described another variant of a luciferase-based imaging approach that capitalizes on the differential fluorescence properties of luciferin versus oxy-luciferin.[171] When excited at 364 nm, luciferin emits fluorescence at 510 to 550 nM, and the emission intensity is reduced upon oxidization. Sorensen and Novak used confocal fluorescence imaging of isolated pancreatic acini bathed in medium containing extracellular luciferase and luciferin.[171] Upon cholinergic stimulation of exocytotic ATP release, the luciferin fluorescence in the immediate vicinity of discrete acinar cells decreased. This interesting approach can be combined with fluorescence imaging of other cellular parameters (e.g., cytosolic Ca^{2+}) that can play roles in the regulation of ATP release.

5.5.3.5 Quinacrine Fluorescence Imaging of ATP-Containing Vesicles in Intact Cells

Quinacrine is a weak base that accumulates in acidic subcellular compartments. When excited by 476-nM light, quinacrine emits fluorescence in the 500 to 540 nm range. Because quinacrine also binds ATP with high affinity, quinacrine fluorescence has been used as a marker of ATP-containing subcellular compartments in intact cells. When combined with confocal fluorescence imaging, localization of quinacrine-labeled puncta may provide an index of likely ATP-enriched secretory granules or vesicles. Likewise, a reduction in the intensity of such quinacrine-labeled granules in response to known ATP-release stimuli has been used to support models of exocytotic secretion of ATP from endothelial cells,[172] pancreatic acini,[171] ocular epithelial cells,[173] and multiple neuronal types.[174]

5.5.3.6 Cell-Based Nucleotide Biosensors—P2X Receptors as ATP Detectors

Cells expressing native or recombinant P2X receptors have used as biosensors of ATP released within the local pericellular space of cells activated by diverse stimuli (Figure 5.6). The basic method, pioneered by Okada and colleagues, is based on the ability of ATP to elicit inward Na^+ current in cells maintained in whole cell voltage-clamp mode.[46,175,176] A patch-clamp microelectrode with the attached biosensor cell is micromanipulated as close as possible to the target cell being studied. The target cell is then exposed to various stimuli known or hypothesized to elicit ATP release; inward currents elicited in the adjacent sensor cell provide an index of local ATP concentration. Exposure of the same sensor cell to known amounts of exogenous ATP is used to calibrate the evoked inward currents. A recent review by Hayashi et al.[176] provides a detailed presentation of the advantages and limitations of this approach.

Like all ligand-gated ion channels, P2X receptor channels are subject to desensitization/inactivation. However, the extent and rate of inactivation varies considerably among the different P2X receptor subtypes. For analysis of ATP release, most studies have used biosensor cells that express the slowly inactivating $P2X_2$ receptors; these have included PC12 pheochromocytoma cells that natively express $P2X_2$ receptors and HEK293 stably transfected with $P2X_2$ receptor cDNA. An important caveat is that some stimuli that trigger ATP release in the target being studied may also elicit ATP release from the sensor cell or may induce other membrane currents in the sensor. While this problem can be evaluated by proper controls, it may restrict the use of this method in some model systems. However, this approach has been successfully used to analyze stimulus-induced ATP release in multiple cell types, including pancreatic beta cells stimulated with insulin secretagogues,[46] neonatal rat cardiomyocytes subjected to hypoxia,[177] and various epithelial cells exposed to hypotonic stress.[178] A particularly elegant application of this method was described by Bell et al.,[175] who used PC12 sensor cells to record ATP release from freshly isolated macula densa cells derived from the thick ascending limb of rabbit nephrons. ATP release was triggered when the macula densa was exposed to increased [NaCl], which is the physiological stimulus for the tubuloglomerular feedback to the mesangial cells to control glomerular filtration. This supports a model wherein increased delivery of luminal NaCl to the distal nephron, as occurs during rapid changes in glomerular permeability or renal blood flow, elicits ATP release from the macula densa for paracrine regulation of P2 and adenosine receptors in cells of the glomerulus.

5.5.3.7 ATP-Sensitive Enzyme Electrodes

Electrochemical or amperometric methods based on carbon fiber electrodes have been extensively used for the online assay of biogenic amine neurotransmitters secreted from single cells in culture or *in situ*. Recently, Dale and colleagues have described an electrochemical method based on ATP-sensitive microelectrodes that can monitor local ATP release in a wide variety of *in vitro* or *in situ* tissue preparations.[179,180] The electrode consists of thin (25 to 100 μm) platinum (Pt) wire, which is coated with an ultrathin layer of protein solution containing glycerol kinase and glycerol-3-phosphate oxidase. In the presence of glycerol, ATP, and oxygen, this provides an enzymatic biosensor surface that rapidly catalyzes the coupled reaction:

$$\text{glycerol} + \text{ATP} \rightarrow \text{glycerol-3-phosphate} + O_2 \rightarrow \text{glycerone phosphate} + H_2O_2$$

When the Pt surface is polarized to +500 mV (relative to an Ag/AgCl reference electrode), the H_2O_2 product is rapidly consumed by the electrochemical reaction: $H_2O_2 \rightarrow O_2 + H^+ + 2e^-$. Thus, the resulting current is proportional to ambient [ATP]. The electrode current is linear in the 0.5 to 50 μM ATP range and the rise time response ($10 \rightarrow 90\%$) is <10 s. An important caveat is that the polarized Pt surface is also sensitive to other redox-reactive extracellular compounds (e.g., ascorbate, serotonin) that may be produced or released in different biological systems. Likewise, changes in extracellular pH will also alter the electrochemical properties of the electrode. Thus, unambiguous application of this technology requires paired control experiments using similar null electrodes that are similar in size and electrochemical properties but which lack the ATP biosensor surface coating.

A major advantage of these ATP-sensitive electrodes is that they can readily be used to monitor local, real-time ATP release within exposed but *in situ* tissues that are very difficult to study by the optical luciferase-based methods. Gourine et al.[180] have recently employed such electrodes to map CO_2-induced changes in extracellular ATP within chemosensitive regions on ventral surfaces of the exposed medulla oblongata in otherwise intact rats. Local changes in extracellular ATP were temporally correlated with phrenic nerve activity and measurements of end-tidal CO_2 content. These impressive studies suggested that ATP can be used as a general mediator of chemosensory signaling in the central nervous system.

5.5.3.8 Atomic Force Microscopy with ATP-Sensitive Cantilever Tips

Schneider et al. have described a very creative application of atomic force microscopy (AFM) methods for the online detection of ATP at the immediate extracellular surface of epithelial cells.[181] This approach was based on earlier AFM studies of various proteins with ATP-sensitive conformations.[182,183] Conventional Si_3Ni_4 AFM cantilever tips were coated with myosin S1 protein. The immobilized myosin S1 ATPase will hydrolyze any local ATP and this hydrolysis will induce a mechanical distortion of the cantilever tip. The exceptional spatial resolution (<1 μm) of AFM-based methods allowed the authors to detect "hotspots" of elevated extracellular ATP as the cantilever tip was scanned over the surfaces of control epithelial cells or epithelial cells engineered to overexpress cystic fibrosis transmembrane regulator (CFTR) channels. The magnitude of the ATP hotspots was increased in the CFTR-overexpressing cells consistent with a possible role of those channels in the regulation of ATP export to interstitial compartments.

5.6 CURRENT PERSPECTIVES ON NUCLEOTIDE RELEASE

5.6.1 Constitutive Release of Nucleotides from Cells in the Basal State

Recent studies from multiple laboratories indicate that many, if not most, cell types constitutively release low amounts of nucleotides even when maintained under basal conditions in the absence of any obvious metabolic or mechanical stress.[26,42,43,47,108,156,161,184–189] These observations echo the early studies by Forrester and Williams, who demonstrated that isolated rat ventricular myocytes release modest amounts of ATP (~0.4 nmol/mg protein) to the extracellular compartment even when incubated in a normal, oxygenated medium.[190] Data from three types of experiments have provided strong support for the existence of physiologically relevant, constitutive nucleotide release.

Several groups have reported that treatment of nominally unstimulated cells with extracellular nucleotide scavenger enzymes, such as potato apyrase or hexokinase, reduced the cytosolic levels of second messengers such as cyclic AMP and inositol trisphosphate.[23,24,27,42,43,188,191,192] This reduction in second messenger levels by nucleotide scavengers indicates that endogenous ATP (and/or UTP) is released from cells in amounts that are sufficient to produce low-level activation of various G protein–coupled P2Y receptors (Figure 5.3).

A second type of experimental support for constitutive ATP release was provided by Lazarowski and colleagues, who employed conventional isotopic tracer methods to measure the steady-state rates of extracellular nucleotide metabolism by four different cell lines.[43] They observed that the four cell types steadily maintained extracellular ATP at the 1 to 10 nM range for many hours under standard, serum-free tissue culture conditions. The cells were then pulsed with extracellular [γ^{32}P]ATP at tracer levels, which did not significantly change the extracellular ATP concentration. Significantly, the [γ^{32}P]ATP tracer was rapidly and completely metabolized. This demonstrated that the steady-state level of extracellular ATP in the cell cultures reflected a constitutive rate of ATP release that was balanced by ATP hydrolysis measured at 20 to 200 fmol/min/cell.

A third group of studies has used luciferase-based methods to directly measure the steady-state levels of extracellular ATP in cultures of various cell types before and after pharmacological inhibition of endogenous ecto-nucleotidases.[38,161,163] For example, adherent murine macrophages were observed to steadily maintain extracellular ATP levels in the 0.3 to 3 nM range in the absence of obvious cell stress or hormonal stimulation.[161] However, this extracellular level of ATP rapidly increased to new steady-state levels of ~10 nM when the culture medium was supplemented with α,β–methylene ATP, a nonhydrolyzable nucleotide that inhibits multiple types of ecto-nucleotidases. Similar findings were noted with 1321N1 astrocytoma cells, which steadily maintained bulk phase [ATP] in the 1 to 3 nM range.[163] Upon exposure to β,γ–methylene ATP (another ecto-nucleotidase inhibitor), the bulk phase [ATP] steadily increased to the 10 to 20 nM range.

5.6.1.1 Constitutive Nucleotide Release: Physiological Relevance

The basal concentrations of released nucleotides may be important in conditioning or "priming" the cells or otherwise establishing their general responsiveness to other stimuli.[40,191,193] Lazarowski et al. have proposed that constitutive ATP release, which occurs at negligible rates compared to intracellular ATP turnover, might provide a basal purinergic "tone" in many tissues.[43] Similarly, Ostrom et al. have argued that constitutively released nucleotides, together with other locally secreted mediators, may contribute to the baseline levels of G protein activity and second messenger synthesis in certain tissues.[27] It should also be stressed that constitutively released nucleotides can be involved in important extracellular processes other than purinergic/pyrimidinergic signaling. As noted previously, NTPs are utilized by the eNPP1 ecto-phosphodiesterase to generate extracellular pyrophosphate, which is a critical modulator of physiological calcification in mineralizing tissues and pathological calcification in various soft tissues such as blood vessels.[87,90,194] In these soft tissues, the local catabolism of constitutively released NTP by eNPP1 may play an unappreciated role in maintaining extracellular pyrophosphate at the steady-state levels required for repression of inappropriate calcification.[195,196]

5.6.1.2 Constitutive Nucleotide Release: Mechanisms

Although most studies of basal nucleotide release have focused on ATP secretion, Lazarowski and colleagues demonstrated that UDP glucose is also constitutively released from most cells types examined thus far.[45] This steady-state externalization of UDP glucose may provide important insights regarding the general mechanisms underlying constitutive nucleotide release. As noted previously, uridine nucleotide sugars play critical roles in driving synthesis of the oligosaccharide precursors used for protein glycosylation within the ER and Golgi. Cellular UDP glucose is largely concentrated within the ER and pre-Golgi structures via the action of a UDP glucose/UMP exchanger. Most of the UDP released following protein glycosylation within the ER/Golgi is hydrolyzed to UMP and returned to the cytosol in exchange for accumulation of new UDP sugar. However, residual amounts of UDP, UMP, and UDP sugars may remain within the Golgi-derived secretory vesicles that constitutively fuse with the plasma membrane. Other studies have shown that ATP is also accumulated within the ER/Golgi to drive the ATP-dependent chaperonin reactions involved in protein folding. Indeed, multiple mechanisms for the transport of ATP, nucleotide sugars, and nucleotide sulfates across ER and Golgi membranes have been characterized in yeast and mammalian cells.[197,198] Thus, it is possible that residual ATP (as well as ADP and AMP) may additionally be present within the constitutive secretory vesicles that deliver folded glycoproteins to extracellular compartments. If so, this may comprise a major pathway for the constitutive or basal ATP release that has been observed in multiple cell types. The use of pharmacological or genetic approaches that target various elements of the constitutive exocytotic machinery will be important for testing its contribution to basal nucleotide release. In this regard, Maroto and Hamill have reported that basal ATP release from single *Xenopus* oocytes was reduced by brefeldin A, which attenuates constitutive exocytosis by collapsing membrane traffic between the *cis*-Golgi and ER.[159] While constitutive exocytosis constitutes an appealing and testable mechanism for basal nucleotide export, it is also possible that facilitative transport proteins of the plasma membrane contribute to the basal release of cytosolic nucleotides, particularly ATP, from some cell types. Transport proteins that are possible candidates as nucleotide exporters are discussed below.

5.6.2 Stimulus-Induced Exocytotic Release of Nucleotides from Excitatory Cell Types

Several neuronal, neuroendocrine, and endocrine cells exhibit a rapid exocytotic secretion of nucleotide-containing vesicles or granules in response to extrinsic depolarizing stimuli that increase cytosolic Ca^{2+}. Platelets also release ATP and ADP via the Ca^{2+}-dependent exocytosis of the dense granules. All of these latter cell types contain morphologically identifiable synaptic vesicles or dense core granules that can be isolated and biochemically characterized. Significant ATP content has been demonstrated in several types of isolated/enriched vesicles or granules including cholinergic vesicles,[199–202] adrenergic vesicles,[203,204] chromaffin granules from adrenal medullary cells,[205–209] dense core granules from platelets,[210–213] and insulin-containing granules from pancreatic beta cells.[214] The copackaging and corelease of ATP and catecholamines from chromaffin granules or platelet dense core granules represent the best-characterized examples of exocytotic nucleotide secretion at the biochemical and physiological levels. Bankston and Guidotti[206]demonstrated that cytosolic ATP can be actively transported into isolated chromaffin granule ghosts via a process dependent on the H^+ electrochemical generated by the bafilomycin-sensitive V-type ATPases present in exocytotic vesicles and granules; this H^+ electrochemical gradient also drives the active compartmentalization of catecholamines. Thus, cytosolic ATP acts as both the substrate for the V-type ATPase that establishes the driving force for nucleotide compartmentalization and a substrate for the presumed nucleotide-permeable transport protein(s) that mediates influx of ATP into the granule/vesicles. The molecular identities of these vesicular nucleotide-permeable transport proteins remain unknown. However, recent studies by Jedlitschky et al. have indicated MRP4 (multidrug resistance protein 4; ABCC4) as a candidate for the nucleotide uptake protein in human platelet dense granules.[215] MRP4 is an ATP-binding cassette protein, present in both cell surface and intracellular membranes, that plays a role in the export of organic anions from cells. MRP4 and the related MRP5 have also been shown to catalyze the efflux of nucleoside analogs used in antiviral therapy or cancer chemotherapy, as well as the efflux of cyclic AMP and cyclic GMP from several cell types.[216] Significantly, Jedlitschky et al. found that MRP4 was highly expressed in the isolated platelet dense granules from normal subjects but not a subject with Hermansky-Pudlak syndrome, a disease state characterized by defective granule function.

5.6.3 Stimulus-Induced Release of Nucleotides from Nonexcitatory Cell Types

5.6.3.1 Overview

Early studies by Forrestor demonstrated that a brief (1 min) exposure of rat cardiac myocytes to anoxic medium (pre-equilibrated with 100% N_2) induced a fourfold increase in ATP release (~1.6 nmol/mg protein) relative to that constitutively released from myocytes in oxygenated medium.[217,218] Subsequently, many different types of stimuli such as mechanical shear,[29,59,60,108,159,164,185,219–229] hypoxia,[53,177,180,230–232] exposure to Ca^{2+}-mobilizing receptor agonists,[126,128,163] hypotonic stress,[22,31,48,49,51,52,55,56,107,108,158,171,173,177,178,188,189,229,233–243] hypertonic stress,[224,244] and membrane hyperpolarization[245] have been shown to increase the rate of nucleotide release from many "nonexcitatory" cell types.

Although exocytosis of ATP-containing vesicles remains a highly plausible possibility (see section 5.6.3.3), a major mechanism proposed to explain the stimulus-induced release of nucleotides from nonexcitatory cells is the facilitated efflux of cytosolic nucleotides through plasma membrane transport proteins. The accelerated release of ATP and other nucleotides in response to diverse stimulus-activated conditions has prompted the search for membrane proteins that can function as "ATP transporters" or "ATP channels." Although proteins belonging to three families of known transporters have been proposed as candidate ATP transporters (see below), the definitive identification of nucleotide transporter proteins remains elusive. This contrasts with recent and significant progress in the molecular characterization of genes encoding several types of transporters for adenosine and other nucleosides.

5.6.3.2 Nucleotide-Permeable Transporters and Channels as Mechanisms for Stimulus-Induced Nucleotide Release from Nonexcitatory Cells

5.6.3.2.1 ABC Transporters

Nonlytic, nonexocytotic ATP release in some cell models can be repressed by drugs known to alter the function of ATP-binding cassette (ABC) family transport proteins.[25] Thus, some putative ATP transporters may belong to the ABC transporter family that includes more than 500 sequenced genes spanning the eukaryotic and prokaryotic kingdoms.[246–248] ABC transporters have been shown or postulated to catalyze transmembrane fluxes of an extremely broad range of organic and inorganic molecules. Three well-characterized ABC transporters, CFTR,[24,49,173,188,219,243,249–256] p-glycoprotein or MDR1,[23,30,51.173,255–257] and MRP4[215,255] have been proposed as playing direct or indirect roles in facilitated efflux of ATP and other nucleotides.

The role of CFTR has been extensively studied by multiple groups who have reported discrepant and contentious observations. Such studies have generally compared rates and extents of nucleotide efflux in sibling cell lines engineered to express or not express CFTR. Some groups have observed that CFTR expression increases the rate of ATP release in response to cyclic AMP-dependent protein kinase, which both phosphorylates and activates CFTR as an anion channel.[49,173,188,219,243,249,250,252–254,258–260] In contrast, other investigators have been unable to demonstrate any significant potentiation.[24,42,178,185,252,261–263] More recent studies have suggested that positive effects of CFTR expression on ATP release are cell specific and variable and thus likely to reflect modulatory effects of CFTR on another differentially expressed ATP transport system.[49,188,252]

Abraham and colleagues suggested that the multidrug resistance (mdr1) gene product, namely P-glycoprotein or MDR1, functions as an ATP conduit.[257] Other ABC transporters related to MDR1 that may be glibenclamide insensitive have also been implicated as ATP release machinery.[23,30,51,173,255,256] It should also be noted that other members of the MDR transporter family (MRP4, MRP5, MRP8) are involved in the extrusion of purine (and pyrimidine) nucleotide analogs such as thiopurine nucleotide monophosphates as well cyclic AMP and cyclic GMP.[215,216,264,265] However, the range of nucleotides that these transporters extrude has not been fully determined. As in the case of CFTR, the presence or absence of MDR proteins in particular cell types may indirectly modulate the activity of other pathways for hypotonic stress-induced ATP release (channel-mediated or exocytotic) by altering the kinetics and magnitude of cell swelling and regulatory volume decrease.[30]

5.6.3.2.2 Connexin Hemichannels

Connexins comprise the second group of known membrane proteins that have been implicated in facilitated ATP release. These proteins play critical roles as subunits of the gap junction channels that facilitate electrical and metabolic coupling between physically connected cells within many tissues. All 13 unique members of the connexin family share the similar structure of four transmembrane domains, intracellular amino- and carboxy termini, and two extracellular loops.[266–268] Six connexin subunits multimerize to form a hexameric hemichannel or connexon. At sites of direct cell–cell contact, the hemichannels from two cells can dock together (via extracellular loop domains) to form the transcellular gap junction channel. Consistent with their role in metabolic coupling, gap junction channels are permeable to cytosolic nucleotides, including ATP.[269–271] Thus, they represent known channels that can readily facilitate the flux of ATP across plasma membranes. Recent studies have suggested that gap junctions composed of different connexin subtypes exhibit graded degrees of permeability to ATP and ADP, with connexin (Cx)-43 channels being 50 times more permeable to these nucleotides than Cx32 channels.[269] The flux of metabolites and ions through gap junctions can be rapidly attenuated by various intracellular signals such as acidification or sustained increases in cytosolic Ca^{2+}.

Although connexin-based channels are generally associated with the transcellular movement of molecules through such gap junction channels, a growing body of observations indicates that connexin hemichannels at nonjunctional membrane sites can also be gated to the open state.[272–274] In this state, hemichannels act as low-resistance conduits for the flux of appropriately sized molecules into and out of the cell. Under normal conditions, the gating of nonjunctional hemichannels in healthy cells is, and must be, a low-frequency event to prevent the collapse of ionic gradients and loss of ATP and other intracellular metabolites that would eventually result in cell death. However, in response to unusual cell stresses, such as reduction in extracellular Ca^{2+} or metabolic inhibition, cells that express native or recombinant connexins become transiently permeable to extracellular fluorescent dyes as large as carboxyfluorescein (M_r 376 Da), lucifer yellow (M_r 522 Da), and calcein (M_r 660 Da), but not to fluorescein-conjugated dextran (M_r 1500 Da).[275,276] Following re-addition of extracellular Ca^{2+}, these water-soluble dyes remain trapped within the cells even after the cells are transferred to dye-free medium; this indicates that the cells rapidly recover normal membrane integrity and have not been irreversibly damaged. Significantly, comparisons using wild-type HeLa cells and HEK293 cells (that do not appreciably express native connexins) versus HeLa and HEK293 that express recombinant Cx43 have demonstrated that the Ca^{2+}-sensitive dye uptake is dependent on expression of that connexin.[275–277] In contrast, repression of native Cx43 expression in Novikoff hepatoma cells transfected with Cx43 antisense constructs was correlated with inhibition of dye uptake.[275] The gating of connexin hemichannels is rapid, reversible, relatively insensitive to reduced temperature, and inhibited by drugs, such as octanol and halothane, known to repress gap junction function. Electrophysiological analyses using HEK293 cells transfected with Cx43 cDNA[275] and single *Xenopus* oocytes injected with Cx50 or Cx46 cRNAs[278] have also indicated that activation of connexin hemichannels by reduced Ca^{2+} reversibly triggers nonselective ionic currents.

Given 1) that gap junctions are highly permeable to ATP and 2) that connexin hemichannels are permeable to dyes (lucifer yellow, calcein) with molecular masses similar to that of ATP, it is possible that hemichannels provide a pathway for the facilitated efflux of ATP from cells that natively express various connexin subtypes. Early reports by Cotrina et al. in the Nedergaard group addressed this possibility as part of studies on intercellular Ca^{2+} wave propagation in astrocytes.[279] Ca^{2+} wave propagation is the process whereby acute changes in cytosolic Ca^{2+} levels within one cell can rapidly be propagated to nearby cells. Connexins and gap junctions have been shown to play critical roles in Ca^{2+} wave propagation in astrocytes and most other cell types.[268,280–283] One underlying mechanism is that gap junction channels between adjacent cells directly facilitate the movement of Ca^{2+} *per se* and Ca^{2+} mobilizing inositol trisphosphate (IP_3) from one cell to another.[281] However, another mechanism for wave propagation is that one cell releases an extracellular signaling molecule that then activates Ca^{2+}-mobilizing receptors in nearby cells. The particular roles of released ATP and Ca^{2+}-mobilizing P2 receptors in wave propagation have now been verified in several cell models including multiple types of astrocytes and glial cells, endothelial cells, osteoblasts, retinal Müller cells, and prostate carcinoma cells.[29,222,228,284–287] In most cases, these analyses of ATP-dependent Ca^{2+} wave propagation involve nonlytic mechanical stimulation of a single cell within an optical field to trigger Ca^{2+} mobilization in that cell. However, rapid Ca^{2+} transients are also observed in surrounding cells after variable delays that are directly proportional to the distance from the mechanically stimulated cell.

Some reports have suggested that the gap junction pathway and the "extracellular ATP" pathway represent parallel and distinct mechanisms for this type of Ca^{2+} wave propagation.[288,289] However, Cotrina et al. observed that connexins are involved in both pathways.[279] They demonstrated that ATP secretion by either C6 glial astrocytes or HeLa cells was increased 5- to 15-fold by expression of Cx43, Cx32, or Cx26, and this was correlated with similar effects of heterologous connexin expression on extent of mechanically induced Ca^{2+} wave propagation. Parallel studies demonstrated that Ca^{2+} wave propagation was markedly attenuated by P2 receptor antagonists and exogenously added apyrase but not inhibitors of direct gap junction coupling. One limitation of these early studies was that the connexin-dependent ATP release (measured by offline luciferase assays) was induced by addition of UTP to activate Ca^{2+}-mobilizing $P2Y_2$ receptors. Significantly, much lower accumulation of extracellular ATP was observed when Ca^{2+}-mobilizing bradykinin or endothelin receptors were stimulated. Although not considered by the authors, the higher efficacy of UTP was likely due to the ability of this nucleotide agonist, but not the peptide agonists, to also repress breakdown of released ATP by endogenous ecto-nucleotidases and to support the ecto-NDPK–catalyzed transphosphorylation of extracellular ADP. Nonetheless, the UTP-induced accumulation of extracellular ATP was strongly enhanced in cells overexpressing the various connexin subtypes and markedly reduced when these Cx-expressing cells were preloaded with Ca^{2+}-chelating BAPTA prior to stimulation.

These seminal findings of Nedergaard and colleagues set the stage for subsequent investigations (by the same group and others) for further analyzing the possible role of connexin hemichannels in stimulus-induced nucleotide release in astrocytes, endothelial cells, and osteocytes. Arcuino et al combined real-time luciferase imaging methods and lowering of extracellular Ca^{2+} concentration as a stimulus for hemichannel gating, spontaneous Ca^{2+} wave propagation, and ATP efflux in monolayers of astrocytes or endothelial cells.[167] Significantly, that study indicated that only a subset of cells within each monolayer act as a "point source" of ATP release, as assayed by highly localized clouds of luciferase-dependent luminescence, in response to the lowering of extracellular Ca^{2+}. The same subset of cells that released ATP also accumulated propidium dyes, which was consistent with the transient gating of connexin hemichannels to the open state. Similarly, Stout et al. combined the use of patch-clamp electrophysiology, propidium dye uptake, and luciferase assays to further support an involvement of Cx43 hemichannels as a pathway for ATP release from C6 glioma cells stimulated by either lowered extracellular Ca^{2+} or mechanical stress.[286]

These latter studies, as well as several others, have also utilized pharmacological blockers of hemichannel gating or conductance to attenuate stimulus-induced ATP release. The commonly used hemichannel inhibitors include 18 beta-glycyrrhetinic acid (18-GA), carbenoxolone (an 18-GA derivative), flufenamic acid, and lanthanides.[268,290,291] A major limitation with this approach is that these reagents lack high selectivity and can interact with channels or transporters other than connexin hemichannels. A more specific group of hemichannel blockers are the so-called connexin mimetic peptides (also known as gap peptides), which are short peptides that correspond to the sequences of extracellular (or intracellular) loops of various connexins.[282,292–294] By interacting with these loop domains, the gap peptides act to disrupt the normal conformation of connexin hexameric complexes with consequent inhibition of either gap junctional or hemichannel function. Leybaert and colleagues have extensively used various connexin mimetics to coordinately repress connexin hemichannel activity and Ca^{2+}-dependent ATP release from various cell types that express either endogenous or transfected connexin subunits.[50,282,295,296] For example, the gap26 peptide (VCYDKSFPISHVR)[43] and gap27 peptide (SRPTEKTIFII)[43] strongly inhibited IP_3- and Ca^{2+}-induced ATP release from Cx-43 expressing endothelial cells, and the gap24 peptide (GHGDPLHLEEVKC)[32] selectively repressed Ca^{2+}-induced ATP release from Cx-32 expressing HCV304 bladder epithelial cells. Gomes et al.[43] similarly used the gap26 peptide to coordinately inhibit mechanical stress-induced Ca^{2+} wave propagation and ATP release in monolayers of bovine corneal endothelial cells.[228]

Recent investigations have demonstrated that connexin hemichannels can be transiently gated to the open state by a variety of stimuli known to trigger ATP release. These include mechanical stress, lowering of extracellular Ca^{2+} concentration, and elevation of cytosolic Ca^{2+}. Although it appears nominally paradoxical that hemichannels can be activated by reduced extracellular Ca^{2+} or increased intracellular Ca^{2+}, a methodical analysis by De Vuyst et al. revealed that removal of extracellular divalent cations triggered a mobilization of intracellular Ca^{2+} stores in ECV304 epithelial cells that was associated with release of ATP.[296] This study also revealed that the coordinated gating of Cx32 hemichannels and ATP release was regulated by elevated cytosolic Ca^{2+} within a relatively narrow window with peak activation at 500 nM and repression at progressively higher $[Ca^{2+}]$.

Thus, a wide range of experimental evidence supports a role for connexin hemichannels in stimulus-induced ATP release in several types of cells. Pannexins, a newly characterized family of connexin-related proteins, have been identified as mechanosensitive membrane channels that may also contribute to ATP release.[297–300] These observations suggest that connexin- or pannexin-based hemichannels may be direct conduits for ATP release or critically regulate other channels/transporters that facilitate the efflux of nucleotides. However, other data indicate that it may be difficult to establish direct cause-and-effect relationships between connexin expression/activity and ATP release. Spray and colleagues have compared the relative contributions of gap junctional communication versus the released ATP/paracrine P2 receptor activation pathway to Ca^{2+} wave propagation in cultures of neonatal cardiac myocytes or neonatal spinal cord astrocytes obtained from wild-type versus Cx43 knockout mice.[288,289,301,302] Cx43 compromises >90% of the connexins expressed in wild-type myocytes and astrocytes. Consistent with this, myocytes from Cx43 null mice exhibited major deficits in gap junction–mediated Ca^{2+} wave propagation.[288] However, Ca^{2+} wave propagation mediated by the paracrine purinergic pathway was similar in both wild-type and Cx43-deficient cells. This indicated that ATP release from the myocytes was not attenuated by the absence of the major connexin component of these cells. In the spinal cord astrocytes, the absence of Cx43 reduced gap junctional coupling by more than 65% but had little effect on the efficiency of mechanically induced Ca^{2+} wave propagation via the paracrine purinergic route.[289] Significantly, deletion of Cx43 from the astrocytes also induced a phenotypic switch in P2Y receptor subtype expression from $P2Y_1$ receptor–dominated profile in wild-type astrocytes to a $P2Y_2$ receptor–dominated profile in the Cx43 null cells. Similar changes in P2Y receptor subtype expression were observed when Cx43 in wild-type was acutely downregulated by antisense oligonucleotides.[302] These various findings with connexin-knockout cells clearly support the existence of mechanisms that reciprocally coordinate Ca^{2+} wave intercellular signaling via the connexin-based and purinergic pathways. However, the observation that purinergic paracrine pathways are not significantly disrupted by deletion of the major connexin subtype within these astrocytes and myocytes raises significant questions as to whether the connexin hemichannels are the actual conduits for ATP release or, rather, act to modulate signaling pathways that regulate other ATP release mechanisms such as exocytosis or flux through anion channels.

5.6.3.2.3 Anion Channels

Given the anionic nature of nucleotides, a variety of anion-selective channels have also been investigated as putative ATP release conduits. These include two types of functionally but not molecularly defined anion channels: 1) the volume-sensitive outwardly rectifying anion channels (VSOAC), which are also known as volume-regulated anion channels (VRAC) or volume-sensitive chloride channels;[31] and 2) the so-called maxi-anion channel.[33] In addition, plasma membrane variants of the molecularly defined mitochondrial voltage-dependent anion channels (plVDAC) comprise a third group of putative ATP channels.[55,237]

A major motivation for considering VRACs and maxi-anion channels as ATP release proteins has been the widely observed correlation between cell volume regulation, nucleotide release, and autocrine P2 receptor activation in many cell types.[22,26,31,48,49,51,52,55,56,107,108,156,158,163,177,178,189,222,225,236–238,241,242,303] Although cells initially swell in response to hypotonic medium, this increase in volume produces both mechanical effects on membrane/cytoskeletal organization and the initiation of signaling cascades that activate efflux of osmotically active intracellular molecules which include K^+, Cl^-, and organic osmolytes such as taurine and inositol. This efflux of these inorganic and organic osmolytes underlies the process of regulatory volume decrease (RVD), which allows cells to recover normal volume.[304] Significantly, hypotonic stress also induces release of ATP in almost all cell types tested, and this release is often accompanied by an autocrine activation of P2 receptors.[49,55,107,188,234,239] In some cell models, this autocrine activation of P2 receptor signaling may contribute to the RVD process because the rate of volume recovery is attenuated in the presence or P2 antagonists or exogenous apyrase added as a scavenger of released nucleotides.

As described in comprehensive reviews, VRAC/VSOAC comprise a ubiquitously expressed type of channel activity that gradually develops within the first few minutes after the initial swelling of cells in response to hypotonic medium.[304,305] It is an outwardly rectifying anion-selective ($I^- > NO_3 > Br > Cl >$ gluconate) current with a single channel conductance of 30 to 70 pS (at +120 mV). Development of VRAC/VSOAC currents requires the presence of cytosolic ATP (or nonmetabolizable ATP analogs) within the patch-clamp pipette but is independent of cytosolic [Ca^{2+}]. VRAC/VSOAC currents are pharmacologically distinguished by their sensitivity to inhibition by a broad range of nonspecific reagents including glibenclamide, stilbenes, phloretin, 1,9-dideoxy-forskolin, tamoxifen, and 5-nitro-2-(3-phenylpropylamino)benzoic acid (NPPB). The gene product(s) that mediate VRAC/VSOAC activity have not yet been unequivocally identified at the molecular level. Initial studies indicated that the gene encoding, the so-called I_{Cln} protein, might be the basis of VRAC/VSOAC channels, but subsequent work has indicated that I_{Cln} is not the channel protein *per se* but may be an indirect regulator of VRAC/VSOAC expression or function. Activation of VRAC/VSOAC currents (as measured electrophysiologically) can also be strongly correlated with an increased efflux (as measured by loss of radioisotope-labeled species) of various organic osmolytes, such as inositol and taurine. Significantly, VRAC/VSOAC-mediated Cl^- currents are also inhibited in the presence of extracellular nucleotides (at millimolar concentrations) and this indicates that nucleotides can enter the permeability pore of the channels. These latter two characteristics of VRAC/VSOAC, permeability to organic metabolites and attenuation of Cl^- current by extracellular nucleotides, has led to the consideration of VRAC/VSOAC as a conduit for release of ATP, particularly in the context of hypotonic stress and RVD responses.

Oike and colleagues have proposed a causal relationship between VRAC/VSOAC activation and hypotonic stress-induced ATP release in bovine aortic and human umbilical vein endothelial cells based on several observations and arguments.[31,242,306,307]

1. Hypotonic stress elicited the temporally correlated activation of VRAC/VSOAC currents and nucleotide release leading to autocrine activation of Ca^{2+}-mobilizing P2 receptors and a consequent stimulation of Ca^{2+}-dependent nitric oxide production.[306]
2. Four chemically disparate blockers (fluoxetine, verapamil, glibenclamide, tamoxifen) of VRAC/VSOAC currents strongly attenuated hypotonic stress-induced ATP release and nitric oxide production.[31]
3. Exposure of patch-clamped endothelial cells to different extracellular nucleotides (ATP, UTP, CTP, GTP, ADP) during peak VRAC/VSOAC

activity produced a voltage-dependent and reversible reduction (20%) of VRAC/VSOAC current.[31]

4. Activation of both VRAC/VSOAC[308,309] and hypotonic stress-induced ATP release[242,307] is markedly attenuated by inhibition of Rho/Rho-kinase pathways and by tyrosine kinase inhibitors.
5. The calculated molecular dimensions of the VRAC/VSOAC permeability pore are consistent with permeability to ATP.[310]

Major challenges to this postulated role of VRAC/VSOAC as a conduit for ATP efflux were raised by Okada and colleagues on the basis of their studies on hypotonic stress-induced ATP release from several epithelial cell models including human intestine 407 cells,[178] murine C127 adenocarcinoma cells,[236,237] and C127 cells engineered to overexpress CFTR channels.[49] In these cells, stimulated ATP release was insensitive to inhibition by glibenclamide or disodium 4-acetamido-4′-isothiocyanato-stilbene-2,2′-disulfonate (SITS) when those reagents were tested at concentrations sufficient for blockade of VRAC/VSOAC currents. In contrast, the ATP release was repressed by Gd^{3+}, which, in contrast did not block VRAC/VSOAC currents. Although overexpression of CFTR greatly increased the magnitude of the hypotonic stress–induced ATP release in the C127 cells, blockade of both CFTR channel activity and VRAC/VSOAC activity by glibenclamide did not attenuate block nucleotide efflux. These data indicated that the ability of hypotonic stress to elicit ATP release could be dissociated from the coincident activation of VRAC/VSOAC. This prompted further studies by Sabirov et al. to identify alternative mechanisms that might underlie hypertonicity-induced ATP release responses in the C127 cells.[236] A salient initial finding was the identification of an alternative anion-permeable conductance in C127 cells that were analyzed using: 1) ATP-free intracellular dialysis medium in the patch-clamp electrode and 2) extracellular saline supplemented with phloretin. As noted previously, both of these latter conditions repress VRAC/VSOAC activity. Like VRAC/VSOAC currents, the alternative current was sensitive to blockade by NPPB and SITS. In contrast to VRAC/VSOAC, it was not blocked by phloretin or glibenclamide, but was blocked by Gd^{3+}. Moreover, when membrane patches were excised from C127 cells and bathed in ATP-free medium, stable single-channel activity was observed. The single-channel conductance was 200 to 400 pS in contrast to the 30 to 70 pS conductance of VRAC/VSOAC single channels, and the voltage–current relationship was linear in contrast to the outwardly rectifying nature of VRAC/VSOAC. Thus, the C127 cell alternative anion current was similar to large conductance anion currents described in earlier studies of N1E115 neuroblastoma cells,[311] rat cortical astrocytes,[312,313] and renal cortical collecting tubule cells.[314,315] Given their nature, these currents have been termed "maxi-anion channels."[304] In addition to their large unitary conductance and pharmacology, the open probability (Po) of maxi-anion channels are characterized by a bell-shaped voltage dependency.

Significantly, Sabirov et al. reported that when the single-channel activity of the C127 excised membranes was studied using Cl^- as the charge carrier, addition of ATP to either the extracellular or cytosolic compartment produced an open-channel block consistent with low affinity (Kd ~13 mM) binding sites for ATP midway through the pore of the presumed maxi-anion channel.[236] Moreover, when ATP was substituted for Cl^- as the charge carrier on the cytosolic face of the excised membrane patch, ATP currents were observed. This indicated that maxi-anion channels can be directly permeated by ATP with a permeability ratio P_{ATP}/P_{Cl} of 0.1. More recent biophysical studies of anion permeability of the single maxi-anion channel indicated that the average radius of the channel's open-state pore is 1.3 nm, which is sufficiently large to accommodate the dimensions of ATP^{4-} (0.57 to 0.65 nm asymmetric radii) or $MgATP^{2-}$ (0.60 nm radius).[316]

Additional studies have supported a role for maxi-anion channels as the ATP release conduits in neonatal rat cardiomyocytes subjected to hypoxic and ischemic stresses as well as hypotonic stress.[177] Each of these three stimuli induced the activation of myocyte single-channel currents with the large unitary conductance (390 pS) characteristic of maxi-anion channels. As in C127 epithelial cells, the myocyte current was selective for anions and permeable to ATP^{4-} and $MgATP^{2-}$. Moreover, the increased ATP release, as assayed by offline luciferase methods, from cardiac myocytes challenged with hypoxia or hypotonicity was similarly inhibited by NPPB, SITS, and Gd^{3+}, but not glibenclamide. (Significantly, appropriate calibration protocols controlled for the known repressive effects of Gd^{3+} on luciferase activity.) Similar investigations by Bell et al. indicated that maxi-anion channel activity and ATP release in rabbit macula densa cells was coordinately stimulated when the extracellular NaCl was increased from low-salt to high-salt conditions that mimic the changes that mediate tubuloglomerular feedback regulation in the kidney.[175] Thus, a role for maxi-anion channels in ATP release is not limited to hypotonic stress stimulation but may also be important for the nucleotide release responses to other physiologically relevant stress stimuli.

Several functional characteristics of the maxi-anion channel, including its large single-channel conductance, permeability to nucleotides, and bell-shaped voltage dependency of open probability, are similar to properties of the well-characterized mitochondrial VDAC porins.[317] The VDAC porins are predominantly expressed in the outer membrane of mitochondria, where they play important roles in regulating the flux of both metabolites and proteins into and out of the mitochondrial intermembrane compartment. Mammals express three different isoforms of VDAC encoded by distinct genes (VDAC1, VDAC2, and VDAC3). Disruption of the vdac2 gene produces embryonic lethality in mice,[318] and knockout of vdac1 produces a conditional embryonic lethality that are among strains of mice.[319] The vdac3 null male mice are infertile due to a defect in sperm motility.[320] Significantly, the presence of anti-VDAC immunoreactive protein has been described in the plasma membranes of several cell types.[321–324] Beuttner et al. reported that alternative splicing of the first exon of murine vdac1 gene could produce a variant protein, pl-VDAC, that trafficked through the Golgi apparatus to the plasma membrane.[325] The possibility that this plasma membrane variant of VDAC might comprise the molecularly defined entity of maxi-anion channels was supported by the observation that antisense oligonucleotides targeted to pl-VDAC reduced maxi-anion channel activity in membrane patches excised from mouse neuroblastoma cells.[324]

Given these findings, Okada et al. directly tested the hypothesis that pl-VDAC may be an ATP release channel in two types of experiments.[55] First, overexpression of pl-VDAC in NIH3T3 fibroblasts produced a 2.4-fold increase in the rate and extent of ATP triggered by mechanical stress stimulation. Second, comparison of mechanical stress-induced ATP release from lung fibroblasts of wild-type versus vdac1 null mice indicated only a modestly reduced (2-fold) rate of ATP release in the latter cells, which lacked both mitochondrial VDAC1 and pl-VDAC. While these data indicated that pl-VDAC may contribute directly or indirectly to ATP release, other ATP export pathways are clearly operative in the vdac1 null cells. Very recent studies by Sabirov et al. examined the possibility that the other VDAC paralogues (VDAC2 and VDAC3) might also contribute to maxi-anion channel activity and stimulate ATP release.[326] Those authors generated mouse embryonic fibroblasts (MEFs) that lacked expression of all three VDAC subtypes by cross-breeding heterozygous $vdac1^{+/-}$ and $vdac3^{+/-}$ mice to generate double-knockout MEFs, which were then treated with short interfering RNA (siRNA) to silence expression of the remaining vdac2 gene. Significantly, the absence of all three VDAC protein subtypes had no effect on maxi-anion channel activity and only slightly diminished hypotonic stress–induced ATP release. This report conclusively indicates that plasma membrane variants of VDAC-type porins do not underlie maxi-anion channel activity or the ATP release mediated (or regulated) by activated maxi-anion channels; the results leave open the issue of the molecular identity of this ATP efflux pathway.[326]

5.6.3.3 Regulated Exocytosis of Nucleotide-Containing Vesicles as a Mechanism for Stimulus-Induced Nucleotide Release from Nonexcitatory Cells

Over the past decade, the majority of studies on stimulus-induced ATP release from nonexcitatory cell types have focused on the possible role of the transporter- or channel-based pathways discussed above. In contrast to the neurons, neuroendocrine cells, and platelets, most nonexcitatory cells lack well-characterized granules or vesicles that contain high concentrations of compartmentalized nucleotides. However, some of these nonexcitatory cell types known to release significant amounts of ATP in response to various acute stimuli do contain specialized types of exocytotic granules, such as mucin granules of mucus-secreting epithelia and the Weibel-Palade bodies within most endothelial cell types.[327] Consistent with the presence of these regulated release granules, various nonexcitatory cell types, including endothelial cells[327] and astrocytes,[328–330] have been shown to express critical components of the Ca^{2+}-regulated exocytotic machinery, such as v- and t-soluble N-ethylmaleimide–sensitive factor attachment protein receptor (SNARES) and N-ethylmaleimide–sensitive factor (NSF). These findings have rekindled speculation that stimulated nucleotide release from many, if not most cell types, involves the Ca^{2+}-sensitive exocytosis of poorly characterized vesicles that contain compartmentalized nucleotides. However, both the relative abundance of these putative vesicles and the absolute nucleotide concentrations within individual vesicles may be much lower than the corresponding values observed in excitatory cells that release nucleotides by well-defined exocytotic mechanisms.

Recent studies using different models of stimulated ATP export (mechanical stress, hypotonic stress, and neurotransmitter receptor activation) have suggested that regulated exocytosis may be a general pathway for nucleotide release from nonexcitatory cells.[52,107,158,189,224,229,331–335] In general, these investigations have used bafilomycin A, proton ionophores, brefeldin A, nocodozole, clostridial toxins, and N-ethylmaleimide (NEM) as pharmacological tools that target particular elements of the regulated exocytotic machinery. Bafilomycin A inhibits the V-type H^{+}-ATPases that maintain the proton gradients required for active accumulation of nucleotides within vesicular compartments.[336] An alternative approach for collapsing the vesicular proton gradients employs proton ionophores such as monensin. Brefeldin A is a fungal metabolite that disrupts the structure and function of the Golgi apparatus as the key intermediate compartment for vesicular trafficking from the endoplasmic reticulum to the plasma membrane.[337] Nocodazole acts to disrupt the microtubules that act as scaffolding and motor proteins for directing vesicular traffic to the plasma membrane. Several clostridial toxins, including tetanus toxin and botulinin B toxin are metalloproteases that cleave the v-SNARE protein synaptobrevin II (VAMP2) and thereby disrupt docking interaction of secretory vesicles with t-SNARE proteins, SNAP-25 and syntaxin, of the plasma membrane.[338–340] Because the SNARE proteins are targets for the NSF (NEM-sensitizing factor) that reverses formation of the SNARE complexes after exocytotic fusion, NEM treatment of cells may also be used to attenuate the turnover of exocytotic processes. As previously discussed, analyses of exocytotic ATP release have also utilized quinacrine fluorescence imaging as an index of ATP-containing vesicular pools that can be depleted upon stimulation. Another fluorescence imaging approach involves correlating ATP release kinetics with accumulation of the FM1-43 dye as a probe of plasma membrane surface area that increases with addition of exocytotic membrane.[341,342]

It is worth highlighting several particularly salient studies that have used these approaches to support a role for exocytotic release of ATP from nonexcitatory cells. Bodin and Burnstock used NEM and monensin to attenuate the ATP release triggered by shear stress stimulation of human endothelial cells.[172] Maroto and Hamill reported that brefeldin A and nocodozole repressed both constitutive and mechanical stress–induced ATP release from single *Xenopus* oocytes.[159] Several studies have similarly indicated that pressure-induced distension of urinary bladder preparations or isolated monolayers of bladder urothelial cells elicits ATP release that is markedly attenuated in the presence of bafilomycin, brefeldin, or botulin toxin.[189,225,343–345] Hypotonic stress–induced ATP release has been correlated with exocytotic vesicle recycling as measured as FM1-43 fluorescence changes in intestine 407 cells and biliary epithelial cells.[52,189] Boudreault and Grygorczyk have recently described an elegant series of experiments that compared the time courses for cell swelling, activation of VRAC-like currents, and ATP release rate in epithelial cells or fibroblasts during sustained hypotonic stimulation.[107] These detailed kinetic studies strongly suggested that ATP was released only transiently during the response to hypotonic stress and that most of the release occurs during the steady-state plateau phase of the cell volume response rather than during the regulatory volume decrease phase when organic and inorganic osmolytes are leaving the cell. Moreover, the transient ATP release, but not the cell swelling *per se*, was almost completely inhibited in cells loaded with Ca^{2+} chelator. Boudreault and Grygorczyk concluded that the Ca^{2+}-regulated exocytosis of a limited pool of ATP-containing vesicles, rather than a channel-mediated efflux pathway, is the most likely explanation for the ATP release response to hypotonic stress.[107]

Subcellular fractionation studies have shown that several types of astroglial cells, including primary astrocytes, contain ATP-enriched vesicles that colocalize with multiple proteins (synaptotagmin-I, synaptobrevin-II, SNAP25) implicated in regulated exocytosis.[328–330,346–350] Coco et al. have reported that mechanical stimulation or pharmacological agonists (phorbol esters) induced a release of ATP from primary neonatal rat astrocytes that could be partially blocked by bafilomycin, an inhibitor of the vesicular H^+-ATPases that set up the proton electrochemical gradients required for active accumulation of neurotransmitters, or by tetanus toxin, which proteolytically inactivates the V-SNARE, synaptobrevin-II.[351] Finally, Haydon and colleagues have very recently described the use of a dominant-negative version of the SNARE domain of synaptobrevin-II that was targeted for selective expression in the astrocytes of transgenic mice.[348,352] Hippocampal slices from these transgenic mice were used to investigate how neuromodulators released from astrocytes via regulated exocytosis can regulate the plasticity of synaptic transmission in nearby neurons. Significantly, these *in situ* studies revealed a critical role for exocytotic release of ATP (from the astrocytes) that was rapidly metabolized into adenosine, which then acted to suppress neurotransmission.

5.7 COLOCALIZATION OF ATP RELEASE SITES, ECTO-NUCLEOTIDASE, AND PURINERGIC RECEPTORS IN PLASMA MEMBRANE SUBDOMAINS

Several studies have indicated that mammalian cells may control the accumulation and signaling of ATP and its metabolites by sequestering complexes of ecto-nucleotidases, P2 receptors and sites of ATP release within specialized cell surface microdomains. Ecto-NTPDases, like CD39, can also be localized in cell surface subdomains. Robson and colleagues identified a potential thioester linkage site for S-acylation within the cytosolic N-terminal region of CD39 expressed in COS-7 and human umbilical vein cells.[69] They showed that this N-terminal region constitutively undergoes covalent lipid modification (palmitoylation) that appears to be important both in the plasma membrane association and in the targeting of CD39 to caveolae. Similarly, Kittel et al. reported the localization of ecto-ATPase activity and 5′-nucleotidase activity in the caveolae of smooth muscle cells of guinea pig vas deferens and the ileum longitudinal muscle strips.[353] Since ecto-5′ nucleotidases like CD73 are GPI-anchored, they are known residents of detergent-resistant and glycolipid-rich membrane subdomains.[106,354] Recent studies demonstrated that P2Y and P2X receptors are associated with cholesterol-rich cell surface microdomains.[355–357]

We have described the accumulation of ATP in the bulk extracellular medium or the cell surface microenvironment of 1321N1 astrocytes stimulated with thrombin.[163] Soluble luciferase was utilized as an indicator of ATP levels within the bulk extracellular compartment, and the chimeric protein A-luciferase adsorbed to antibodies against CD14, a GPI-anchored plasma membrane protein, was used as a spatially localized probe of ATP levels at the immediate extracellular surface. Significant accumulation of ATP in the bulk extracellular compartment of thrombin-stimulated cells was observed only when endogenous ecto-ATPase activity was pharmacologically inhibited by the poorly metabolizable analog β,γ–methylene ATP. In contrast, accumulation of submicromolar ATP in the cell surface microenvironment was readily measured even in the absence of ecto-ATPase inhibition, suggesting that the spatially colocalized luciferase could effectively compete with endogenous ecto-ATPases for released ATP. Such observations support the intriguing idea that mammalian cells may express a nucleotide-signaling metabolism complex within specialized plasma membrane subdomains. In this way, the activity of P2 nucleotide receptors, ecto-nucleotidases, and ATP release sites may provide efficient temporal and spatial coordination of purinergic signaling within cell surface microenvironments (Figure 5.4).

5.8 INTRACELLULAR SIGNALING PATHWAYS INVOLVED IN THE REGULATION OF NUCLEOTIDE RELEASE FROM NONEXCITATORY CELLS

Figure 5.7 provides an overview of the established or proposed intracellular signaling pathways involved in the regulation of nucleotide release from nonexcitatory cells. Regardless of whether stimulated ATP release in nonexcitable cells is mediated by exocytotic or channel-mediated pathways, most studies indicate that increased cytosolic [Ca^{2+}] is a key second messenger for signal transduction from the extrinsic signal to the ATP release mechanism. An important signaling role for Ca^{2+} has been reported for ATP/nucleotide release triggered by mechanical stress, hypotonic stress, and occupancy of receptors for various neurotransmitters, local mediators, or hormones.[24,48,50,107,163,225,303,334,358–362] In general, most experimental support for this role of cytosolic [Ca^{2+}] is based on: 1) the ability of permeant intracellular Ca^{2+} chelators such as BAPTA to attenuate stimulus-induced increases in ATP export; 2) the attenuation of nucleotide release when extracellular [Ca^{2+}] is reduced to micromolar levels; and 3) the usually strong temporal correlation between stimulus-induced changes in the appearance of extracellular ATP and the increase in intracellular [Ca^{2+}].

Other second messengers and signaling cascades that may be involved in ATP release mechanisms have been studied in a more limited number of cell types. These include lipid-regulated signals, such as protein kinase C and phosphatidylinositol-3-kinase (PI3K).[189,306,363,364] Based on the inhibitory effects of wortmannin on PI3K, this signaling enzyme has been suggested as a putative component of the signal transduction pathway(s) involved in mechanical or hypotonic stress–activated ATP release from epithelial or endothelial cells.[306,364] ATP release from epithelial and endothelial cells by hypo-osmotic stress may also involve the activation of the rho/rho-activated kinase (ROCK) signal transduction pathway.[242,307,365] It is known that activation of rho family GTPases can lead to significant effects on cytoskeletal components or trafficking of membrane vesicles. For example, rhoA activation can lead to the tyrosine phosphorylation of integrin-associated proteins like focal adhesion kinase and paxillin.[242] Integrin-dependent ATP release has also been proposed as a mechanism in *Xenopus* oocytes.[159]

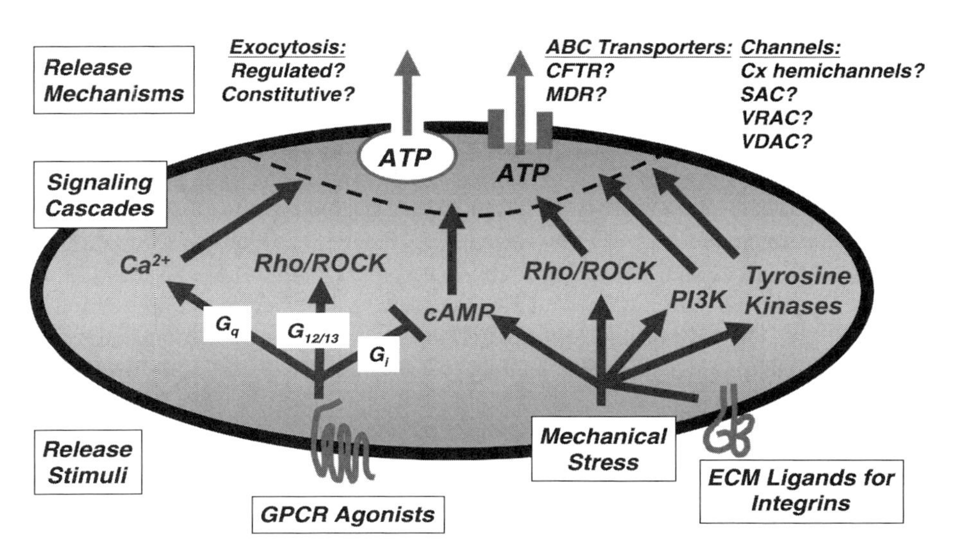

FIGURE 5.7 (See color insert following page 80) Intracellular signaling pathways involved in the regulation of nucleotide release. Mechanisms for release of ATP (and other nucleotides) can include constitutive or regulated exocytosis or nucleotides concentrated within intracellular granules or vesicles. Alternatively, cytosolic ATP (or other cytosolic nucleotides) can flow down its concentration gradient across the plasma membrane during the transient activation of various nucleotide-permeable channels or transporters. These ATP-conductive transport mechanisms may include ABC family transporters such as the cystic fibrosis transmembrane regulator (CFTR) anion channel or various multidrug-resistance (MDR) transporters. Other ATP-conductive pathways that have been studied include connexin (Cx) hemichannels, stretch-activated anion channels (SAC), volume-regulated anion channels (VRAC), and plasma membrane variants of the mitochondrial voltage-dependent anion channels (VDAC). Extrinsic stimuli for increased ATP release may include: (1) neurotransmitters, hormones, or local mediators that target diverse G protein–coupled receptors (GPCR); (2) different types of mechanical stress (see Figure 5.3); and (3) perturbed interaction (possibly via mechanical stress) of cell surface integrins with components of the extracellular matrix (ECM). Most of these extrinsic stimuli are coupled to exocytotic or conductive ATP release via intracellular signaling pathways based on transient or sustained increases in cytosolic [Ca^{2+}]. However, a growing body of data suggests that the central Ca^{2+}-based regulatory pathway acts synergistically with other signaling pathways that—depending on tissue or cell type—may include rho family GTPases, rho-operated protein kinases (ROCK), phosphatidylinositol-3-kinases (PI3K), and protein tyrosine kinases.

However, it also possible that rho family GTPases may play roles in the regulation of ATP release pathways by ROCK-independent mechanisms. Our group has observed that rapid ATP release from 1321N1 astrocytes can be triggered by agonists for multiple G protein–coupled receptors that activate phospholipase C and Ca^{2+} mobilization, including carbachol and thrombin.[163] While Ca^{2+} mobilization *per se* acted as a necessary signal for the release triggered by these receptors, it was not a sufficient signal because Ca^{2+} ionophores elicited only minor ATP release. Moreover, carbachol was fivefold less efficacious than thrombin as an ATP secretagogue despite being equally efficacious as a Ca^{2+}-mobilizing agonist. In recent unpublished studies we have observed that the higher efficacy of thrombin (and lyso-phosphatidic acid, or LPA) versus carbachol as ATP secretagogues was correlated with their relative ability to significantly activate the small GTPase rhoA as a likely consequence of a $G_{12/13}$-to-rhoGEF signaling cascade. Consistent with this, ATP release triggered by thrombin or LPA, but not carbachol, was markedly attenuated in astrocytes pretreated with *Clostridium difficile* toxin B. This clostridial toxin glycosylates all members of the rho family GTPase family (rho, rac, and CDC42) and prevents their interactions with downstream effectors.[366,367] However, ROCK inhibitors did not recapitulate the effects of the clostridial toxin, indicating that rho-GTPases other than rho *per se*, and/or that rho effectors other than ROCK, may be involved in the regulation of ATP release. Thus, rapid and efficient ATP release from nonexcitatory cells may involve an integration of parallel signaling cascades involving Ca^{2+} mobilization and multiple rho family GTPases.

5.9 NUCLEOTIDE RELEASE PATHWAYS IN BONE AND OTHER MINERALIZING TISSUES

Several recent studies have utilized the various approaches described in this chapter for the analysis of nucleotide release mechanisms from cells involved in the regulation of bone mineralization or cartilage formation.[59,220,222,229,368–373] D'Andrea and colleagues have utilized the HOBIT cell line (human osteoblast-like initial transformant) to demonstrate that both mechanical shear stress and hypotonic stress trigger a rapid ATP release from these osteoblast-type cells.[222,229,370] Moreover, this ATP release was neither decreased by pharmacological inhibitors of connexin hemichannels nor increased by overexpression of Cx43. Follow-up studies by the same group reported that HOBIT cells contain quinacrine-labeled vesicles consistent with a potential exocytotic pool of releasable ATP.[229] This quinacrine-labeling was depleted by monensin pretreatment, which also decreased the amount of ATP released in response to hypotonic stress. Significantly, hypotonic stress also triggered Ca^{2+} mobilization in the HOBIT cells, which was temporally correlated with ATP release. Treatment of the cells with Ca^{2+} ionophore also triggered a massive ATP release, which was prevented by preloading the cells with Ca^{2+} chelator. The authors concluded that Ca^{2+}-dependent exocytosis is a likely mechanism for mechanical stress–induced ATP release from osteoblasts.

Buckley et al. have extensively analyzed constitutive ATP release and extracellular metabolism in SaOS-2 cells, another human osteoblast line.[374] Those studies revealed that osteoblasts constitutively release significant amounts of ATP that is catabolised to maintain nanomolar levels of ATP and ADP in the extracellular culture medium. Those authors also reported that osteoblasts express a strong ecto-NDPK activity that acts to generate additional extracellular ATP in the presence of exogenously added ADP and other nucleotide triphosphates.

Other types of mechanical stimulation, such as fluid shear stress, also trigger ATP release from cells involved in the regulation of mineralization. Genetos et al. used a flow chamber and offline luciferase assays to monitor ATP release from the MC3T3-E1 murine pre-osteoblast cells during calibrated increases in fluid shear.[59] While these osteoblasts constitutively released ATP to achieve steady-state levels of 6 nM extracellular ATP in the low-flow state, application of 12 dynes/cm^2 fluid shear produced a 10-fold increase in extracellular ATP accumulation. This fluid shear–induced ATP release was not reduced in the presence of pharmacological inhibitors of connexin hemichannels but was markedly attenuated when the osteoblasts were pretreated with NEM, monensin, or brefeldin. Moreover, like human HOBIT cells, these MC3T3-E1 cells contained quinacrine-labeled vesicles. Thus, these studies support the findings of the D'Andrea group regarding the apparent importance of exocytotic mechanisms in osteoblast ATP release responses to diverse types of mechanical stress.

Genetos et al. also reported that fluid shear stimulation of the MC3T3-E1 cells induced a release of prostaglandin E_2 (PGE_2), which acts as a critical autocrine/paracrine regulator of load-induced bone formation.[59] Inclusion of apyrase in the extracellular medium during application of the fluid shear stress markedly repressed this release of PGE_2. This indicated that mechanical stress triggered an autocrine/paracrine cascade that involved localized ATP release and stimulation of osteoblast P2 receptors that are coupled to PGE_2 generation. Recent studies by Li et al. have used primary cultures of murine calvarial osteoblasts from wild-type versus $P2X_7$ receptor knockout mice to further analyze the role of fluid shear–induced ATP release and to identify the specific P2 receptor subtype(s) that drive the autocrine regulation of PGE_2 release.[373] Significantly, osteoblasts from both the wild-type mice and $P2X_7$ receptor null animals showed similar 10- to 30-fold increases in ATP release rate when challenged by fluid shear stress. Despite the similar increases in release of endogenous ATP, only the wild-type osteoblasts, but not the $P2X_7$ receptor null cells, secreted increased amounts of PGE_2 in response to the fluid shear stimulation. The authors concluded that autocrine activation of the $P2X_7$ receptor by locally released ATP is required for the mechanically induced secretion of PGE_2.

In summary, these various studies highlight the critical roles of local nucleotide release and induction of P2 receptor signaling cascades in the regulation of osteogenesis. They also indicate that osteoblasts may be an excellent model system for future mechanistic studies aimed at characterization of the nucleotide-containing vesicles involved in the regulated exocytotic release of ATP from nonexcitatory cell types.

REFERENCES

1. Burnstock, G., Purinergic nerves, *Pharmacol. Rev.*, 24, 509, 1972.
2. White, T.D., Direct detection of depolarisation-induced release of ATP from a synaptosomal preparation, *Nature*, 267, 67, 1977.
3. Burnstock, G. et al., Direct evidence for ATP release from non-adrenergic, non-cholinergic ("purinergic") nerves in the guinea-pig taenia coli and bladder, *Eur. J. Pharmacol.*, 49, 145, 1978.
4. Holmsen, H., Biochemistry of the platelet release reaction, *Ciba Found. Symp.*, 35, 175, 1975.
5. Macfarlane, D.E. et al., The role of thrombin in ADP-induced platelet aggregation and release: a critical evaluation, *Br. J. Haematol.*, 30, 457, 1975.
6. Holmsen, H. et al., Short communication: possible association of newly absorbed serotonin with nonmetabolic, granule-located adenine nucleotides in human blood platelets, *Blood*, 45, 413, 1975.

7. Ralevic, V. and Burnstock, G., Receptors for purines and pyrimidines, *Pharmacol. Rev.*, 50, 413, 1998.
8. Shaver, S.R., P2Y receptors: biological advances and therapeutic opportunities, *Curr. Opin. Drug Discov. Devel.*, 4, 665, 2001.
9. Lazarowski, E.R. et al., Mechanisms of release of nucleotides and integration of their action as P2X- and P2Y-receptor activating molecules, *Mol. Pharmacol.*, 64, 785, 2003.
10. North, R.A., Molecular physiology of P2X receptors, *Physiol. Rev.*, 82, 1013, 2002.
11. Girdler, G. and Khakh, B.S., ATP-gated P2X channels, *Curr. Biol.*, 14, R6, 2004.
12. Fredholm, B.B. et al., International Union of Pharmacology. XXV. Nomenclature and classification of adenosine receptors, *Pharmacol. Rev.*, 53, 527, 2001.
13. Linden, J., Molecular approach to adenosine receptors: receptor-mediated mechanisms of tissue protection, *Ann. Rev. Pharmacol. Toxicol.*, 41, 775, 2001.
14. Zimmermann, H., 5′-Nucleotidase: molecular structure and functional aspects, *Biochem. J.*, 285, 345, 1992.
15. Zimmermann, H., Extracellular metabolism of ATP and other nucleotides, *N. Schmied. Arch. Pharmacol.*, 362, 299, 2000.
16. Zimmermann, H., Two novel families of ectonucleotidases: molecular structures, catalytic properties and a search for function, *Trends Pharmacol. Sci.*, 20, 231, 1999.
17. Robson, S.C. et al., Ectonucleotidases of CD39 family modulate vascular inflammation and thrombosis in transplantation, *Semin. Thromb. Hemost.*, 31, 217, 2005.
18. Stefan, C. et al., NPP-type ectophosphodiesterases: unity in diversity, *Trends Biochem. Sci.*, 30, 542, 2005.
19. Milner, P. et al., Endothelial cells cultured from human umbilical vein release ATP, substance P and acetylcholine in response to increased flow, *Proc. Biol. Sci.*, 241, 245, 1990.
20. Milner, P. et al., Rapid release of endothelin and ATP from isolated aortic endothelial cells exposed to increased flow, *Biochem. Biophys. Res. Commun.*, 170, 649, 1990.
21. Osipchuk, Y. and Cahalan, M., Cell-to-cell spread of calcium signals mediated by ATP receptors in mast cells, *Nature*, 359, 241, 1992.
22. Wang, Y. et al., Autocrine signaling through ATP release represents a novel mechanism for cell volume regulation, *Proc. Natl. Acad. Sci. USA*, 93, 12020, 1996.
23. Roman, R.M. et al., Hepatocellular ATP-binding cassette protein expression enhances ATP release and autocrine regulation of cell volume, *J. Biol. Chem.*, 272, 21970, 1997.
24. Watt, W.C. et al., Cystic fibrosis transmembrane regulator-independent release of ATP. Its implications for the regulation of $P2Y_2$ receptors in airway epithelia, *J. Biol. Chem.*, 273, 14053, 1998.
25. Schwiebert, E.M., ABC transporter-facilitated ATP conductive transport, *Am. J. Physiol.*, 276, C1, 1999.
26. Feranchak, A.P. et al., Volume-sensitive purinergic signaling in human hepatocytes, *J. Hepatol.*, 33, 174, 2000.
27. Ostrom, R.S. et al., Cellular release of and response to ATP as key determinants of the set-point of signal transduction pathways, *J. Biol. Chem.*, 275, 11735, 2000.
28. Schwiebert, E.M. and Zsembery, A., Extracellular ATP as a signaling molecule for epithelial cells, *Biochim. Biophys. Acta*, 1615, 7, 2003.
29. Sauer, H. et al., Mechanical strain-induced Ca^{2+} waves are propagated via ATP release and purinergic receptor activation, *Am. J. Physiol.*, 279, C295, 2000.
30. Roman, R.M. et al., Evidence for multidrug resistance-1 P-glycoprotein-dependent regulation of cellular ATP permeability, *J. Membr. Biol.*, 183, 165, 2001.

31. Hisadome, K. et al., Volume-regulated anion channels serve as an auto/paracrine nucleotide release pathway in aortic endothelial cells, *J. Gen. Physiol.*, 119, 511, 2002.
32. Bennett, M.V. et al., New roles for astrocytes: gap junction hemichannels have something to communicate, *Trends Neurosci.*, 26, 610, 2003.
33. Sabirov, R.Z. and Okada, Y., ATP-conducting maxi-anion channel: a new player in stress-sensory transduction, *Jpn. J. Physiol.*, 54, 7, 2004.
34. Crane, J.K. et al., Two pathways for ATP release from host cells in enteropathogenic *Escherichia coli* infection, *Am. J. Physiol.*, 289, G407, 2005.
35. Buxton, I.L. et al., Evidence supporting the Nucleotide Axis Hypothesis: ATP release and metabolism by coronary endothelium, *Am. J. Physiol.*, 281, H1657, 2001.
36. Yegutkin, G.G. et al., Extracellular ATP formation on vascular endothelial cells is mediated by ecto-nucleotide kinase activities via phosphotransfer reactions, *FASEB J.*, 15, 251, 2001.
37. Yegutkin, G.G. et al., The evidence for two opposite, ATP-generating and ATP-consuming, extracellular pathways on endothelial and lymphoid cells, *Biochem. J.*, 367, 121, 2002.
38. Joseph, S.M. et al., Methylene ATP analogs as modulators of extracellular ATP metabolism and accumulation, *Br. J. Pharmacol.*, 142, 1002, 2004.
39. Burrell, H.E. et al., Human keratinocytes release ATP and utilize three mechanisms for nucleotide interconversion at the cell surface, *J. Biol. Chem.*, 280, 29667, 2005.
40. Lazarowski, E.R. et al., Identification of an ecto-nucleoside diphosphokinase and its contribution to interconversion of P2 receptor agonists, *J. Biol. Chem.*, 272, 20402, 1997.
41. Lazarowski, E.R. and Harden, T.K., Quantitation of extracellular UTP using a sensitive enzymatic assay, *Br. J. Pharmacol.*, 127, 1272, 1999.
42. Donaldson, S.H. et al., Basal nucleotide levels, release, and metabolism in normal and cystic fibrosis airways, *Mol. Med.*, 6, 969, 2000.
43. Lazarowski, E.R. et al., Constitutive release of ATP and evidence for major contribution of ecto-nucleotide pyrophosphatase and nucleoside diphosphokinase to extracellular nucleotide concentrations, *J. Biol. Chem.*, 275, 31061, 2000.
44. Muller, C.E., P2-pyrimidinergic receptors and their ligands, *Curr. Pharm. Des.*, 8, 2353, 2002.
45. Lazarowski, E.R. et al., Release of cellular UDP-glucose as a potential extracellular signaling molecule, *Mol. Pharmacol.*, 63, 1190, 2003.
46. Hazama, A. et al., Cell surface measurements of ATP release from single pancreatic beta cells using a novel biosensor technique, *Pflugers Arch.*, 437, 31, 1998.
47. Roman, R.M. et al., Endogenous ATP release regulates Cl^- secretion in cultured human and rat biliary epithelial cells, *Am. J. Physiol.*, 276, G1391, 1999.
48. Van der Wijk, T. et al., Osmotic cell swelling-induced ATP release mediates the activation of extracellular signal–regulated protein kinase (Erk)-1/2 but not the activation of osmo-sensitive anion channels, *Biochem. J.*, 343, 579, 1999.
49. Hazama, A. et al., Swelling-activated, cystic fibrosis transmembrane conductance regulator–augmented ATP release and Cl^- conductances in murine C127 cells, *J. Physiol.*, 523, 1, 2000.
50. Braet, K. et al., Photoliberating inositol-1,4,5-trisphosphate triggers ATP release that is blocked by the connexin mimetic peptide gap 26, *Cell Calcium*, 33, 37, 2003.
51. Darby, M. et al., ATP released from astrocytes during swelling activates chloride channels, *J. Neurophysiol.*, 89, 1870, 2003.
52. van der Wijk, T. et al., Increased vesicle recycling in response to osmotic cell swelling. Cause and consequence of hypotonicity-provoked ATP release, *J. Biol. Chem.*, 278, 40020, 2003.

53. Buttigieg, J. and Nurse, C.A., Detection of hypoxia-evoked ATP release from chemoreceptor cells of the rat carotid body, *Biochem. Biophys. Res. Commun.*, 322, 82, 2004.
54. Kawakami, M. et al., Hypo-osmotic potentiation of acetylcholine-stimulated ciliary beat frequency through ATP release in rat tracheal ciliary cells, *Exp. Physiol.*, 89, 739, 2004.
55. Okada, S.F. et al., Voltage-dependent anion channel-1 (VDAC-1) contributes to ATP release and cell volume regulation in murine cells, *J. Gen. Physiol.*, 124, 513, 2004.
56. Selzner, N. et al., Water induces autocrine stimulation of tumor cell killing through ATP release and P2 receptor binding, *Cell Death Differ.*, 11, S172, 2004.
57. Solini, A. et al., Enhanced $P2X_7$ activity in human fibroblasts from diabetic patients: a possible pathogenetic mechanism for vascular damage in diabetes, *Arterioscler. Thromb. Vasc. Biol.*, 24, 1240, 2004.
58. Bianco, F. et al., Astrocyte-derived ATP induces vesicle shedding and IL-1β release from microglia, *J. Immunol.*, 174, 7268, 2005.
59. Genetos, D.C. et al., Fluid shear–induced ATP secretion mediates prostaglandin release in MC3T3-E1 osteoblasts, *J. Bone Miner. Res.*, 20, 41, 2005.
60. Patel, A.S. et al., Paracrine stimulation of surfactant secretion by extracellular ATP in response to mechanical deformation, *Am. J. Physiol.*, 289, L489, 2005.
61. Wang, L. et al., ADP acting on $P2Y_{13}$ receptors is a negative feedback pathway for ATP release from human red blood cells, *Circ. Res.*, 96, 189, 2005.
62. Picher, M. et al., Ecto 5′-nucleotidase and nonspecific alkaline phosphatase. Two AMP-hydrolyzing ectoenzymes with distinct roles in human airways, *J. Biol. Chem.*, 278, 13468, 2003.
63. Kansas, G.S. et al., Expression, distribution, and biochemistry of human CD39. Role in activation-associated homotypic adhesion of lymphocytes, *J. Immunol.*, 146, 2235, 1991.
64. Kaczmarek, E. et al., Identification and characterization of CD39/vascular ATP diphosphohydrolase, *J. Biol. Chem.*, 271, 33116, 1996.
65. Wang, T.F. and Guidotti, G., CD39 is an ecto-(Ca^{2+},Mg^{2+})-apyrase, *J. Biol. Chem.*, 271, 9898, 1996.
66. Chadwick, B.P. and Frischauf, A.M., Cloning and mapping of a human and mouse gene with homology to ecto-ATPase genes, *Mamm. Genome*, 8, 668, 1997.
67. Marcus, A.J. et al., The endothelial cell ecto-ADPase responsible for inhibition of platelet function is CD39, *J. Clin. Invest.*, 99, 1351, 1997.
68. Enjyoji, K. et al., Targeted disruption of cd39/ATP diphosphohydrolase results in disordered hemostasis and thromboregulation, *Nat. Med.*, 5, 1010, 1999.
69. Koziak, K. et al., Analysis of CD39/ATP diphosphohydrolase (ATPDase) expression in endothelial cells, platelets and leukocytes, *Thromb. Haemost.*, 82, 1538, 1999.
70. Wang, T.F. et al., The transmembrane domains of ectoapyrase (CD39) affect its enzymatic activity and quaternary structure, *J. Biol. Chem.*, 273, 24814, 1998.
71. Chadwick, B.P. and Frischauf, A.M., The CD39-like gene family: identification of three new human members (CD39L2, CD39L3, and CD39L4), their murine homologues, and a member of the gene family from *Drosophila melanogaster*, *Genomics*, 50, 357, 1998.
72. Smith, T.M. and Kirley, T.L., Cloning, sequencing, and expression of a human brain ecto-apyrase related to both the ecto-ATPases and CD39 ecto-apyrases1, *Biochim. Biophys. Acta*, 1386, 65, 1998.
73. Mateo, J. et al., Functional expression of a cDNA encoding a human ecto-ATPase, *Br. J. Pharmacol.*, 128, 396, 1999.

74. Hicks-Berger, C.A. and Kirley, T.L., Expression and characterization of human ecto-ATPase and chimeras with CD39 ecto-apyrase, *IUBMB Life*, 50, 43, 2000.
75. Yang, F. et al., Site-directed mutagenesis of human nucleoside triphosphate diphosphohydrolase 3: the importance of residues in the apyrase conserved regions, *Biochemistry*, 40, 3943, 2001.
76. Wang, T.F. and Guidotti, G., Golgi localization and functional expression of human uridine diphosphatase, *J. Biol. Chem.*, 273, 11392, 1998.
77. Yeung, G. et al., CD39L2, a gene encoding a human nucleoside diphosphatase, predominantly expressed in the heart, *Biochemistry*, 39, 12916, 2000.
78. Murphy-Piedmonte, D.M. et al., Bacterial expression, folding, purification and characterization of soluble NTPDase5 (CD39L4) ecto-nucleotidase, *Biochim. Biophys. Acta*, 1747, 251, 2005.
79. Ivanenkov, V.V. et al., Bacterial expression, characterization, and disulfide bond determination of soluble human NTPDase6 (CD39L2) nucleotidase: implications for structure and function, *Biochemistry*, 42, 11726, 2003.
80. Hicks-Berger, C.A. et al., Expression and characterization of soluble and membrane-bound human nucleoside triphosphate diphosphohydrolase 6 (CD39L2), *J. Biol. Chem.*, 275, 34041, 2000.
81. Mulero, J.J. et al., CD39-L4 is a secreted human apyrase, specific for the hydrolysis of nucleoside diphosphates, *J. Biol. Chem.*, 274, 20064, 1999.
82. Braun, N. et al., Sequencing, functional expression and characterization of rat NTPDase6, a nucleoside diphosphatase and novel member of the ecto-nucleoside triphosphate diphosphohydrolase family, *Biochem. J.*, 351, 639, 2000.
83. Sevigny, J. et al., Differential catalytic properties and vascular topography of murine nucleoside triphosphate diphosphohydrolase 1 (NTPDase1) and NTPDase2 have implications for thromboregulation, *Blood*, 99, 2801, 2002.
84. Failer, B.U. et al., Determination of native oligomeric state and substrate specificity of rat NTPDase1 and NTPDase2 after heterologous expression in Xenopus oocytes, *Eur. J. Biochem.*, 270, 1802, 2003.
85. Stefan, C. et al., Differential regulation of the expression of nucleotide pyrophosphatases/phosphodiesterases in rat liver, *Biochim. Biophys. Acta*, 1450, 45, 1999.
86. Hessle, L. et al., Tissue-nonspecific alkaline phosphatase and plasma cell membrane glycoprotein-1 are central antagonistic regulators of bone mineralization, *Proc. Natl. Acad. Sci. USA*, 99, 9445, 2002.
87. Rutsch, F. et al., PC-1 nucleoside triphosphate pyrophosphohydrolase deficiency in idiopathic infantile arterial calcification, *Am. J. Pathol.*, 158, 543, 2001.
88. Johnson, K. et al., Linked deficiencies in extracellular PP(i) and osteopontin mediate pathologic calcification associated with defective PC-1 and ANK expression, *J. Bone Miner. Res.*, 18, 994, 2003.
89. Rutsch, F. et al., Mutations in ENPP1 are associated with "idiopathic" infantile arterial calcification, *Nat. Genet.*, 34, 379, 2003.
90. Johnson, K. et al., Chondrogenesis mediated by PPi depletion promotes spontaneous aortic calcification in NPP1$^{-/-}$ mice, *Arterioscler. Thromb. Vasc. Biol.*, 25, 686, 2005.
91. Maddux, B.A. et al., Membrane glycoprotein PC-1 and insulin resistance in non-insulin–dependent diabetes mellitus, *Nature*, 373, 448, 1995.
92. Goldfine, I.D. et al., Role of PC-1 in the etiology of insulin resistance, *Ann. NY Acad. Sci.*, 892, 204, 1999.
93. Meyre, D. et al., Variants of ENPP1 are associated with childhood and adult obesity and increase the risk of glucose intolerance and type 2 diabetes, *Nat. Genet.*, 37, 863, 2005.

94. Maddux, B.A. et al., Overexpression of the insulin receptor inhibitor, PC-1/NPP1, induces insulin resistance and hyperglycemia, *Am. J. Physiol.* [Epub ahead of print, 8/11/05] 2005.
95. Gijsbers, R. et al., The hydrolysis of lysophospholipids and nucleotides by autotaxin (NPP2) involves a single catalytic site, *FEBS Lett.*, 538, 60, 2003.
96. Cimpean, A. et al., Substrate-specifying determinants of the nucleotide pyrophosphatases/phosphodiesterases NPP1 and NPP2, *Biochem. J.*, 381, 71, 2004.
97. Bollen, M. et al., Nucleotide pyrophosphatases/phosphodiesterases on the move, *Crit. Rev. Biochem. Mol. Biol.*, 35, 393, 2000.
98. Meerson, N.R. et al., Intracellular traffic of the ecto-nucleotide pyrophosphatase/phosphodiesterase NPP3 to the apical plasma membrane of MDCK and Caco-2 cells: apical targeting occurs in the absence of N-glycosylation, *J. Cell Sci.*, 113, 4193, 2000.
99. Yano, Y. et al., Expression and localization of ecto-nucleotide pyrophosphatase/phosphodiesterase I-3 (E-NPP3/CD203c/PD-I beta/B10/gp130RB13-6) in human colon carcinoma, *Int. J. Mol. Med.*, 12, 763, 2003.
100. Buhring, H.J. et al., The basophil-specific ectoenzyme E-NPP3 (CD203c) as a marker for cell activation and allergy diagnosis, *Int. Arch. Allergy Immunol.*, 133, 317, 2004.
101. Vaingankar, S.M. et al., Subcellular targeting and function of osteoblast nucleotide pyrophosphatase phosphodiesterase 1, *Am. J. Physiol.*, 286, C1177, 2004.
102. Yano, Y. et al., Expression and localization of ecto-nucleotide pyrophosphatase/phosphodiesterase I-1 (E-NPP1/PC-1) and -3 (E-NPP3/CD203c/PD-Ibeta/B10/gp130(RB13-6)) in inflammatory and neoplastic bile duct diseases, *Cancer Lett.*, 207, 139, 2004.
103. Jansen, S. et al., Proteolytic maturation and activation of autotaxin (NPP2), a secreted metastasis-enhancing lysophospholipase D, *J. Cell Sci.*, 118, 3081, 2005.
104. Resta, R. et al., Ecto-enzyme and signaling functions of lymphocyte CD73, *Immunol. Rev.*, 161, 95, 1998.
105. Resta, R. et al., Murine ecto-5′-nucleotidase (CD73): cDNA cloning and tissue distribution, *Gene*, 133, 171, 1993.
106. Resta, R. et al., Glycosyl phosphatidylinositol membrane anchor is not required for T cell activation through CD73, *J. Immunol.*, 153, 1046, 1994.
107. Boudreault, F. and Grygorczyk, R., Cell swelling-induced ATP release is tightly dependent on intracellular calcium elevations, *J. Physiol.*, 561, 499, 2004.
108. Guyot, A. and Hanrahan, J.W., ATP release from human airway epithelial cells studied using a capillary cell culture system, *J. Physiol.*, 545, 199, 2002.
109. Plock, N. and Kloft, C., Microdialysis—theoretical background and recent implementation in applied life-sciences, *Eur. J. Pharm. Sci.*, 25, 1, 2005.
110. Matsuka, Y. et al., Concurrent release of ATP and substance P within guinea pig trigeminal ganglia *in vivo*, *Brain Res.*, 915, 248, 2001.
111. Melani, A. et al., ATP extracellular concentrations are increased in the rat striatum during *in vivo* ischemia, *Neurochem. Int.*, 47, 442, 2005.
112. Neubert, J.K. et al., Microdialysis in trigeminal ganglia, *Brain Res. Brain Res. Protoc.*, 10, 102, 2002.
113. Nishiyama, A. et al., Relation between renal interstitial ATP concentrations and autoregulation-mediated changes in renal vascular resistance, *Circ. Res.*, 86, 656, 2000.
114. Headrick, J.P. et al., Interstitial adenosine and cellular metabolism during beta-adrenergic stimulation of the *in situ* rabbit heart, *Cardiovasc. Res.*, 31, 699, 1996.
115. Harrison, G.J. et al., Extracellular adenosine levels and cellular energy metabolism in ischemically preconditioned rat heart, *Cardiovasc. Res.*, 40, 74, 1998.

116. Kuzmin, A.I. et al., Interstitial ATP level and degradation in control and postmyocardial infarcted rats, *Am. J. Physiol.*, 275, C766, 1998.
117. Kuzmin, A.I. et al., Effects of preconditioning on myocardial interstitial levels of ATP and its catabolites during regional ischemia and reperfusion in the rat, *Basic Res. Cardiol.*, 95, 127, 2000.
118. Sato, T. et al., Nicorandil increases adenosine 5′-monophosphate–primed interstitial adenosine via activation of ecto-5′-nucleotidase in rat hearts, *Heart Vessels*, 15, 81, 2000.
119. Ninomiya, H. et al., Complementary role of extracellular ATP and adenosine in ischemic preconditioning in the rat heart, *Am. J. Physiol.*, 282, H1810, 2002.
120. Obata, T., Adenosine production and its interaction with protection of ischemic and reperfusion injury of the myocardium, *Life Sci.*, 71, 2083, 2002.
121. Gourine, A.V. et al., Interstitial purine metabolites in hearts with LV remodeling, *Am. J. Physiol.*, 286, H677, 2004.
122. Zakaria, M. and Brown, P.R., High-performance liquid chromatography of nucleotides, nucleosides and bases, *J. Chromatogr.*, 226, 267, 1981.
123. Stevens, P. et al., Synthesis, storage and drug-induced release of atp-8-3h in the perfused bovine adrenal gland, *Pharmacology*, 13, 40, 1975.
124. Pearson, J.D. and Gordon, J.L., Vascular endothelial and smooth muscle cells in culture selectively release adenine nucleotides, *Nature*, 281, 384, 1979.
125. Westfall, D.P. et al., Release of endogenous ATP from rat caudal artery, *Blood Vessels*, 24, 125, 1987.
126. Yang, S. et al., Purinergic axis in cardiac blood vessels. Agonist-mediated release of ATP from cardiac endothelial cells, *Circ. Res.*, 74, 401, 1994.
127. McConalogue, K. et al., Direct measurement of the release of ATP and its major metabolites from the nerve fibres of the guinea-pig taenia coli, *Clin. Exp. Pharmacol. Physiol.*, 23, 807, 1996.
128. Shinozuka, K. et al., *In vitro* studies of release of adenine nucleotides and adenosine from rat vascular endothelium in response to alpha 1-adrenoceptor stimulation, *Br. J. Pharmacol.*, 113, 1203, 1994.
129. Shinozuka, K. et al., Differences in purinoceptor modulation of norepinephrine release between caudal arteries of normotensive and hypertensive rats, *J. Pharmacol. Exp. Ther.*, 272, 1193, 1995.
130. Hashimoto, M. et al., Source of ATP, ADP, AMP and adenosine released from isolated rat caudal artery exposed to noradrenaline, *J. Smooth Muscle Res.*, 33, 127, 1997.
131. Shinozuka, K. et al., Regional differences in noradrenaline-induced release of adenosine triphosphate from rat vascular endothelium, *J. Smooth Muscle Res.*, 33, 135, 1997.
132. Kawamoto, Y. et al., Determination of ATP and its metabolites released from rat caudal artery by isocratic ion-pair reversed-phase high-performance liquid chromatography, *Anal. Biochem.*, 262, 33, 1998.
133. Lazarowski, E.R. et al., Nucleotide release provides a mechanism for airway surface liquid homeostasis, *J. Biol. Chem.*, 279, 36855, 2004.
134. Denburg, J.L. and McElroy, W.D., Catalytic subunit of firefly luciferase, *Biochemistry*, 9, 4619, 1970.
135. Lee, R.T. et al., Substrate-binding properties of firefly luciferase. II. ATP-binding site, *Arch. Biochem. Biophys.*, 141, 38, 1970.
136. Denburg, J.L. et al., Substrate-binding properties of firefly luciferase. I. Luciferin-binding site, *Arch. Biochem. Biophys.*, 134, 381, 1969.
137. Kimmich, G.A. et al., Assay of picomole amounts of ATP, ADP, and AMP using the luciferase enzyme system, *Anal. Biochem.*, 69, 187, 1975.

138. McElroy, W.D. and DeLuca, M.A., Firefly and bacterial luminescence: basic science and applications, *J. Appl. Biochem.*, 5, 197, 1983.
139. Leach, F.R. and Webster, J.J., Commercially available firefly luciferase reagents, *Meth. Enzymol.*, 133, 51, 1986.
140. Ford, S.R. et al., Enhancement of firefly luciferase activity by cytidine nucleotides, *Anal. Biochem.*, 204, 283, 1992.
141. Ford, S.R. et al., Bioluminescent determination of 0.1 picomole amounts of guanine nucleotides, *J. Biolumin. Chemilumin.*, 9, 251, 1994.
142. Ford, S.R. et al., Effect of periodate-oxidized ATP and other nucleotides on firefly luciferase, *Arch. Biochem. Biophys.*, 314, 261, 1994.
143. Ford, S.R. et al., Use of firefly luciferase for ATP measurement: other nucleotides enhance turnover, *J. Biolumin. Chemilumin.*, 11, 149, 1996.
144. Holmsen, H. et al., Determination of ATP and ADP in blood platelets: a modification of the firefly luciferase assay for plasma, *Anal. Biochem.*, 46, 489, 1972.
145. Borst, M.M. and Schrader, J., Adenine nucleotide release from isolated perfused guinea pig hearts and extracellular formation of adenosine, *Circ. Res.*, 68, 797, 1991.
146. Brovko, L. et al., Bioluminescent assay of bacterial intracellular AMP, ADP, and ATP with the use of a coimmobilized three-enzyme reagent (adenylate kinase, pyruvate kinase, and firefly luciferase), *Anal. Biochem.*, 220, 410, 1994.
147. Gorman, M.W. et al., Measurement of adenine nucleotides in plasma, *Luminescence*, 18, 173, 2003.
148. Ishii, S. et al., A novel method for determination of ATP, ADP, and AMP contents of a single pancreatic islet before transplantation, *Transplant Proc.*, 36, 1191, 2004.
149. White, T.D., Release of ATP from a synaptosomal preparation by elevated extracellular K+ and by veratridine, *J. Neurochem.*, 30, 329, 1978.
150. White, T. et al., Tetrodotoxin-resistant release of ATP from guinea-pig taenia coli and vas deferens during electrical field stimulation in the presence of luciferin-luciferase, *Can. J. Physiol. Pharmacol.*, 59, 1094, 1981.
151. White, T.D. et al., Direct and continuous detection of ATP secretion from primary monolayer cultures of bovine adrenal chromaffin cells, *J. Neurochem.*, 49, 1266, 1987.
152. Rojas, E. et al., Real-time measurements of acetylcholine-induced release of ATP from bovine medullary chromaffin cells, *FEBS Lett.*, 185, 323, 1985.
153. Cena, V. and Rojas, E., Kinetic characteristics of calcium-dependent, cholinergic receptor controlled ATP secretion from adrenal medullary chromaffin cells, *Biochim. Biophys. Acta*, 1023, 213, 1990.
154. Fiedler, J.L. et al., Quantitative analysis of depolarization-induced ATP release from mouse brain synaptosomes: external calcium dependent and independent processes, *J. Membr. Biol.*, 127, 21, 1992.
155. Taylor, A.L. et al., Bioluminescence detection of ATP release mechanisms in epithelia, *Am. J. Physiol.*, 275, C1391, 1998.
156. Schwiebert, L.M. et al., Extracellular ATP signaling and P2X nucleotide receptors in monolayers of primary human vascular endothelial cells, *Am. J. Physiol.*, 282, C289, 2002.
157. Baggett, B. et al., Thermostability of firefly luciferases affects efficiency of detection by *in vivo* bioluminescence, *Mol. Imaging*, 3, 324, 2004.
158. Boudreault, F. and Grygorczyk, R., Cell swelling-induced ATP release and gadolinium-sensitive channels, *Am. J. Physiol.*, 282, C219, 2002.
159. Maroto, R. and Hamill, O.P., Brefeldin A block of integrin-dependent mechanosensitive ATP release from *Xenopus* oocytes reveals a novel mechanism of mechanotransduction, *J. Biol. Chem.*, 276, 23867, 2001.

160. Webster, J.J. et al., Buffer effects on ATP analysis by firefly luciferase, *Anal. Biochem.*, 106, 7, 1980.
161. Beigi, R.D. and Dubyak, G.R., Endotoxin activation of macrophages does not induce ATP release and autocrine stimulation of P2 nucleotide receptors, *J. Immunol.*, 165, 7189, 2000.
162. Beigi, R. et al., Detection of local ATP release from activated platelets using cell surface–attached firefly luciferase, *Am. J. Physiol.*, 276, C267, 1999.
163. Joseph, S.M. et al., Colocalization of ATP release sites and ecto-ATPase activity at the extracellular surface of human astrocytes, *J. Biol. Chem.*, 278, 23331, 2003.
164. Godecke, S. et al., Do rat cardiac myocytes release ATP on contraction?, *Am. J. Physiol.*, 289, C609, 2005.
165. Pellegatti, P. et al., A novel recombinant plasma membrane–targeted luciferase reveals a new pathway for ATP secretion, *Mol. Biol. Cell.*, 16, 3659, 2005.
166. Wang, Z. et al., Direct observation of calcium-independent intercellular ATP signaling in astrocytes, *Anal. Chem.*, 72, 2001, 2000.
167. Arcuino, G. et al., Intercellular calcium signaling mediated by point-source burst release of ATP, *Proc. Natl. Acad. Sci. USA*, 99, 9840, 2002.
168. Gruenhagen, J.A. et al., Monitoring real-time release of ATP from the molluscan central nervous system, *J. Neurosci. Meth.*, 139, 145, 2004.
169. Newman, E.A., Glial cell inhibition of neurons by release of ATP, *J. Neurosci.*, 23, 1659, 2003.
170. Wang, X. et al., $P2X_7$ receptor inhibition improves recovery after spinal cord injury, *Nat. Med.*, 10, 821, 2004.
171. Sorensen, C.E. and Novak, I., Visualization of ATP release in pancreatic acini in response to cholinergic stimulus. Use of fluorescent probes and confocal microscopy, *J. Biol. Chem.*, 276, 32925, 2001.
172. Bodin, P. and Burnstock, G., Evidence that release of adenosine triphosphate from endothelial cells during increased shear stress is vesicular. *J. Cardiovasc. Pharmacol.*, 38, 900, 2001.
173. Mitchell, C.H. et al., A release mechanism for stored ATP in ocular ciliary epithelial cells, *Proc. Natl. Acad. Sci. USA*, 95, 7174, 1998.
174. White, P.N. et al., Quinacrine staining of marginal cells in the stria vascularis of the guinea-pig cochlea: a possible source of extracellular ATP?, *Hear. Res.*, 90, 97, 1995.
175. Bell, P.D. et al., Macula densa cell signaling involves ATP release through a maxi anion channel, *Proc. Natl. Acad. Sci. USA*, 100, 4322, 2003.
176. Hayashi, S. et al., Detecting ATP release by a biosensor method, *Sci. STKE*, 2004, pl14, 2004.
177. Dutta, A.K. et al., Role of ATP-conductive anion channel in ATP release from neonatal rat cardiomyocytes in ischaemic or hypoxic conditions, *J. Physiol.*, 559, 799, 2004.
178. Hazama, A. et al., Swelling-induced, CFTR-independent ATP release from a human epithelial cell line: lack of correlation with volume-sensitive Cl^- channels, *J. Gen. Physiol.*, 114, 525, 1999.
179. Llaudet, E. et al., Microelectrode biosensor for real-time measurement of ATP in biological tissue, *Anal. Chem.*, 77, 3267, 2005.
180. Gourine, A.V. et al., Release of ATP in the ventral medulla during hypoxia in rats: role in hypoxic ventilatory response, *J. Neurosci.*, 25, 1211, 2005.
181. Schneider, S.W. et al., Continuous detection of extracellular ATP on living cells by using atomic force microscopy, *Proc. Natl. Acad. Sci. USA*, 96, 12180, 1999.

182. Ehrenhofer, U. et al., The atomic force microscope detects ATP-sensitive protein clusters in the plasma membrane of transformed MDCK cells, *Cell Biol. Int.*, 21, 737, 1997.
183. Rakowska, A. et al., ATP-induced shape change of nuclear pores visualized with the atomic force microscope, *J. Membr. Biol.*, 163, 129, 1998.
184. Imai, M. et al., Suppression of ATP diphosphohydrolase/CD39 in human vascular endothelial cells, *Biochemistry*, 38, 13473, 1999.
185. Grygorczyk, R. and Hanrahan, J.W., CFTR-independent ATP release from epithelial cells triggered by mechanical stimuli, *Am. J. Physiol.*, 272, C1058, 1997.
186. Roman, R.M. et al., Evidence for Gd^{3+} inhibition of membrane ATP permeability and purinergic signaling, *Am. J. Physiol.*, 277, G1222, 1999.
187. Salter, K.D. et al., Domain-specific purinergic signaling in polarized rat cholangiocytes, *Am. J. Physiol.*, 278, G492, 2000.
188. Braunstein, G.M. et al., Cystic fibrosis transmembrane conductance regulator facilitates ATP release by stimulating a separate ATP release channel for autocrine control of cell volume regulation, *J. Biol. Chem.*, 276, 6621, 2001.
189. Gatof, D. et al., Vesicular exocytosis contributes to volume-sensitive ATP release in biliary cells, *Am. J. Physiol.*, 286, G538, 2004.
190. Forrester, T. and Williams, C.A. ,Effect of pH and Ca^{2+} on ATP release from isolated adult heart cells, *J. Physiol.*, 272, 44P, 1977.
191. Lazarowski, E.R. et al., Direct demonstration of mechanically induced release of cellular UTP and its implication for uridine nucleotide receptor activation, *J. Biol. Chem.*, 272, 24348, 1997.
192. Madara, J.L. et al., 5′-adenosine monophosphate is the neutrophil-derived paracrine factor that elicits chloride secretion from T84 intestinal epithelial cell monolayers, *J. Clin. Invest.*, 91, 2320, 1993.
193. Baricordi, O.R. et al., Increased proliferation rate of lymphoid cells transfected with the $P2X_7$ ATP receptor, *J. Biol. Chem.*, 274, 33206, 1999.
194. Rutsch, F. and Terkeltaub, R., Deficiencies of physiologic calcification inhibitors and low-grade inflammation in arterial calcification: lessons for cartilage calcification, *Joint Bone Spine*, 72, 110, 2005.
195. Abedin, M. et al., Vascular calcification: mechanisms and clinical ramifications, *Arterioscler. Thromb. Vasc. Biol.* 24, 1161, 2004.
196. Towler, D.A., Inorganic pyrophosphate: a paracrine regulator of vascular calcification and smooth muscle phenotype, *Arterioscler. Thromb. Vasc. Biol.*, 25, 651, 2005.
197. Hirschberg, C.B. et al., Transporters of nucleotide sugars, ATP, and nucleotide sulfate in the endoplasmic reticulum and Golgi apparatus, *Ann. Rev. Biochem.*, 67, 49, 1998.
198. Berninsone, P.M. and Hirschberg, C.B., Nucleotide sugar transporters of the Golgi apparatus, *Curr. Opin. Struct. Biol.*, 10, 542, 2000.
199. Unsworth, C.D. and Johnson, R.G., Acetylcholine and ATP are coreleased from the electromotor nerve terminals of Narcine brasiliensis by an exocytotic mechanism, *Proc. Natl. Acad. Sci. USA*, 87, 553, 1990.
200. Morel, N. and Meunier, F.M., Simultaneous release of acetylcholine and ATP from stimulated cholinergic synaptosomes, *J. Neurochem.*, 36, 1766, 1981.
201. Richardson, P.J. and Brown, S.J., ATP release from affinity-purified rat cholinergic nerve terminals, *J. Neurochem.*, 48, 622, 1987.
202. Schweitzer, E., Coordinated release of ATP and ACh from cholinergic synaptosomes and its inhibition by calmodulin antagonists, *J. Neurosci.*, 7, 2948, 1987.
203. Stjarne, L. et al., Spatiotemporal pattern of quantal release of ATP and noradrenaline from sympathetic nerves: consequences for neuromuscular transmission, *Adv. Sec. Messenger Phosphoprotein Res.*, 29, 461, 1994.

204. Palaty, V., Swelling of the small adrenergic storage vesicles and the loss of vesicular ATP induced by cocaine in the isolated rat tail artery, *N. Schmied. Arch. Pharmacol.*, 343, 149, 1991.
205. Baumgartner, H. et al., Isolated chromaffin granules maintenance of ATP content during incubation at 31 degrees C, *Eur. J. Pharmacol.*, 22, 102, 1973.
206. Bankston, L.A. and Guidotti, G., Characterization of ATP transport into chromaffin granule ghosts, Synergy of ATP and serotonin accumulation in chromaffin granule ghosts, *J. Biol. Chem.*, 271, 17132, 1996.
207. Weber, A. et al., Specificity and properties of the nucleotide carrier in chromaffin granules from bovine adrenal medulla, *Biochem. J.*, 210, 789, 1983.
208. Bevington, A. et al., Phosphorus-31 nuclear magnetic resonance studies of pig adrenal glands, *Neuroscience*, 11, 281, 1984.
209. Costa, J.L. et al., Effects of osmotic dehydration on the nucleotides of isolated chromaffin granules: evaluation by 31p nuclear magnetic resonance, *Res. Commun. Chem. Pathol. Pharmacol.*, 45, 389, 1984.
210. Holmsen, H. and Weiss, H.J., Further evidence for a deficient storage pool of adenine nucleotides in platelets from some patients with thrombocytopathia—"storage pool disease," *Blood*, 39, 197, 1972.
211. Daniel, J.L. et al., Radioactive labeling of the adenine nucleotide pool of cells as a method to distinguish among intracellular compartments. Studies on human platelets, *Biochim. Biophys. Acta*, 632, 444, 1980.
212. Meyers, K.M. et al., Comparative study of platelet dense granule constituents, *Am. J. Physiol.*, 243, R454, 1982.
213. Ugurbil, K. et al., 31P NMR studies of nucleotide storage in the dense granules of pig platelets, *Biochemistry*, 23, 409, 1984.
214. Obermuller, S. et al., Selective nucleotide-release from dense-core granules in insulin-secreting cells, *J. Cell. Sci.*, 118, 4271, 2005.
215. Jedlitschky, G. et al., The nucleotide transporter MRP4 (ABCC4) is highly expressed in human platelets and present in dense granules, indicating a role in mediator storage, *Blood*, 104, 3603, 2004.
216. Ritter, C.A. et al., Cellular export of drugs and signaling molecules by the ATP-binding cassette transporters MRP4 (ABCC4) and MRP5 (ABCC5), *Drug Metab. Rev.*, 37, 253, 2005.
217. Forrester, T. and Williams, C.A., Release of adenosine triphosphate from isolated adult heart cells in response to hypoxia, *J. Physiol.*, 268, 371, 1977.
218. Williams, C.A. and Forrester, T., Possible source of adenosine triphosphate released from rat myocytes in response to hypoxia and acidosis, *Cardiovasc. Res.*, 17, 301, 1983.
219. Sprague, R.S. et al., Deformation-induced ATP release from red blood cells requires CFTR activity, *Am. J. Physiol.*, 275, H1726, 1998.
220. Graff, R.D. et al., ATP release by mechanically loaded porcine chondrons in pellet culture, *Arthritis Rheum.*, 43, 1571, 2000.
221. John, K. and Barakat, A.I., Modulation of ATP/ADP concentration at the endothelial surface by shear stress: effect of flow-induced ATP release, *Ann. Biomed. Eng.*, 29, 740, 2001.
222. Romanello, M. et al., Mechanically induced ATP release from human osteoblastic cells, *Biochem. Biophys. Res. Commun.*, 289, 1275, 2001.
223. Srinivas, S.P. et al., Shear-induced ATP release by cultured rabbit corneal epithelial cells, *Adv. Exp. Med. Biol.*, 506, 677, 2002.

224. Aleu, J. et al., Release of ATP induced by hypertonic solutions in *Xenopus* oocytes, *J. Physiol.*, 547, 209, 2003.
225. Birder, L.A. et al., Feline interstitial cystitis results in mechanical hypersensitivity and altered ATP release from bladder urothelium, *Am. J. Physiol.*, 285, F423, 2003.
226. Yamazaki, S. et al., Annulus cells release ATP in response to vibratory loading *in vitro*, *J. Cell. Biochem.*, 90, 812, 2003.
227. Furuya, K. et al., Characteristics of subepithelial fibroblasts as a mechano-sensor in the intestine: cell-shape–dependent ATP release and $P2Y_1$ signaling, *J. Cell Sci.*, 118, 3289, 2005.
228. Gomes, P. et al., ATP release through connexin hemichannels in corneal endothelial cells, *Invest. Ophthalmol. Vis. Sci.*, 46, 1208, 2005.
229. Romanello, M. et al., Autocrine/paracrine stimulation of purinergic receptors in osteoblasts: contribution of vesicular ATP release, *Biochem. Biophys. Res. Commun.*, 331, 1429, 2005.
230. Bergfeld, G.R. and Forrester, T., Release of ATP from human erythrocytes in response to a brief period of hypoxia and hypercapnia, *Cardiovasc. Res.*, 26, 40, 1992.
231. Bodin, P. et al., Chronic hypoxia changes the ratio of endothelin to ATP release from rat aortic endothelial cells exposed to high flow, *Proc. Biol. Sci.*, 247, 131, 1992.
232. Bodin, P. and Burnstock, G., Synergistic effect of acute hypoxia on flow-induced release of ATP from cultured endothelial cells, *Experientia*, 51, 256, 1995.
233. Wilson, P.D. et al., ATP release mechanisms in primary cultures of epithelia derived from the cysts of polycystic kidneys, *J. Am. Soc. Nephrol.*, 10, 218, 1999.
234. Dezaki, K. et al., Receptor-mediated facilitation of cell volume regulation by swelling-induced ATP release in human epithelial cells, *Jpn. J. Physiol.*, 50, 235, 2000.
235. Mitchell, C.H., Release of ATP by a human retinal pigment epithelial cell line: potential for autocrine stimulation through subretinal space, *J. Physiol.*, 534, 193, 2001.
236. Sabirov, R.Z. et al., Volume-dependent ATP-conductive large-conductance anion channel as a pathway for swelling-induced ATP release, *J. Gen. Physiol.*, 118, 251, 2001.
237. Dutta, A.K. et al., Regulation of an ATP-conductive large-conductance anion channel and swelling-induced ATP release by arachidonic acid, *J. Physiol.*, 542, 803, 2002.
238. Jans, D. et al., Hypotonic treatment evokes biphasic ATP release across the basolateral membrane of cultured renal epithelia (A6), *J. Physiol.*, 545, 543, 2002.
239. Phillis, J.W. and O'Regan, M.H., Evidence for swelling-induced adenosine and adenine nucleotide release in rat cerebral cortex exposed to monocarboxylate-containing or hypotonic artificial cerebrospinal fluids, *Neurochem. Int.*, 40, 629, 2002.
240. Goldstein, L. et al., ATP release from hypotonically stressed skate RBC: potential role in osmolyte channel regulation, *J. Exp. Zoolog. A Comp. Exp. Biol.*, 296, 160, 2003.
241. Gheorghiu, M. and Van Driessche, W., Modeling of basolateral ATP release induced by hypotonic treatment in A6 cells, *Eur. Biophys. J.*, 33, 412, 2004.
242. Hirakawa, M. et al., Sequential activation of RhoA and FAK/paxillin leads to ATP release and actin reorganization in human endothelium, *J. Physiol.*, 558, 479, 2004.
243. Reigada, D. and Mitchell, C.H., Release of ATP from retinal pigment epithelial cells involves both CFTR and vesicular transport, *Am. J. Physiol.*, 288, C132, 2005.
244. Loomis, W.H. et al., Hypertonic stress increases T cell interleukin-2 expression through a mechanism that involves ATP release, P2 receptor, and p38 MAPK activation, *J. Biol. Chem.*, 278, 4590, 2003.
245. Bodas, E. et al., ATP crossing the cell plasma membrane generates an ionic current in xenopus oocytes, *J. Biol. Chem.*, 275, 20268, 2000.
246. Dean, M. et al., The human ATP-binding cassette (ABC) transporter superfamily, *Genome Res.*, 11, 1156, 2001.

247. Dean, M. and Annilo, T., Evolution of the ATP-binding cassette (ABC) transporter superfamily in vertebrates, *Annu. Rev. Genomics Hum. Genet.*, 6, 123, 2005.
248. Uitto, J., The gene family of ABC transporters—novel mutations, new phenotypes, *Trends Mol. Med.*, 11, 341, 2005.
249. Schwiebert, E.M. et al., CFTR regulates outwardly rectifying chloride channels through an autocrine mechanism involving ATP, *Cell*, 81, 1063, 1995.
250. Prat, A.G. et al., Cellular ATP release by the cystic fibrosis transmembrane conductance regulator, *Am. J. Physiol.*, 270, C538, 1996.
251. Abraham, E.H. et al., Cystic fibrosis transmembrane conductance regulator and adenosine triphosphate, *Science*, 275, 1324, 1997.
252. Jiang, Q. et al., Cystic fibrosis transmembrane conductance regulator–associated ATP release is controlled by a chloride sensor, *J. Cell. Biol.*, 143, 645, 1998.
253. Olearczyk, J.J. et al., Receptor-mediated activation of the heterotrimeric G-protein Gs results in ATP release from erythrocytes, *Med. Sci. Monit.*, 7, 669, 2001.
254. Sprague, R.S. et al., Participation of cAMP in a signal-transduction pathway relating erythrocyte deformation to ATP release, *Am. J. Physiol.*, 281, C1158, 2001.
255. Ballerini, P. et al., Glial cells express multiple ATP binding cassette proteins which are involved in ATP release, *Neuroreport*, 13, 1789, 2002.
256. Naumann, N. et al., P-glycoprotein expression increases ATP release in respiratory cystic fibrosis cells, *J. Cyst. Fibros.*, 4, 157, 2005.
257. Abraham, E.H. et al., The multidrug resistance (mdr1) gene product functions as an ATP channel, *Proc. Natl. Acad. Sci. USA*, 90, 312, 1993.
258. Reisin, I.L. et al., The cystic fibrosis transmembrane conductance regulator is a dual ATP and chloride channel, *J. Biol. Chem.*, 269, 20584, 1994.
259. Cantiello, H.F. et al., Electrodiffusional ATP movement through the cystic fibrosis transmembrane conductance regulator, *Am. J. Physiol.*, 274, C799, 1998.
260. Konig, J. et al., No evidence for inhibition of ENaC through CFTR-mediated release of ATP, *Biochim. Biophys. Acta*, 1565, 17, 2002.
261. Huang, P. et al., Compartmentalized autocrine signaling to cystic fibrosis transmembrane conductance regulator at the apical membrane of airway epithelial cells, *Proc. Natl. Acad. Sci. USA*, 98, 14120, 2001.
262. Tarran, R. et al., Normal and cystic fibrosis airway surface liquid homeostasis. The effects of phasic shear stress and viral infections, *J. Biol. Chem.*, 280, 35751, 2005.
263. Tarran, R. et al., Regulation of normal and cystic fibrosis airway surface liquid volume by phasic shear stress, *Annu. Rev. Physiol.*, 2005.
264. Rius, M. et al., Cotransport of reduced glutathione with bile salts by MRP4 (ABCC4) localized to the basolateral hepatocyte membrane, *Hepatology*, 38, 374, 2003.
265. Nies, A.T. et al., Expression and immunolocalization of the multidrug resistance proteins, MRP1-MRP6 (ABCC1-ABCC6), in human brain, *Neuroscience*, 129, 349, 2004.
266. Bennett, M.V. et al., Structure–function studies of voltage sensitivity of connexins, the family of gap junction forming proteins, *Jpn. J. Physiol.*, 43, S301, 1993.
267. Bennett, M.V. et al., The connexins and their family tree, *Soc. Gen. Physiol. Ser.*, 49, 223, 1994.
268. Saez, J.C. et al., Plasma membrane channels formed by connexins: their regulation and functions, *Physiol. Rev.*, 83, 1359, 2003.
269. Goldberg, G.S. et al., Selective transfer of endogenous metabolites through gap junctions composed of different connexins, *Nat. Cell Biol.*, 1, 457, 1999.
270. Goldberg, G.S. et al., Gap junctions between cells expressing connexin 43 or 32 show inverse permselectivity to adenosine and ATP, *J. Biol. Chem.*, 277, 36725, 2002.

271. Alexander, D.B. and Goldberg, G.S., Transfer of biologically important molecules between cells through gap junction channels, *Curr. Med. Chem.*, 10, 2045, 2003.
272. Ebihara, L., New roles for connexons, *News Physiol. Sci.*, 18, 100, 2003.
273. Saez, J.C. et al., Gap junction hemichannels in astrocytes of the CNS, *Acta Physiol. Scand.*, 179, 9, 2003.
274. Saez, J.C. et al., Connexin-based gap junction hemichannels: gating mechanisms, *Biochim. Biophys. Acta.*, 1711, 215, 2005.
275. Li, H. et al., Properties and regulation of gap junctional hemichannels in the plasma membranes of cultured cells, *J. Cell Biol.*, 134, 1019, 1996.
276. Quist, A.P. et al., Physiological role of gap-junctional hemichannels. Extracellular calcium–dependent isosmotic volume regulation, *J. Cell Biol.*, 148, 1063, 2000.
277. John, S.A. et al., Connexin-43 hemichannels opened by metabolic inhibition, *J. Biol. Chem.*, 274, 236, 1999.
278. Trexler, E.B. et al., Voltage gating and permeation in a gap junction hemichannel, *Proc. Natl. Acad. Sci. USA*, 93, 5836, 1996.
279. Cotrina, M.L. et al., Connexins regulate calcium signaling by controlling ATP release, *Proc. Natl. Acad. Sci. USA*, 95, 15735, 1998.
280. Scemes, E., Components of astrocytic intercellular calcium signaling, *Mol. Neurobiol.*, 22, 167, 2000.
281. Braet, K. et al., Astrocyte-endothelial cell calcium signals conveyed by two signalling pathways, *Eur. J. Neurosci.*, 13, 79, 2001.
282. Leybaert, L. et al., Connexin channels, connexin mimetic peptides and ATP release, *Cell Commun. Adhes.*, 10, 251, 2003.
283. Leybaert, L. et al., Calcium signal communication between glial and vascular brain cells, *Acta Neurol. Belg.*, 104, 51, 2004.
284. Schwiebert, E.M., Extracellular ATP–mediated propagation of Ca^{2+} waves. Focus on "mechanical strain–induced Ca^{2+} waves are propagated via ATP release and purinergic receptor activation," *Am. J. Physiol.*, 279, C281, 2000.
285. Verderio, C. and Matteoli, M., ATP mediates calcium signaling between astrocytes and microglial cells: modulation by IFN-γ, *J. Immunol.*, 166, 6383, 2001.
286. Stout, C.E. et al., Intercellular calcium signaling in astrocytes via ATP release through connexin hemichannels, *J. Biol. Chem.*, 277, 10482, 2002.
287. Yao, J. et al., ATP-dependent mechanism for coordination of intercellular Ca^{2+} signaling and renin secretion in rat juxtaglomerular cells, *Circ. Res.*, 93, 338, 2003.
288. Suadicani, S.O. et al., Slow intercellular Ca^{2+} signaling in wild-type and Cx43-null neonatal mouse cardiac myocytes, *Am. J. Physiol.*, 279, H3076, 2000.
289. Scemes, E. et al., Intercellular communication in spinal cord astrocytes: fine tuning between gap junctions and P2 nucleotide receptors in calcium wave propagation, *J. Neurosci.*, 20, 1435, 2000.
290. Bennett, M.V. et al., Gap junctions: new tools, new answers, new questions, *Neuron*, 6, 305, 1991.
291. Spray, D.C., Molecular physiology of gap junction channels, *Clin. Exp. Pharmacol. Physiol.*, 23, 1038, 1996.
292. Boitano, S. and Evans, W.H., Connexin mimetic peptides reversibly inhibit Ca^{2+} signaling through gap junctions in airway cells, *Am. J. Physiol.*, 279, L623, 2000.
293. Evans, W.H. and Boitano, S., Connexin mimetic peptides: specific inhibitors of gap-junctional intercellular communication, *Biochem. Soc. Trans.*, 29, 606, 2001.
294. Chaytor, A.T. et al., Connexin-mimetic peptides dissociate electrotonic EDHF-type signalling via myoendothelial and smooth muscle gap junctions in the rabbit iliac artery, *Br. J. Pharmacol.*, 144, 108, 2005.

295. Braet, K. et al., Pharmacological sensitivity of ATP release triggered by photoliberation of inositol-1,4,5-trisphosphate and zero extracellular calcium in brain endothelial cells, *J. Cell. Physiol.*, 197, 205, 2003.
296. De Vuyst, E. et al., Intracellular calcium changes trigger connexin 32 hemichannel opening, *EMBO J.*, 25, 34, 2006.
297. Bruzzone, R. et al., Pannexins, a family of gap junction proteins expressed in brain, *Proc. Natl. Acad. Sci. USA*, 100, 13644, 2003.
298. Bao, L. et al., Pannexin membrane channels are mechanosensitive conduits for ATP, *FEBS Lett.*, 572, 65, 2004.
299. Baranova, A. et al., The mammalian pannexin family is homologous to the invertebrate innexin gap junction proteins, *Genomics*, 83, 706, 2004.
300. Locovei, S. et al., Activation of pannexin 1 channels by ATP through P2Y receptors and by cytoplasmic calcium, *FEBS Lett.*, 580, 239, 2006.
301. Scemes, E. et al., Calcium waves between astrocytes from Cx43 knockout mice, *Glia*, 24, 65, 1998.
302. Suadicani, S.O. et al., Acute downregulation of Cx43 alters P2Y receptor expression levels in mouse spinal cord astrocytes, *Glia*, 42, 160, 2003.
303. Takemura, H. et al., Hypotonicity-induced ATP release is potentiated by intracellular Ca^{2+} and cyclic AMP in cultured human bronchial cells, *Jpn. J. Physiol.*, 53, 319, 2003.
304. Strange, K. et al., Cellular and molecular physiology of volume-sensitive anion channels, *Am. J. Physiol.*, 270, C711, 1996.
305. Nilius, B. and Droogmans, G., Amazing chloride channels: an overview, *Acta Physiol. Scand.*, 177, 119, 2003.
306. Kimura, C. et al., Hypotonic stress-induced NO production in endothelium depends on endogenous ATP, *Biochem. Biophys. Res. Commun.*, 274, 736, 2000.
307. Koyama, T. et al., Involvement of Rho-kinase and tyrosine kinase in hypotonic stress-induced ATP release in bovine aortic endothelial cells, *J. Physiol.*, 532, 759, 2001.
308. Carton, I. et al., RhoA exerts a permissive effect on volume-regulated anion channels in vascular endothelial cells, *Am. J. Physiol.*, 283, C115, 2002.
309. Nilius, B. et al., Role of Rho and Rho kinase in the activation of volume-regulated anion channels in bovine endothelial cells, *J. Physiol.*, 516, 67, 1999.
310. Droogmans, G. et al., Sulphonic acid derivatives as probes of pore properties of volume-regulated anion channels in endothelial cells, *Br. J. Pharmacol.*, 128, 35, 1999.
311. Falke, L.C. and Misler, S., Activity of ion channels during volume regulation by clonal N1E115 neuroblastoma cells, *Proc. Natl. Acad. Sci. USA*, 86, 3919, 1989.
312. Jalonen, T., Single-channel characteristics of the large-conductance anion channel in rat cortical astrocytes in primary culture, *Glia*, 9, 227, 1993.
313. Kimelberg, H.K. et al., Functional consequences of astrocytic swelling, *Prog. Brain Res.*, 94, 57, 1992.
314. Schwiebert, E.M. et al., Adenosine regulates a chloride channel via protein kinase C and a G protein in a rabbit cortical collecting duct cell line, *J. Clin. Invest.*, 89, 834, 1992.
315. Schwiebert, E.M. et al., Actin-based cytoskeleton regulates a chloride channel and cell volume in a renal cortical collecting duct cell line, *J. Biol. Chem.*, 269, 7081, 1994.
316. Sabirov, R.Z. and Okada, Y., Wide nanoscopic pore of maxi-anion channel suits its function as an ATP-conductive pathway, *Biophys. J.*, 87, 1672, 2004.
317. Shoshan-Barmatz, V. and Gincel, D., The voltage-dependent anion channel: characterization, modulation, and role in mitochondrial function in cell life and death, *Cell Biochem. Biophys.*, 39, 279, 2003.

318. Cheng, E.H. et al., VDAC2 inhibits BAK activation and mitochondrial apoptosis, *Science*, 301, 513, 2003.
319. Anflous, K. et al., Altered mitochondrial sensitivity for ADP and maintenance of creatine-stimulated respiration in oxidative striated muscles from VDAC1-deficient mice, *J. Biol. Chem.*, 276, 1954, 2001.
320. Sampson, M.J. et al., Immotile sperm and infertility in mice lacking mitochondrial voltage-dependent anion channel type 3, *J. Biol. Chem.*, 276, 39206, 2001.
321. Dermietzel, R. et al., Cloning and *in situ* localization of a brain-derived porin that constitutes a large-conductance anion channel in astrocytic plasma membranes, *Proc. Natl. Acad. Sci. USA*, 91, 499, 1994.
322. Bathori, G. et al., Porin is present in the plasma membrane where it is concentrated in caveolae and caveolae-related domains, *J. Biol. Chem.*, 274, 29607, 1999.
323. Bahamonde, M.I. et al., Plasma membrane voltage-dependent anion channel mediates antiestrogen-activated maxi Cl^- currents in C1300 neuroblastoma cells, *J. Biol. Chem.*, 278, 33284, 2003.
324. Bahamonde, M.I. and Valverde, M.A., Voltage-dependent anion channel localises to the plasma membrane and peripheral but not perinuclear mitochondria, *Pflugers Arch.*, 446, 309, 2003.
325. Buettner, R. et al., Evidence for secretory pathway localization of a voltage-dependent anion channel isoform, *Proc. Natl. Acad. Sci. USA*, 97, 3201, 2000.
326. Sabirov, R.Z. et al., Genetic demonstration that the plasma membrane maxi-anion channel and voltage-dependent anion channels (VDACs) are unrelated proteins, *J. Biol. Chem.*, 2005.
327. Lowenstein, C.J. et al., Regulation of Weibel-Palade body exocytosis, *Trends Cardiovasc. Med.*, 15, 302, 2005.
328. Volknandt, W., Vesicular release mechanisms in astrocytic signalling, *Neurochem. Int.*, 41, 301, 2002.
329. Volknandt, W. et al., Expression and allocation of proteins of the exo-endocytotic machinery in U373 glioma cells: similarities to long-term cultured astrocytes, *Cell. Mol. Neurobiol.*, 22, 153, 2002.
330. Wilhelm, A. et al., Localization of SNARE proteins and secretory organelle proteins in astrocytes *in vitro* and *in situ*, *Neurosci. Res.*, 48, 249, 2004.
331. Bodin, P. and Burnstock, G., Purinergic signalling: ATP release, *Neurochem. Res.*, 26, 959, 2001.
332. Abdipranoto, A. et al., Mechanisms of secretion of ATP from cortical astrocytes triggered by uridine triphosphate, *Neuroreport*, 14, 2177, 2003.
333. Baratta, R. et al., Evidence for genetic epistasis in human insulin resistance: the combined effect of PC-1 (K121Q) and PPARgamma2 (P12A) polymorphisms, *J. Mol. Med.*, 81, 718, 2003.
334. Anderson, C.M. et al., ATP-induced ATP release from astrocytes, *J. Neurochem.*, 88, 246, 2004.
335. Chan, W.H., Effect of resveratrol on high glucose-induced stress in human leukemia K562 cells, *J. Cell. Biochem.*, 94, 1267, 2005.
336. Drose, S. and Altendorf, K., Bafilomycins and concanamycins as inhibitors of V-ATPases and P-ATPases, *J. Exp. Biol.*, 200, 1, 1997.
337. Chardin, P. and McCormick, F., Brefeldin A: the advantage of being uncompetitive, *Cell*, 97, 153, 1999.
338. Rossetto, O. et al., Tetanus and botulinum neurotoxins: turning bad guys into good by research, *Toxicon*, 39, 27, 2001.

339. Humeau, Y. et al., How botulinum and tetanus neurotoxins block neurotransmitter release, *Biochimie*, 82, 427, 2000.
340. Poulain, B. et al., *In vitro* physiological studies on clostridial neurotoxins. Biological models and procedures for extracellular and intracellular application of toxins, *Meth. Mol. Biol.*, 145, 259, 2000.
341. Betz, W.J. et al., Imaging exocytosis and endocytosis, *Curr. Opin. Neurobiol.*, 6, 365, 1996.
342. Cochilla, A.J. et al., Monitoring secretory membrane with FM1-43 fluorescence, *Annu. Rev. Neurosci.*, 22, 1, 1999.
343. Knight, G.E. et al., ATP is released from guinea pig ureter epithelium on distension, *Am. J. Physiol.*, 282, F281, 2002.
344. Khera, M. et al., Botulinum toxin A inhibits ATP release from bladder urothelium after chronic spinal cord injury, *Neurochem. Int.*, 45, 987, 2004.
345. Sun, Y. and Chai, T.C., Augmented extracellular ATP signaling in bladder urothelial cells from patients with interstitial cystitis, *Am. J. Physiol.*, 290, C27, 2006.
346. Bezzi, P. et al., Astrocytes contain a vesicular compartment that is competent for regulated exocytosis of glutamate, *Nat. Neurosci.*, 7, 613, 2004.
347. Kreft, M. et al., Properties of Ca^{2+}-dependent exocytosis in cultured astrocytes, *Glia*, 46, 437, 2004.
348. Zhang, Q. et al., Fusion-related release of glutamate from astrocytes, *J. Biol. Chem.*, 279, 12724, 2004.
349. Zhang, Q. et al., Synaptotagmin IV regulates glial glutamate release, *Proc. Natl. Acad. Sci. USA*, 101, 9441, 2004.
350. Crippa, D. et al., Synaptobrevin2-expressing vesicles in astrocytes: insights into molecular characterization, dynamics and exocytosis, *J. Physiol.*, 2005.
351. Coco, S. et al., Storage and release of ATP from astrocytes in culture, *J. Biol. Chem.*, 278, 1354, 2003.
352. Pascual, O. et al., Astrocytic purinergic signaling coordinates synaptic networks, *Science*, 310, 113, 2005.
353. Kittel, A. and Bacsy, E., Presynaptic ecto- and postsynaptic endo-calcium-adenosine-triphosphatases in synaptosomes: doubts about biochemical interpretation of localization, *Int. J. Dev. Neurosci.*, 12, 207, 1994.
354. Strohmeier, G.R. et al., Surface expression, polarization, and functional significance of CD73 in human intestinal epithelia, *J. Clin. Invest.*, 99, 2588, 1997.
355. Vacca, F. et al., $P2X_3$ receptor localizes into lipid rafts in neuronal cells, *J. Neurosci. Res.*, 76, 653, 2004.
356. Quinton, T.M. et al., Lipid rafts are required in Gαi signaling downstream of the $P2Y_{12}$ receptor during ADP-mediated platelet activation, *J. Thromb. Haemost.*, 3, 1036, 2005.
357. Vial, C. and Evans, R.J., Disruption of lipid rafts inhibits $P2X_1$ receptor–mediated currents and arterial vasoconstriction, *J. Biol. Chem.*, 280, 30705, 2005.
358. Bal-Price, A. et al., Nitric oxide induces rapid, calcium-dependent release of vesicular glutamate and ATP from cultured rat astrocytes, *Glia*, 40, 312, 2002.
359. Brain, K.L. et al., Intermittent ATP release from nerve terminals elicits focal smooth muscle Ca^{2+} transients in mouse vas deferens, *J. Physiol.*, 541, 849, 2002.
360. Katsuragi, T. et al., Inositol(1,4,5)trisphosphate signal triggers a receptor-mediated ATP release, *Biochem. Biophys. Res. Commun.*, 293, 686, 2002.
361. Liu, G.J. et al., Secretion of ATP from Schwann cells in response to uridine triphosphate, *Eur. J. Neurosci.*, 21, 151, 2005.

362. Migita, K. et al., Adenosine induces ATP release via an inositol 1,4,5-trisphosphate signaling pathway in MDCK cells, *Biochem. Biophys. Res. Commun.*, 328, 1211, 2005.
363. Feranchak, A.P. et al., Phosphatidylinositol 3-kinase contributes to cell volume regulation through effects on ATP release, *J. Biol. Chem.*, 273, 14906, 1998.
364. Feranchak, A.P. et al., The lipid products of phosphoinositide 3-kinase contribute to regulation of cholangiocyte ATP and chloride transport, *J. Biol. Chem.*, 274, 30979, 1999.
365. Grygorczyk, R. and Guyot, A., Osmotic swelling-induced ATP release: a new role for tyrosine and Rho-kinases? *J. Physiol.*, 532, 582, 2001.
366. Just, I. and Gerhard, R., Large clostridial cytotoxins, *Rev. Physiol. Biochem. Pharmacol.*, 152, 23, 2004.
367. Aktories, K. and Just, I., Clostridial Rho-inhibiting protein toxins, *Curr. Top. Microbiol. Immunol.*, 291, 113, 2005.
368. Jørgensen, N.R. et al., ATP- and gap junction–dependent intercellular calcium signaling in osteoblastic cells, *J. Cell Biol.*, 139, 497, 1997.
369. Bowler, W.B. et al., Extracellular nucleotide signaling: a mechanism for integrating local and systemic responses in the activation of bone remodeling, *Bone*, 28, 507, 2001.
370. Romanello, M. et al., Mechanosensitivity and intercellular communication in HOBIT osteoblastic cells: a possible role for gap junction hemichannels, *Biorheology*, 40, 119, 2003.
371. Hayton, M.J. et al., Involvement of adenosine 5′-triphosphate in ultrasound-induced fracture repair, *Ultrasound Med. Biol.*, 31, 1131, 2005.
372. Jørgensen, N.R., Short-range intercellular calcium signaling in bone, *APMIS Suppl.*, 5, 2005.
373. Li, J. et al., The $P2X_7$ nucleotide receptor mediates skeletal mechanotransduction, *J. Biol. Chem.*, 280, 42952, 2005.
374. Buckley, K.A. et al., Release and interconversion of P2 receptor agonists by human osteoblast-like cells, *FASEB J.*, 17, 1401, 2003.

6 Purinergic Signaling in Inflammation and Immunomodulation

Francesco Di Virgilio and Irma Lemaire

CONTENTS

6.1 INTRODUCTION

The most obvious link between bone and the immune system lies at the origin of the immune cells themselves, as they form in the bone marrow from hematopoietic precursors. Less appreciated is that immune cell activity also shapes the skeletal scaffolding by closely interacting with osteoblasts and osteoclasts.[1] Osteoclasts, and to a lesser extent osteoblasts, secrete a large number of cytokines that affect immune cell recruitment and activation. Thus, bone is not only a key structure for mechanical support of the body but also a key player in immunity. The massive remodeling that bone undergoes during several inflammatory and neoplastic diseases is due to the subversion of normal mechanisms of bone deposition and resorption caused by soluble mediators released by immune cells and by the direct action of activated T lymphocytes.[2,3] Immune cells affect bone metabolism by multiple means, including direct stimulation of osteoblasts and osteoclasts, secretion of cytokines and metabolites of arachidonic acid, and mobilization of bone cell precursors. But at the same time, immune cells also secrete factors that inhibit development and activation of bone-resorbing cells. A clear example of this dynamic relationship between bone and the immune system is afforded by the opposite roles on osteoclast activation of the T lymphocyte cytokines RANK ligand (receptor activator of NF-κB ligand) and interferon-γ (INF-γ).[4–6] Both cytokines are secreted by activated T cells, but while RANKL promotes osteoclast differentiation, INF-γ exerts a strong inhibitory effect, and thus it possibly functions as a feedback system to prevent excess bone loss in the course of inflammatory diseases.

The many complex emerging interrelationships between bone and the immune system have far-reaching implications for investigators active in both fields and have led to the definition of "osteoimmunology" as a novel field aimed at the exploration of the exciting interface between these two disciplines.

6.2 P2 RECEPTORS AS "DANGER SENSORS"

Although extracellular actions of ATP have been known since the late 1920s,[7] its possible involvement in immunomodulation has been seriously considered only during the last 10 years, and even now many immunologists raise their eyebrows when they hear about extracellular nucleotides and purinergic receptors. But, if we think of ATP for a moment with an unbiased attitude, we immediately realize that there is no better extracellular messenger. ATP is a very simple molecule, evolutionarily very old and tightly associated to the very first manifestations of life. Due to its charge, ATP is highly hydrophilic and therefore can easily diffuse through the aqueous environment in which life thrives. Large stores of ATP are available within each cell; thus, there is a virtually inexhaustible reservoir of this nucleotide. With a few exceptions in some anatomic sites where it can reach several hundred nanomolars,[8] the ATP concentration in the extracellular space is close to nil,[9–11] thus even a minute cellular release of ATP would cause a large increase in the extracellular concentration and a high signal-to-noise ratio.[12] It is not surprising therefore that living organisms have quickly learned to exploit ATP as a messenger by developing specific antennae for this and other nucleotides (the P2 receptors) and enzyme systems to quickly degrade it (ecto-nucleotidases).[13–15] Sometimes the objection is raised that ATP, as the universal intermediate of intracellular energy transactions, is too precious a molecule to be wasted as a messenger. This is clearly a very naïve objection because all the other well accepted extracellular messengers (hormones, neurotransmitters, growth factors, cytokines, etc.) are synthesized at the expense of ATP and are consequently much more energy intensive. In addition, if we think that the intracellular ATP concentration is in the millimolar range (5 to 10 mM), that due to its negative charges ATP cannot passively diffuse across the plasma membrane, and that virtually all living cells possess specific and sensitive sensors that reveal the presence of nucleotides in the extracellular space (the P2 receptors),[16] then ATP (and probably other nucleotides such as UTP and UDP) have all the prerequisites of the perfect signal to alert the immune system of cell or tissue damage (danger signals).

Every pathologist knows well that the most potent stimulus for inflammation is tissue damage; accordingly, it is generally thought that cell injury is the final common pathway through which disease conditions as different as cancer and infections, or noxious agents such as UV light or chemical irritants, trigger inflammation. Immune cells resident in the connective tissue (mast cells, macrophages, dendritic cells) react

TABLE 6.1
Intracellular molecules as candidates for the role of constitutive danger signals

Ligand (Danger Signal)	Receptor (Danger Sensor)
Heat shock proteins (HSP60, HSP70, HSP90)	CD91, TLR4, TLR4/TLR2, LOX-1
Uric acid crystals	?
DNA	TLR9
ssRNA	TLR3, TLR7, TLR8
Nucleotides	$P2Y_{1,2,4,6,11}$, $P2X_{4,7}$
Sugar metabolites (UDP glucose)	$P2Y_{14}$

Note: TLR = Toll-like receptors; ssRNA = single strand RNA.

to the presence of intracellular molecules that are normally segregated within the cytoplasm and that, upon cell injury, are released into the extracellular space (Table 6.1).[17,18] Depending on the extent of cell damage and the modality of cell death, release of these signals of cell distress will be more or less massive, thus leading to a varying degree of immune cell activation. The role of intracellular components released by cells undergoing stress has been greatly re-evaluated during the last few years, mainly thanks to the work and the theoretical elaborations of Polly Matzinger.[17,19] Thus, we are becoming familiar with the concept that we do not react (or we do not *only* react) to foreign ("non-self") molecules, but rather to molecules that damage our tissues. In other words, it seems that our organism reacts to all those factors that are actual or potential causes of damage, rather than to mere foreign agents. To this aim, our cells are equipped with a network of receptors specialized in the detection in the pericellular environment of molecules normally confined within the intracellular space. The molecules that alert the body of an impending danger have been collectively named "danger signals," and their receptors "danger sensors."[17,19] A key feature of all danger signals is that they have a strong activating and differentiating activity on dendritic cells and, by this mechanism, also a potent immunoregulatory function. Thus, we may define a danger signal as something that is released by injured or dying cells, is sensed as a sign of tissue distress, and causes dendritic cell activation and differentiation. Nucleotides fit almost perfectly with this definition, and accordingly nucleotide receptors fit with the definition of danger signals.

ATP is stored in the cytoplasm at a concentration varying in the 5 to 10 mM range, while UTP is about 10-fold less concentrated. It is also known that in some cell types ATP is accumulated within intracellular granules (i.e., platelets or chromaffin cells),[20–22] less clear is whether it is also compartmentalized in the aqueous cytoplasm. In contrast, the extracellular ATP concentration is extremely low (1 to 100 nM) under normal conditions. This creates a very large concentration gradient that allows substantial ATP levels to build up in the extracellular space via comparatively small cellular ATP releases, thus generating a high signal-to-noise ratio. Furthermore, thanks to its negative charges (from 2 to 4, depending on the divalent cation concentration and pH of the solvent), ATP can easily diffuse through the extracellular environment and thus efficiently spread the alarm. In addition, the ATP alert signal can be terminated quickly by ubiquitous ecto-nucleotidases, thus preventing overstimulation of the immune system or even desensitization.[15] Last but not least, virtually all cells express receptors for extracellular nucleotides with a wide range of affinities that translate the ATP signal into numerous specific cellular responses.[16]

Fifteen P2 receptors have been so far identified (seven P2X and eight P2Y receptors), most of which are expressed to varying degree by immune cells.[23,24] While ATP is the preferred agonist (with EC_{50} ranging from 1 μM to 500 μM, depending on the given receptor subtype) at all P2X receptors, among P2Y receptors, $P2Y_{11}$ is the only true ATP-selective receptor, as at $P2Y_2$ receptors ATP and UTP are equipotent, and at the remaining P2Y receptors other nucleotides are preferred ligands (ADP at $P2Y_1$, $P2Y_{12}$, and $P2Y_{13}$; UTP at $P2Y_4$; UDP at $P2Y_6$; and UDP-glucose at $P2Y_{14}$).[14,25] The high number of subtypes and the widely differing affinities make the purinergic system a very flexible apparatus for the decoding of signals originating from stressed tissues. Low concentrations of extracellular ATP mainly stimulate P2Y receptors and, as such, exert a modulatory activity on immune cell differentiation and cytokine secretion. The effect of chronic (24 h) stimulation with low (10 to 250 μM) ATP doses has been most thoroughly investigated in *in vitro* human dendritic cell cultures. In the presence of ATP, typical plasma membrane markers of dendritic cell maturation (CD54, CD80, CD83, CD86) are upregulated.[26–28] Furthermore, ATP also potentiates the expression of CD83, CD86, and CD54 stimulated by bacterial endotoxin (lipopolysaccharide, LPS) or CD40L, two potent dendritic cell activating agents. Maturation of dendritic cells is normally accompanied by a decrease in endocytic activity, which translates into a reduced ability of antigen uptake. In keeping with its maturating ability, chronic exposure to ATP also causes a substantial decrease in fluid phase endocytosis. Finally, antigen-presenting function of dendritic cells is strongly enhanced.[26]

It has been proposed that P2 receptors are also involved in the early phases of dendritic cell activation by nonnucleotide stimulatory factors such as extracellular mRNA.[29] It is known that exogenous mRNA delivers a strong activation signal to dendritic cells and stimulates a potent primary immune response. Data obtained by Ni and coworkers suggest that mRNA interacts on the dendritic cell plasma membrane with a G protein–coupled receptor that can be desensitized by ATP and ADP but not by UTP.[29] However, although these data are suggestive of an involvement of P2Y receptors, they do not unequivocally prove that exogenous mRNA directly interacts with P2Y receptors, as the putative mRNA-binding receptor might be down-modulated or cross-desensitized by the concomitant activation of an as yet unidentified P2Y receptor.

A typical feature of mature dendritic cells is the ability to synthesize and release cytokines and chemokines with key regulatory functions on inflammation and T cell differentiation.[30] However, ATP-stimulated dendritic cells did not substantially release cytokines or chemokines, showing that the differentiation pathways promoted by this nucleotide is incomplete. However, ATP had a striking effect on the pattern of cytokine and chemokine secretion stimulated by the conventional maturation agents LPS and CD40L. By themselves LPS and CD40 cause a large secretion of interleukin (IL)-1α, -1β, tumor necrosis factor-α (TNF-α), IL-6, IL-12, IL-10, and the IL-1 receptor antagonist (IL-1ra). Preincubation in the presence of ATP had negligible effect on secretion of IL-10 and IL-1ra but drastically inhibited release of the other cytokines.[26] Inhibition of IL-12 secretion is of particular relevance since this cytokine is the most important factor that drives differentiation of naïve T cells toward the Th1 phenotype. Thus, and not surprisingly, dendritic cells matured in the presence of ATP promote a Th2 polarization.[26,31] Treatment with low ATP concentrations also modifies the pattern of chemokine receptor expression and chemokine release. Receptors for inflammatory chemokines such as CC chemokine receptor 5 (CCR5) are down-modulated, whereas those for lymphoid chemokines, such as CXC chemokine receptor 4 (CXCR4) and CCR7, are upregulated.[31] The pattern of chemokine release is modified accordingly: constitutive secretion of the lymphoid chemokine CCl22 is increased, and the LPS-stimulated secretion of the inflammatory chemokines CXCL10 and CCL5 is inhibited. This shift in chemokine release and chemokine receptor expression, along with the inhibition of IL-12 secretion, favors recruitment of Th2 rather than Th1 lymphocytes. In summary, chronic exposure to low ATP concentrations enhances lymphocyte localization to the lymph nodes and favors a Th2 rather than a Th1 immune response. Ability to activate dendritic cells and modulate their differentiation is a key requirement of danger signals, fully satisfied by extracellular ATP[32] (Figure 6.1).

The type of response triggered by ATP changes drastically when immune cells are acutely exposed to high (millimolar) extracellular ATP levels, as may happen as a consequence of acute tissue injury or massive bacterial invasion. Under these conditions, release of pro-inflammatory cytokines is potently stimulated and inflammation is initiated.[24,33] This response also fits well with the danger signal paradigm, since primary, or constitutive, danger signals should also be able to induce release of additional mediators in order to activate body defensive mechanisms. Extracellular ATP is well known to cause release of several important cytokines that have a key immunoregulatory role such as IL-1β, IL-1α, IL-1ra, IL-6, IL-8, IL-10, IL-18, and TNF-α.[27,34–42] This might be of particular relevance for bone tissue, as it is well known that TNF-α and IL-1 markedly upregulate expression of RANKL,[43] a molecule expressed by activated osteoblasts that has a crucial role in osteoclastogenesis. RANKL, by binding to its receptor named RANK, regulates osteoclast development and osteoclast survival, fusion, and pit-formation activity.[4] Cytokines also modulate release of the osteoclast inhibitory factor osteoprotegerin (OPG), a decoy receptor that binds to RANKL, prevents its interaction with RANK, and negatively regulates osteoclast formation and function.[44,45] Secretion of OPG is stimulated by IL-1, IL-6, IL-11, TNF-α, and transforming growth factor β (TGFβ).[46] Nucleotides can directly trigger IL-6 secretion from human osteoblasts,[47] a finding that further emphasizes the local immunoregulatory function of nucleotides in the bone. Both P2X and P2Y receptors are implicated in cytokine release: IL-1β and IL-1α are secreted in response to $P2X_7$ receptor activation via a mechanism that has been extensively investigated, especially in the case of IL-1β. The key step appears to be the drastic depletion of intracellular K^+, which causes caspase 1 activation, pro-IL-1β processing, and exteriorization.[35,38,48] The pathway for caspase 1 activation triggered by ATP appears to

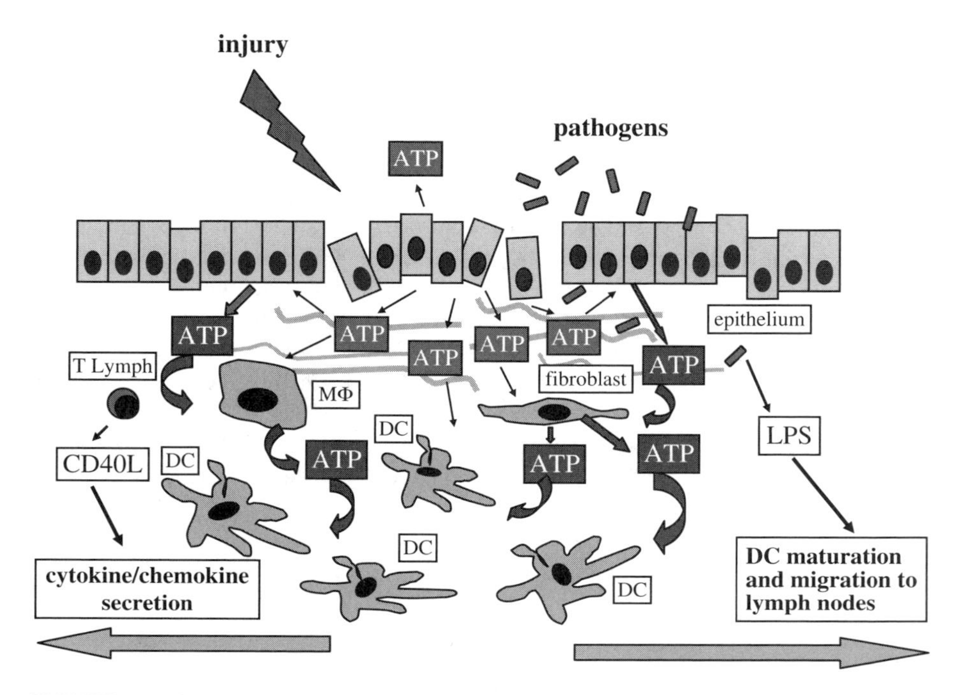

FIGURE 6.1 (See color insert following page 80) Cell damage caused by physical trauma or pathogens causes an initial wave of ATP release (red) from injured cells. This first wave of ATP release diffuses through the pericellular space and activates adjacent cells (e.g., epithelial cells, macrophages, fibroblasts). P2 receptors expressed on the plasma membrane of the bystander cells will be activated and trigger the release of several bioactive agents, among which ATP itself (ATP-induced ATP release, dark blue). Accumulation of ATP into the extracellular milieu will prime Langerhans/dendritic cells (DC) and modify their responses to bacterial endotoxin (LPS) or T lymphocyte–derived CD40 ligand (CD40L). The combined action of these agents drives DC maturation and the secretion of cytokines and chemokines. (Reprinted from Di Virgilio, F., *Purinergic Signalling*, 1, 205–209, 2006. With permission of Springer The Netherlands.)

follow the physiological route involving the "inflammasome".[49] Recent studies have also provided an answer to the long-standing question of the pathway involved in IL-1β transmembrane transport. It is now clear that most secreted mature IL-1β is exteriorized by a vesicular pathway that may involve cytokine packaging in the lumen of microvesicles budding from the plasma membrane[39] as well as classical secretory exocytosis of specialized lysosomes.[50] These microvesicles diffuse in the pericellular space and deliver their cargo to neighboring cells. It is, however, as yet unclear how the IL-1β is released from within the vesicles once they reach their targets. More recently it has been shown that secretion of type-1 IL-1ra from macrophages and endothelial cells is also triggered by ATP via $P2X_7$ receptor stimulation.[51] As previously shown for IL-1β release from human macrophages and mouse microglial cells, IL-1ra secretion is triggered not only by challenge with exogenously added ATP, but also by autocrine/paracrine stimulation by locally released ATP. Not surprisingly, LPS was found to be a strong stimulus for ATP release. ATP-stimulated

IL-18 release is probably also mediated by $P2X_7$ receptors,[52,53] but the role of nucleotides in the release of this cytokine has not been investigated as extensively as that of IL-1β.

Extracellular ATP also triggers release of TNF-α from microglia and dendritic cells. The receptor implicated might also be a $P2X_7$ receptor, but the ability of rather low (10 to 100 μM) ATP concentrations, subthreshold for $P2X_7$ activation, to trigger release of this cytokine in dendritic cells suggests that P2Y receptors might also be involved.[54–56] In mouse peritoneal macrophages, chronic incubation in the presence of ATP suppressed TNF-α release stimulated by LPS,[57] a finding reminiscent of the inhibitory activity of extracellular ATP on the release of inflammatory cytokines in dendritic cells activated with LPS or CD40L.[26,27] Autocrine/paracrine purinergic stimulation also sustains IL-10 release from LPS-activated microglia, probably via P2Y receptors, as this effect can be mimicked by exogenous ADP in addition to ATP.[58] Of particular interest is the ability of extracellular nucleotides to induce synthesis and release of IL-8 from human astrocytes, macrophages, and eosinophils.[59,60] Receptors subtypes involved in astrocyte or eosinophil activation appear to be different; in astrocytes, ATP and ADP were equipotent, whereas 2′– and 3′–*O*-(4-benzoyl-benzoyl)-ATP (BzATP) was less effective than ATP, and UTP was ineffective.[59] On the contrary, in eosinophils the best stimulus was UDP, followed by ATP, while ADP and UTP were ineffective. A role of the $P2Y_6$ receptor (UDP selective) in IL-8 release is supported by data from THP-1 cells.[61]

The transcription factor NF-6B has a central role in inflammation and immunity as it controls transcription of several genes encoding key inflammatory mediators such as TNF-α and IL-6 or adhesion molecules such as ICAM-1.[62] In further support of the role of ATP as a danger signal, this nucleotide potently stimulates NF-κB activation in microglia,[63] osteoclasts,[64] T lymphocytes,[65] monocytes,[66] and endothelial cells.[67] The mechanism has been thoroughly investigated by Shulze-Osthoff and coworkers, who showed that, in contrast to other traditional pro-inflammatory stimuli, ATP-dependent NF-κB activation was rather delayed and required caspase 1 activity in addition to production of reactive oxygen species and proteasome function.[63] In addition to NF-κB, NFAT is also activated by extracellular ATP.[68] For both transcription factors the receptor involved is $P2X_7$.

In a living organism, the paradigmatic "signal of danger" is pain, an unavoidable accompanying symptom of inflammation. P2X receptors are increasingly recognized as key participants in the sensory transduction of inflammatory, chronic, and neuropathic pain. It is hypothesized that tonic ATP release from injured tissues or from stimulated nerve terminals triggers sustained activation of $P2X_3$ homomers or $P2X_{2/3}$ heteromers, which then signal pain sensation.[69] Inoue and coworkers recently showed that pharmacological blockade of $P2X_4$ receptors abolishes tactile allodynia after peripheral nerve injury in the rat.[70] In addition, after nerve injury, microglia are turned into an activated state, a feature of which is a dramatic overexpression of the $P2X_4$ receptors. As $P2X_7$ is the P2X receptor with the best established role in inflammation, it is not surprising that this receptor may also mediate inflammatory pain. It is of great interest that pain induced by adjuvant injection or by nerve ligation is completely absent in $P2X_7$ receptor knockout mice.[71] Not surprisingly, in these animals IL-1β secretion following adjuvant inoculation was also defective. These very recent reports highlight the importance of the "extracellular ATP/P2 receptor" system as a ubiquitous mechanism that senses and transduces danger from the cellular to the tissue level.

6.3 ROLE OF P2 RECEPTORS IN IMMUNE CELL CHEMOTAXIS

Cyclic AMP has been known for years as a potent chemoattractant for the amoeba *Dyctiostelium discoideum* acting at a family of seven membrane-spanning, G protein–coupled receptors with chemotactic and differentiating activity. In eukaryotic cells, a chemotactic activity for extracellular nucleotides has been postulated by several authors,[72,73] but formal proof was provided only recently by showing that neutrophils or HL-60 cells chemotact in response to UTP, ATP, and ATPγS,[74] and rat mast cells chemotact in response to ADP, ATP, and UTP.[75] Rather interestingly, J774 mouse macrophages respond to ADP but not to UTP. Likewise, primary mouse microglia are potently attracted by ATP or ADP but not UTP gradients.[76] In all these cases, chemotaxis was inhibited by pertussis toxin.

A very interesting case is that of human dendritic cells. These cells are normally in a quiescent state (immature), but in the presence of stimulating agents (bacteria, viruses, endotoxin, or tissue-derived danger or distress signals) undergo a complex process of maturation, as described in section 6.2 of this chapter. Common P2 receptor agonists such as ATP or the pharmacological analogue 2-methylthio ATP (2-MeSATP), or selective P2Y receptor agonists such as ADP or UTP, are potent chemotactic stimuli for immature dendritic cells.[77] In contrast, mature dendritic cells did not chemotact in response to any of the nucleotides used. Signal transduction is also impaired in mature compared to immature dendritic cells: while immature dendritic cells undergo an increase in $[Ca^{2+}]_i$ and actin polymerization in response to several nucleotides, mature dendritic cells respond with a $[Ca^{2+}]_i$ rise only to ATP, a common P2 agonist, but not ADP (a $P2Y_1$, $P2Y_{12}$, and $P2Y_{13}$ receptor agonist) or UTP (a $P2Y_2$, $P2Y_4$, and $P2Y_6$ receptor agonist). Furthermore, no nucleotide is capable of triggering actin polymerization in mature dendritic cells. Nevertheless, immature and mature dendritic cells show no substantial difference in the pattern of P2 receptor expression. Therefore, it appears that during maturation, dendritic cells uncouple selected P2Y receptors from the downhill signaling cascade. In the setting of inflammation, this selective negative modulation of P2Y responses might have a deep physiological significance, because immature dendritic cells should be able to chemotact toward the ATP-rich inflammatory site and, once they have captured the antigen, should be able to complete their maturation and depart toward lymphoid organs under the stimulation of chemokines. In support of this model, the chemotactic response to nucleotides is bell shaped, thus enabling dendritic cells to receive a "stop signal" when they reach the site of maximal ATP accumulation. Optimal ATP dose for chemotaxis is 100 nM, which is a concentration subthreshold for effector responses or for expression of differentiation markers.[26,55] The precise identity of the P2Y receptor responsible for leukocyte chemotaxis is as yet unknown. Published experiments at the moment appear to rule out $P2Y_{11}$ receptors, the only

ATP-selective receptor in the P2Y subfamily, but leaves open the choice among at least the following candidates: $P2Y_1$, $P2Y_2$, $P2Y_4$, $P2Y_6$, $P2Y_{12}$, and $P2Y_{13}$ receptors. The $P2Y_6$ receptor might have a more defined role because, besides driving chemotaxis of immature dendritic cells by itself, it also mediates release of the chemokine CXCL8 (formerly known as IL-8) from mature dendritic cells.[78] IL-8 is a potent neutrophil chemotactic factor; thus, by increasing dendritic cells' ability to recruit neutrophils, the $P2Y_6$ receptor has a possible important role in the amplification of inflammation. A more general role in chemotaxis of $P2Y_2$, $P2Y_4$, and $P2Y_6$ receptors is supported by the strong chemotactic activity of UTP, but not ATP, on vascular endothelial cells.[79] Even in smooth muscle cells, where chemotaxis is driven by several nucleotides, UTP is the most potent.[80] In this cell type, UTP increases osteopontin expression dose dependently, and blockade of this ligand or of its integrin receptors inhibits UTP-induced migration.

Robson and coworkers[81] provided an elegant indirect demonstration of the importance of nucleotides in chemotaxis by showing that monocytes/macrophages, endothelial cells, and pericytes from mice deleted of the major ecto-ATPase (CD39) were unable to repopulate implanted Matrigel plugs containing angiogenic factors. This suggests that aberrant regulation of P2 receptors due to elevated pericellular ATP levels disrupts chemotaxis. The strong *in vitro* chemotactic activity of uracyl nucleotides suggests an important, albeit as yet ill-defined, *in vivo* role for these molecules, especially because extracellular UDP and UTP are not easily quantitated.

Burnstock and coworkers first reported the cellular release of UTP, showing that changes in perfusion rates of the rabbit ear artery caused an accumulation of extracellular UTP.[82] Later, an enzymatic assay for UTP was developed by Lazarowski and coworkers based on the generation of UDP glucose by UDP glucose pyrophosphorylase in the presence of UTP.[83] This has allowed measurement of extracellular UTP concentrations of 1 to 5 nM in the medium bathing quiescent cell cultures. These UTP levels are thought to be about one third of the extracellular ATP concentration.[8] While it is reasonable to hypothesize that many cells are able to release UTP, this has been conclusively shown only for a few cell types: endothelial cells, platelets, astrocytoma cells, C6 glioma, airway, and epithelial cells.[82,84] It is calculated that this may result in the accumulation of 20 to 100 nM UTP into the pericellular space.[85] As to the mechanism of release, it is thought that UTP and ATP are released through a common pathway that does not discriminate between the two nucleotides, since the time course of release is the same and the relative extracellular concentrations reflect the relative intracellular levels.[83]

6.4 PARTICIPATION OF P2 RECEPTORS IN ANTIGEN HANDLING AND PRESENTATION

It is a common observation that P2 receptors are expressed to a high level by dendritic cells and, in general, by mononuclear phagocytes. This has raised the intriguing question of the possible participation of these receptors to antigen handling and presentation. In order to start an immune response, antigens must first be endocytosed by antigen-presenting cells (APCs), processed, and then exposed on the plasma membrane in association with molecules belonging to class I or class II major histocompatibility complex (MHC). Once the antigen/MHC has been properly exposed on the plasma membrane, the APCs interact with specific target T cells expressing the cognate T cell receptor (TCR) and initiate a complex chain of reactions that initiate the immune response. The P2 receptors might, in theory, participate in this complex chain of events at different steps of antigen handling: endocytosis, processing within the endosomes or in the proteasome, or association to MHC molecules. A role in APC target cell interaction can also be postulated.

Very few studies have addressed the modulation of phagocyte endocytosis via P2 receptors. Neutrophil phagocytosis was reported to be enhanced by exposure to platelet supernatant, very likely thanks to the high content of ATP, since the potentiating effect was fully blocked by apyrase or by treatment with suramin.[86] More relevant to antigen presentation is the finding that dendritic cell endocytosis is transiently stimulated by exposure to low (100 μM) ATP doses.[28] These are the same doses that induce dendritic cell maturation, and therefore it makes physiological sense that the increase in endocytic activity is transient since antigen uptake must be shut down once the dendritic cells enter the maturation pathway. Higher ATP concentrations, on the other hand, inhibit endocytosis both in dendritic cells and in macrophages[28] (F. Di Virgilio, unpublished observations). Although no systematic analysis has been carried out, it is very likely that the endocytosis-potentiating effect is mediated by P2Y receptors. As described in section 6.2, P2Y receptors have a striking effect on dendritic cell maturation, and this also bears on the ability of these cells to handle and present the antigen. However, very few studies have directly addressed the issue of the modulation of antigen presentation by P2 receptors. Several P2 receptor knockout mice are now available, but a thorough analysis of the immune response in these animals has not been carried out yet. In the one study that addressed directly the issue of the possible participation of P2 receptors in antigen presentation, P2X rather than P2Y receptors were implicated, as this study shows that antigen-pulsed mouse dendritic cells lacking the $P2X_7$ receptor have a reduced ability to stimulate antigen-specific Th lymphocytes.[87] These defective cells showed no obvious immunological deficits, but a strikingly reduced secretion of IL-1β during their interaction with the specific Th lymphocytes was observed. Furthermore, blockade of the $P2X_7$ receptor with oxidized ATP (oATP) also drastically reduced antigen-presenting ability. Inhibition of Th lymphocyte stimulation was reverted by addition of exogenous IL-1β to dendritic cell-lymphocyte cocultures. However, as the role of IL-1β in the overall process of antigen presentation and Th lymphocyte activation is as yet poorly understood, these results are difficult to interpret.

P2 receptors might also participate in the overall process of antigen handling and presentation by directing dendritic cell homing to regional lymph nodes. It is currently understood that maturing dendritic cells migrate from peripheral tissues to the lymph nodes under the action of the coordinate stimulation by inflammatory and lymphoid chemokines. During this process, receptors for inflammatory chemokines (CCR1, CCR2, CCR5, and CXCR1) are down-modulated, whereas those for lymphoid chemokines (CXCR4, CCR4, CCR7) are induced. Stimulation with ATP upregulates CXCR4 and CCR7 and reduces CCR5. Accordingly, ATP-treated dendritic cells increase migration to lymphoid chemokines such as CXCL12 and

CCL19 and lose responsiveness to an inflammatory chemokine such as CCL4. Finally, the pattern of chemokines released by dendritic cells is also drastically affected by ATP: constitutive production of CCL22 is enhanced, and LPS-induced secretion of CXCL10 and CCL5 is down-modulated. Overall, this complex alteration of chemokine receptor expression and chemokine release induced by ATP facilitates dendritic cell migration to secondary lymphoid organs and dendritic cell-Th interaction, thus ultimately favoring antigen presentation. However, very few studies have explored purinergic modulation of dendritic cell responses, leaving this field largely unexplored.

6.5 PURINERGIC SIGNALING IN ANTIGEN DESTRUCTION/ELIMINATION

6.5.1 Purinergic Signaling and Elimination of Infectious Agents

The ultimate goal of the concerted action of innate and adaptive immunity is to eliminate any invading pathogens or foreign molecules that represent a threat to the host. In particular, innate responses are key in confining and eliminating the invaders, thus preventing further access to tissues.[88] This process requires the participation of parallel responses carried out by a variety of soluble factors and cells that involve purinergic as well as other inter- and intracellular signaling events.[89–91]

Following interaction with the host, some pathogens can be readily eliminated at outer extra-epithelial sites by various effector molecules, including secreted enzymes and antimicrobial peptides that kill pathogens. The latter are natural small cationic peptides,[92] generally 12 to 50 amino acids in length, that have a net positive charge and a molecular structure well suited to interacting with membranes, especially negatively charged bacterial membranes. Cationic peptides have a wide range of antimicrobial activities that can include action against most Gram-negative and Gram-positive bacteria, fungi, enveloped viruses, and eukaryotic parasites.[93] Recent studies have provided both direct and indirect evidence that purinergic signaling is involved in mediating their actions (Figure 6.2A). Thus Histatin-5, a small histidine-rich basic peptide of salivary acinar cell origin,[94] kills *Candida albicans*, the most prevalent human fungal pathogen, via a mechanism that involves a nonlytic release of ATP from the yeast.[95] Conductive transport of ATP has been observed under certain conditions in various cell types and is thought to implicate transmembrane efflux through ATP-specific channels and ATP-binding cassette proteins.[96,97] Furthermore, release of ATP in response to Histatin-5 precedes the signal for cytotoxicity, which was shown to be mediated by extracellular ATP acting on $P2X_7$-like receptors.[95]

Alpha-defensin 1 (HNP-1), a cationic peptide present in saliva of patients with oral diseases,[98] is secreted by epithelial cells and human neutrophils in candidiasis.[99] The killing effect of HNP-1 on *C. albicans*[100] shares very similar features to Histatin-5 cytotoxic action with respect to magnitude of induction of nonlytic ATP efflux, depletion of intracellular ATP pools, and purinergic inhibitor profile.[95] Although these two classes of molecules are structurally different, their killing pathways appear

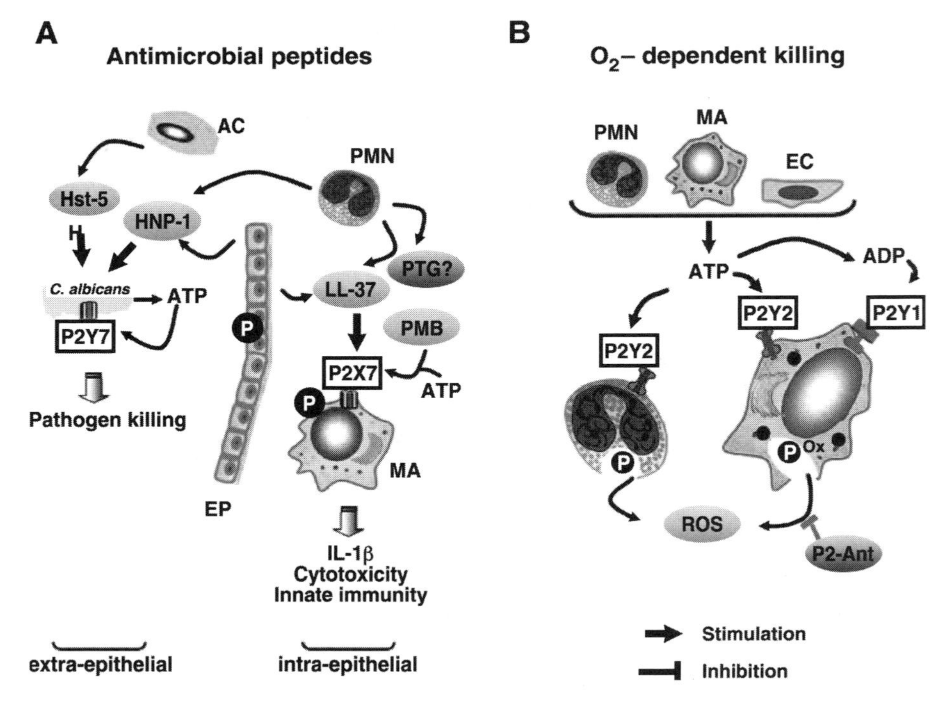

FIGURE 6.2 (See color insert following page 80) Schematic representation of the potential implication of P2 receptors in extracellular killing of pathogens. (A) P2 receptors and antimicrobial peptides. (B) P2 receptors and O_2-dependent killing. AC = salivary acinar cells; P = pathogen; EP = epithelium; PMN = neutrophil; MA = monocyte/macrophage; EC = endothelial cells; Ox = NADPH oxidase; P2-Ant = P2 antagonist; Hst-5 = Histatin-5; HNP-1 = alpha-defensin 1; PTG = protegrin; LL-37 = human cathelicidin peptide; ROS = reactive oxygen species.

to converge through purinergic signaling. As found for Histatin-5, the candidacidal activity of HNP-1 may also be related to the known cytotoxic effects of extracellular ATP through activation of membrane P2 nucleotide receptors.[33] Collectively, these studies support the concept that extracellular ATP plays a role in the extracellular killing of some *C. albicans* by antimicrobial peptides through activation of $P2X_7$ receptors. This mechanism of action may also be involved in response to other pathogens such as *Saccharomyces cerevisiae* and *Aspergillus niger*, which display nonlytic release of ATP and express plasma membrane ATP-specific transporters.[101,102] These observations also raise the possibility that other host defensins may kill their targets through $P2X_7$ receptor activation.

Some microorganisms are able to escape first-line defense mechanisms, giving them access to the epithelial cells and invading underlying tissues. Binding of the pathogen with the epithelium triggers a series of alarm signals, including purine nucleotides that result in the secretion of chemokines, which are involved in the recruitment of phagocytic cells (see sections 6.1 to 6.3). As shown in Figure 6.2A, a number of recent studies suggest a role for $P2X_7$ receptors in modulating effects of cationic peptides other than their prototypical antimicrobial activities. It is known that defensins and cathelicidin released either constitutively or following cell activation by epithelial cells and neutrophils can alert additional mechanisms of defense aimed at amplifying the inflammatory reaction and regulating adaptive immunity.[103,104] Indeed, cationic peptides are chemotactic for T cells, immature dendritic cells, and monocytes and exert a number of innate and adaptive immune effects.[105] Notably, the human cathelicidin-derived peptide LL-37 reportedly promotes IL-1β processing by LPS-primed monocytes via direct activation of $P2X_7$ receptors.[106] In connection with this, it is worth mentioning that lung transfer of a genetic construct capable of expressing supraphysiological levels of LL37 was found to augment the innate immune response and alleviate the pulmonary bacterial infection associated with cystic fibrosis in an animal model.[107] Another natural cationic peptide, polymyxin B, has recently been shown to potentiate significantly ATP-mediated $P2X_7$ receptor stimulation in various cell types including $P2X_7$ receptor–transfected HEK293 cells, chronic lymphocytic leukemia cells (CLL), mouse J774 macrophages,[108] and rat alveolar macrophages (I. Lemaire et al., unpublished observations). In particular, polymyxin B enhanced ATP-mediated Ca^{2+} influx, plasma membrane permeabilization, and cell cytotoxicity, as well as $P2X_7$ receptor–dependent IL-1ß processing in LPS-primed macrophages.[108] Other work has brought to light an intriguing similarity between the action of the protegrin (PTG) antimicrobial peptides and that of extracellular ATP. Both protegrin-1 (PTG-1) and protegrin-3 (PTG-3), two antimicrobial peptides derived by proteolysis of cathelicidin precursor molecule, have been shown to act as a second stimulus in LPS-primed monocytes to release mature IL-1β. Although the PTG-mediated release of IL-1β engaged a mechanism similar to that initiated by extracellular ATP acting via $P2X_7$ receptors,[35,109,110] its effects were not blocked by the specific $P2X_7$ receptor antagonist KN-62,[111] leading to the suggestion that PTG-1 does not act through the $P2X_7$ receptor.[112] It is noteworthy that KN-62 does not prevent the potentiating effect of polymyxin B (PMB) on $P2X_7$ receptor activity.[108] Rather, PMB relieved the inhibitory effect of this antagonist on HEK-$P2X_7$ receptor–transfected cells. As suggested for PMB, the possibility remains that PTG-1 may act near or at the KN-62 binding site on $P2X_7$ receptors or may bind nonspecifically to KN-62. Alternatively, PTG-1 may interact directly with a common effector molecule downstream from P2X receptor.

There is a wealth of evidence to indicate that purinergic signaling regulates other mechanisms of bacterial killing by phagocytic cells, including oxygen-dependent killing of extracellular bacteria. This process, often referred to as the "respiratory burst," involves the production of reactive oxygen species (ROS) by NADPH oxidase, such as superoxide anion (O_2^-), hydrogen peroxide (H_2O_2), and hydroxyl radical (OH), which are strong bactericidal agents. Further reaction of H_2O_2 with Cl^- catalyzed by myeloperoxidase (MPO) yields highly toxic HOCl. While 50 to 70% of ROS generated by NADPH oxidase are produced within closed intracellular compartments (phagosomes, endosomes, intracellular granules) following phagocytosis, the remaining 30 to 50% of the ROS are delivered to the extracellular space either by direct oxidase activation at the plasma membrane or by oxidase recruitment to the developing phagosomes before they are sealed and internalized.

As illustrated in Figure 6.2B, ATP levels can be increased during infection due to either cell death or regulated release as a signal between cells. Both adenine nucleotides and adenosine can be released from neutrophils, macrophages, and endothelial cells. There is ample evidence that extracellular ATP and ADP can prime neutrophils for enhanced ROS production when the cells receive a subsequent oxidase-activating stimulus.[113] In addition, ATP induces superoxide release in guinea pig peritoneal macrophages,[114] rat alveolar macrophages,[115] and human macrophages.[116] In rat alveolar macrophages, the main resident phagocytes in the lungs, ADP is more potent than ATP as a stimulus of the respiratory burst. This is associated with a significant increase in inositol (1,4,5)-triphosphate (IP_3) production and a transient increase in intracellular calcium, suggesting the involvement of P2Y receptor(s),[117,118] possibly $P2Y_1$ receptors. Indeed, ADP is a potent agonist at this receptor subtype,[119] which has been shown to be present on rat alveolar macrophages.[120] In human macrophages, there is evidence that ATP-dependent generation of ROS is also mediated through a P2Y receptor, most likely $P2Y_2$ (originally known as P2U), since both ATP and UTP have the same efficiency.[116] In neutrophils, the P2 receptor subtypes mediating these effects have not been fully identified. It is noteworthy that human neutrophils do not express significant levels of $P2Y_1$ receptor[121] and that only $P2Y_2$ receptors have been shown to be functionally expressed,[122] therefore raising the possibility that ATP-dependent ROS production in neutrophils may be mediated via this receptor. More direct evidence that purinergic signaling regulates the generation of extracellular oxygen radicals during infection has been provided by the demonstration that ATP is released by infected macrophages and that inhibitors of P2 receptors blocked reactive oxygen intermediates production in mouse macrophages infected with *Escherichia coli* or stimulated with LPS, the prototypical bacterial antigen.[123] Experiments with macrophages from $P2X_7$ receptor knockout mice ruled out the participation of $P2X_7$ receptors in such responses. In view of the known cytotoxic action of the $P2X_7$ receptor, a $P2X_7$ receptor–independent mechanism would be more in line with the participation of other P2 receptor(s) acting on viable and fully functional macrophages. Rather surprisingly, oATP has been found to inhibit ROS production in *E. coli*–infected macrophages from $P2X_7$ knockout mice. However, this is not totally unexpected since there is evidence that oATP exerts $P2X_7$ receptor–independent effects.[124,125] Oxidized ATP can partially antagonize ADP activation of $P2Y_1$[124] but not $P2Y_2$ receptors,[126] thus pointing again at the $P2Y_1$ receptor as a potential candidate in mediating oxygen production by rodent macrophages.

Besides their participation in extracellular killing, purinergic nucleotides are also implicated in intracellular destruction of pathogens. Upon recognition of invading microorganisms through various pattern recognition receptors (PRR) such as Toll-like receptors (TLRs), Fc receptors, and lectin-like receptors (mannose receptor), neutrophils and macrophages engulf pathogens by the process of phagocytosis. This involves internalization of pathogens within endocytic vesicules called "phagosomes" and is associated with the oxidative burst and the generation of oxygen radicals within the phagocytic vacuole. The newly formed phagosomes then fuse with lysosomes, which release their contents of hydrolytic enzymes, proteases, and antimicrobial proteins that damage and destroy the endocytosed bacteria.

Evidence exists that extracellular ATP exerts a profound bactericidal effect *in vitro* against highly virulent strains of *Mycobacterium tuberculosis*, an obligate intracellular pathogen, while ADP, AMP, and UTP are not effective.[127] This activity does not result from a direct effect of ATP on bacterial viability and requires stimulation of macrophage $P2X_7$ receptors. The induction of macrophage apoptosis through $P2X_7$ receptor activation was originally identified as an important feature of ATP-induced pathway of mycobacterial killing.[127,128] However, later studies demonstrated that ATP-mediated cell death and associated mycobacterial killing could be functionally uncoupled. Thus, ATP-induced killing of virulent *M. tuberculosis* within macrophages required phospholipase D (PLD),[129] and inhibitors of PLD antagonized ATP-mediated killing of intracellular mycobacteria but had no effect on macrophage death.[130] Stimulation of the $P2X_7$ receptor has also been linked to activation of PLD,[131] which was previously shown to regulate the intracellular trafficking of immune complexes to lysosomes,[132] suggesting that both $P2X_7$ receptors and PLD may be involved in the maturation of phagosomes to microbicidal phagolysosomes. In support of this, there is evidence that $P2X_7$ receptor mRNA levels are elevated in bovine macrophages infected *in vitro* with *Mycobacterium bovis*[133] and that ATP induces the progressive colocalization of mycobacteria-containing phagosomes with lysosomes.[134] Additional experiments using macrophages obtained from $P2X_7$ knockout mice revealed unequivocally that $P2X_7$ receptors are the major purinergic receptor involved in the *in vitro* mycobacterial killing through acidification of and promotion of phagolysosome formation.[135] This in turn would expose mycobacteria to the detrimental effect of low pH[136] and to the actions of lysosomal proteases and antimicrobial proteins. Neither reactive oxygen nor nitrogen radicals were found to be involved in the $P2X_7$ receptor effector pathway mediating death of intracellular mycobacteria as demonstrated by the observations that ATP-mediated killing was unaffected in $p47^{phox}$ null or inducible NO synthase–deficient mice unable to produce reactive oxygen and nitrogen radicals, respectively.[135] The participation of Nramp1 (natural resistance-associated macrophage protein gene-1), reported to play a major role in defense against various intracellular infections and in human resistance to tuberculosis,[137] was also ruled out in the latter study because ATP-mediated killing was unaffected in $Nramp^{susceptible}$ cells. Studies in human populations harboring polymorphisms of the $P2X_7$ receptor also demonstrated that loss-of-function polymorphism in the $P2X_7$ receptor of human macrophages is associated with failure of ATP to induce *in vitro* mycobacterial killing by these cells.[138] Therefore, it has been proposed that extracellular ATP acting through $P2X_7$ receptor–dependent activation of PLD would favor maturation of mycobacteria containing phagosomes and, hence, effective killing of mycobacteria (Figure 6.3A). Additional evidence also implicates the $P2X_7$ receptor in mediating many other intracellular pathogens *in vitro*, notably *Chlamydia psittaci*[139] and *Chlamydia trachomatis*.[140] $P2X_7$ receptor–dependent killing of *Chlamydia* and *M. tuberculosis* share similar features and a prominent role of $P2X_7$ receptor signaling pathway in host protection against intracellular pathogens has been postulated. However, this assumption, based mainly on *in vitro* experimentation, has been recently challenged by the findings that deficiency in $P2X_7$ receptors did not affect priming of naïve T cells and antimycobacterial response in mice infected by low-dose aerosol *M. tuberculosis*.[141] Although this may be due to compensatory mechanisms, which develop in the absence of the $P2X_7$ receptor, these results demand a more thorough investigation of the role of $P2X_7$

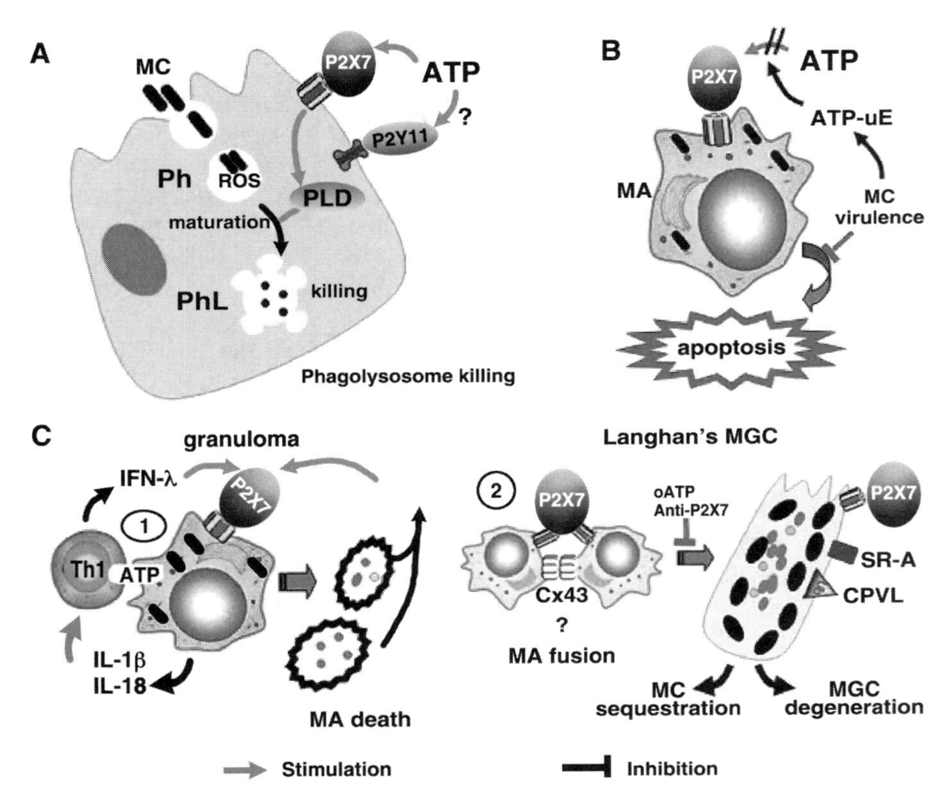

FIGURE 6.3 (See color insert following page 80) Schematic diagram illustrating the potential role of P2 receptors in intracellular killing of pathogens. (A) P2 receptors and endocytic phagolysosome killing. MC = mycobacteria; Ph = phagosome; PhL = phagolysosome. (B) $P2X_7$-mediated apoptosis. ATP-uE = ATP-utilizing enzymes. (C) Infective granulomas and multinucleated giant cells. (1) ATP-mediated signaling in macrophage–lymphocyte coordinated activation of mycobactericidal activity. (2) $P2X_7$ receptor and MGC formation. Cx-43 = connexin-43; SR-A = type-A scavenger receptor; CPVL = carboxypeptidase-related vitellogenin-like molecule; MA = macrophage; oATP = oxidized ATP; Th1 = T-helper 1; PLD = phospholipase D.

and P2 receptors during infection with intracellular pathogens. It is possible that low-dose infection used in the latter study stimulates release of low doses of ATP that are likely to act on P2Y receptors. As mentioned in sections 6.2 and 6.4, P2Y receptors have a stimulating effect on dendritic cell maturation and are likely to mediate dendritic cell endocytosis. It is noteworthy that macrophages from $P2X_7$ receptor knockout mice have been found to display mycobactericidal activity, albeit at a low level.

The small mycobacterial activity seen in $P2X_7$ receptor knockout mice could be due to the involvement of additional purinergic receptor(s). Macrophages express other P2 receptors that can also respond to ATP with varying sensitivities, including $P2X_4$, $P2X_1$, and $P2X_5$ as well as $P2Y_1$, $P2Y_2$, $P2Y_4$, and $P2Y_{11}$ receptors. Based on the observation that UTP did not induce tuberculocidal activity, Kusner and colleagues[129] ruled out the participation of $P2Y_2$ receptors in mediating the

$P2X_7$-independent killing effect of ATP. In agreement with this, work by Stober et al.[134] demonstrated that mobilization of intracellular Ca^{2+} via an ATP-activated P2Y receptor other than $P2Y_2$ may also operate to mediate the bactericidal effects of ATP. These investigators presented evidence for the existence of ATP-activated $P2Y_{11}$ receptors on human macrophages and suggested that it could be involved as an alternative mycobactericidal mechanism in the absence of $P2X_7$ receptor activation.

Apoptosis of infected macrophages may also contribute to eliminate cells infected with pathogens that can evade intracellular phagosolysome-mediated killing mechanisms. In this respect, $P2X_7$ receptor–mediated apoptosis may represent a complementary killing pathway for intracellular pathogens, which, like virulent *M. tuberculosis*, can inhibit TNF-α dependent apoptosis of infected macrophages[142] (Figure 6.3B). Inhibition of apoptosis of infected cells has been proposed as a potential virulence mechanism of many intracellular pathogens including bacteria,[143,144] protozoa,[145] and fungi.[145–147] A corollary to $P2X_7$ receptor–dependent antimycobacterial response is that secretion by *M. bovis* of ATP-utilizing enzymes[148] that sequester ATP from macrophage $P2X_7$ receptors may represent yet another mechanism linked to mycobacteria survival in macrophages.

One of the hallmarks of an effective immune response against *M. tuberculosis* is the formation of granulomas, which is considered a necessary step in the antimycobacterial response. It is characterized by a defined granuloma delineated by activated macrophages and T lymphocytes. Human immunity to mycobacteria is thought to involve an interplay between infected macrophages and specific CD4-positive T lymphocytes. Macrophages produce cytokines such as IL-12, IL-18, IL-23, and IL-27 that activate T helper–1 lymphocytes, which in turn further stimulate the macrophages through the release of interferon-γ (IFN-γ) and TNF-α. These two classes of effector cells interact in a positive feedback manner to contain the infection within T cell–activated macrophages, which is the principal effector cell to kill the pathogen. During these interactions macrophages are known to establish close contact with each other and with T helper lymphocytes. As depicted in Figure 6.3C, purinergic signaling is likely involved in such cell–cell communication. A precedent for this has been established by the demonstration that hepatocytes can communicate via nucleotide release.[149] ATP efflux is restricted to abrupt point-source bursts,[150] and transient activation of connexin hemichannels is thought to mediate this type of ATP release.[151] In addition, the centers of granulomas are often marked by high numbers of dying cells and the presence of such areas with relatively high cell death may serve as a significant source of ATP, sufficient to trigger purinergic signaling and drive P2 receptor–mediated effects. As discussed previously, $P2X_7$ receptor activation can promote the production of IL-1β[35,109,110] and IL-18[152–154], which activate T lymphocytes, and IFN-γ potentiates $P2X_7$ receptor function in macrophages.[155,156] Moreover, IL-12 and IL-27 expression is thought to be regulated by the $P2Y_{11}$ receptor, whereas ATP/ADP acting on a P2 receptor with affinity to ADP, such as $P2Y_1$, $P2Y_2$, $P2Y_{12}$, or $P2Y_{13}$ receptors, reportedly upregulate IL-23.[157] It is likely that P2 receptor–mediated signaling is involved in macrophage–lymphocyte coordinated activation of protective responses against mycobacteria. It should be stressed that in human macrophages, T $helper_1$-associated cytokines cannot induce mycobactericidal activity *in vitro*,[158] whereas extracellular ATP can, as discussed previously. Moreover, infected macrophages are thought to become progressively unresponsive to further activation by Th1-derived cytokines as the pathogen undergoes intracellular replication.[159] This, coupled to the observation that, *in vivo*, macrophages are able to contain mycobacteria in the vast majority of infected individuals, points to ATP-mediated signaling as a potentially relevant effector pathway in regulating mycobacteriostatic and/or mycobactericidal activity within human macrophages.

An intriguing aspect of macrophage response during granulomatous reactions associated with *M. tuberculosis* is the formation of multinucleated giant cells (MGC) of the Langhan's-type, which display nuclei organized at the periphery of the cell in a ring-like fashion. As yet, the mechanisms involved in MGC formation are not understood. There is evidence that they arise as a result of macrophage fusion[160] and may be a manifestation of specific cellular immunity since lymphocyte-derived cytokines such as IFN-γ, IL-3, IL-4, or granulocyte macrophage colony-stimulating factor (GMCSF) promote their formation.[161] Quite interestingly, purinergic signaling through $P2X_7$ receptors has been implicated in the formation of MGC (Figure 6.3C). Falzoni and colleagues[155] first reported that formation of concanavalin-A–induced MGC in human monocyte populations was inhibited by oATP, and these observations were later corroborated by others.[162] Further work demonstrated that fusion occurred only between macrophages displaying high expression of $P2X_7$ receptors and not among macrophages with low expression.[163] Moreover, blockade of $P2X_7$ receptors by a specific monoclonal antibody directed against the outer domain of human $P2X_7$ receptors prevented fusion *in vitro*.[164] In agreement with this, the formation of MGC was found to be significantly higher in HEK293 cells stably transfected with the rat or the human $P2X_7$ receptor than in mock-transfected cells. Increased expression and activation of $P2X_7$ receptors may be important for an efficient and sustained promotion of MGC formation during granulomatous reactions. Other molecules are likely to participate in this process. Among these, cell surface adhesion molecules such as CD11a (LFA-1),[165] and ICAM-1,[166] as well as cytokines, including IFN-γ, IL-3, IL-4,[167,168] IL-13,[169] MCSF, and GMCSF,[170,171] have all been shown to be involved in MGC formation. In particular, membrane-bound molecules of both macrophages and mycobacteria (including MDP, a component of cell walls of *M. Tuberculosis*) appear to be involved in the MGC formation.[162,172] Interestingly, in the latter study, giant cell formation could be inhibited by antibodies against IFN-γ, which increases $P2X_7$ receptor activity,[155,173] and the β2-integrin, which reportedly interacts with the $P2X_7$ receptor C-terminal domain.[174]

MGC formation results from a coordinated series of events involving recognition, adhesion, activation, and fusion of partner cells. As suggested by Di Virgilio et al.,[175] the pore-forming $P2X_7$ receptor might be involved in the final cell membrane fusion step in a manner analogous to the "fusion pore" model of biological membrane fusion in which a small dynamic fusion pore forms in the common bilayer following hemifusion and expands to allow complete membrane merging.[176]

To date, there is no direct evidence that links $P2X_7$ receptor expression or activation to MGC formation in granulomas associated with *M. tuberculosis* or other intracellular pathogens. However, indirect evidence for this has been provided by the observations that enhanced MGC formation in monocyte cultures of patients with tuberculosis (TB) is inhibited by oATP.[162] The precise role and functional significance of MGC still remains obscure. With respect to mycobacterial infection, growth restriction of *M. tuberculosis* has been associated with the presence of MGC. There is evidence that dense growth of *M. tuberculosis* is initially confined to the center of MGC and that physical sequestration of *M. tuberculosis* by MGC may limit cell-to-cell evasion. In this respect, MGC from TB patients have been found to express type-A scavenger receptor,[177] a pattern recognition membrane molecule thought to be involved in non-opsonic uptake of mycobacteria by macrophages[178] and to play a role in TNF-α production.[179] Surface expression of type-A scavenger receptor would be consistent with a role in cell adhesion/retention[180] as well as in phagocytosis of microorganisms[181] and apoptotic cells.[182] MGC from TB patients also expresses carboxypeptidase-related vitellogenin-like molecule, whose possible functions include antigen processing, microbial degradation, and modulation of extracellular polypeptide mediators of inflammation after secretion.[183] Therefore, MGC may contribute to the formation of a cellular barrier to impede the spread of pathogens from an initial focus of infection. Physical containment by MGC has also been shown for other intracellular pathogens such as *Cryptococcus neoformans*, where sequestration of the yeast precedes full induction of fungicidal mechanisms,[184] and for *Schistosoma mansoni*,[185] which was found to be trapped within MGC in the lung. Such a mechanism of cell containment apparently involves cytoskeletal interactions,[186] and the $P2X_7$ receptor, which reportedly interacts with cytoskeletal elements such as β–actin and α-actinin,[174] may participate in this process. At later stages, MGC containing central mass of mycobacteria were observed to undergo degeneration,[169] suggesting that MGC may represent a mechanism for disposal of infected macrophages. This could be related to the presence of functional $P2X_7$ receptors on MGC that would trigger a $P2X_7$-dependent suicide pathway. Consistent with such an assumption, monocytes from patients with TB and other granulomatous diseases were found to be more susceptible than cells from healthy controls to $P2X_7$ receptor–mediated cytotoxicity.[162] In accordance with the functional attributes of $P2X_7$ receptors, it is tempting to speculate that during infection with intracellular pathogens, $P2X_7$-mediated MGC formation and $P2X_7$-induced cytotoxicity are intricately linked to limitation of infection and disposal of infected macrophages.

6.5.2 Purinergic Signaling and Destruction of Noninfectious Agents

The presence of nonbacterial antigens such as foreign particles or agents and, in some cases, self-molecules, can be recognized as danger signals by the host and can trigger a set of responses similar to those recruited during host defense against pathogens. The overall process aimed at the elimination of the offending antigen(s) gives rise to what is referred to as "sterile inflammation." The response to noninfectious agents is often characterized by ineffective removal of antigen(s) and leads to uncontrolled production of oxygen and nitrogen radicals as well as pro-inflammatory cytokines by phagocytic cells, inappropriate stimulation of adaptive immunity, and tissue damage. These reactions form the basis of many chronic granulomatous disorders such as sarcoidosis, foreign body granulomas, and osteolytic granulomas. Accumulating evidence suggests a role for extracellular nucleotides and purinergic signaling in some granulomatous reactions not clearly related to an infective disease (Figure 6.4).

Sarcoidosis is a systemic disease characterized by non-caseating granulomas, which most commonly affects the lungs. Although the cause of sarcoidosis is unknown, there is evidence that it results from exposure of genetically susceptible hosts to either noninfectious environmental factors (pesticides, insecticides, pure pollen, silica, talc, metal dusts, and man-made mineral fibers) or infectious agents such as *M. tuberculosis*, *Proproni bacterium acnes*, or *P. granulosum*.[187,188] In addition, cell-wall deficient bacteria that adapt to live within macrophages and are extremely hard to kill have been linked to sarcoidosis.[189,190] Sarcoidosis and TB are very similar diseases with respect to histopathology. Macrophages are prominent in both groups and share several features, including the formation of Langhan's-type MGC with comparable expression of SR-A and angiotensin-converting enzyme.[177] Langhan's giant cells containing indigestible particles are surrounded by a lymphocytic collar indicative of a Th1 immune response to the inciting agent. As mentioned

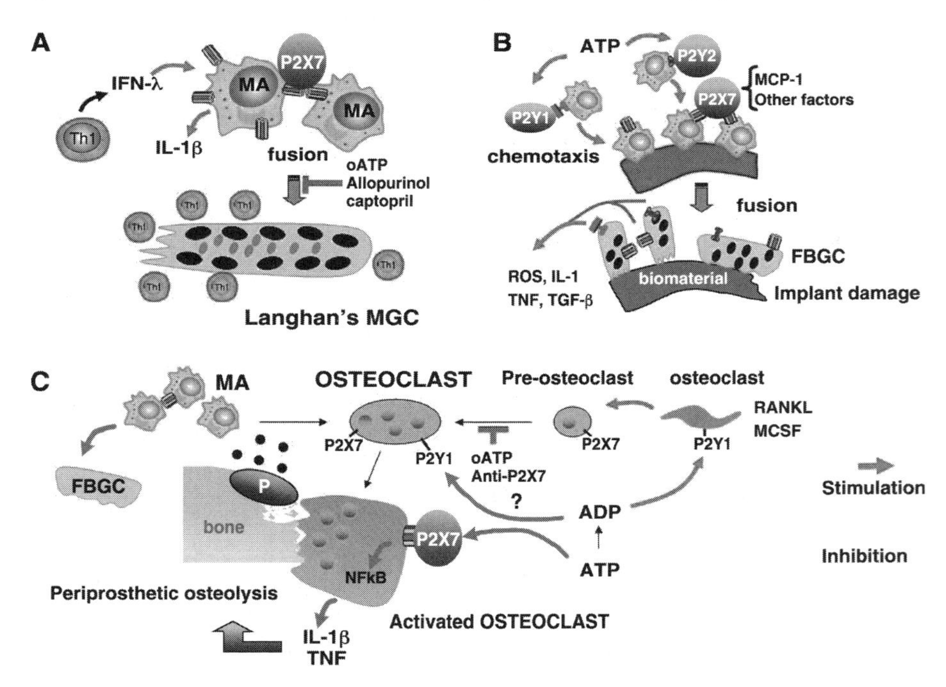

FIGURE 6.4 (See color insert following page 80) General scheme illustrating the potential involvement of P2 receptors in the elimination of noninfectious agents. (A) Sarcoidosis and Langhan's MGC. (B) Foreign body reaction to implanted biomaterials and FBGC. (C) Periprosthetic bone destruction and osteoclasts. P = prosthesis; MCP-1 = monocyte chemoattractant protein–1; FBGC = foreign body giant cell.

previously, Th1 immune response is associated with the production of IFN-γ, a cytokine known to increase $P2X_7$ receptor function in macrophages *in vitro*.[155,156] Mizuno et al.[162] demonstrated that monocytes from sarcoidosis patients were more susceptible to the cytotoxic effect of BzATP, a potent $P2X_7$ receptor agonist,[191] suggesting enhanced expression or activity of $P2X_7$ receptors in these cells (Figure 6.4A). Consistent with this, a separate study[192] showed that culture supernatants of alveolar macrophages from sarcoidosis patients contained higher levels of IL-1β, a cytokine known to be released following $P2X_7$ receptor activation.[193] Moreover, in the Mizuno study,[162] increased $P2X_7$ receptor–dependent activity of monocytes and macrophages from sarcoidosis patients was associated with a greater ability to generate Langhan's-type MGC upon *in vitro* culture. Conversely, treatment with the potent $P2X_7$ receptor antagonist, oATP, abolished MGC formation. Additional evidence linking $P2X_7$ receptors to MGC formation in sarcoidosis was provided by recent observations that allopurinol and captopril, two agents that have therapeutic effects on sarcoidosis,[194–196] also inhibited MGC formation from monocytes concomitant with a downregulation of $P2X_7$ receptors.[197]

The foreign body reaction (FBR) is a sterile inflammation that develops in response to implantation of almost all biomaterials, prostheses, and medical devices and can be detrimental to their function. Shortly after implantation, phagocytic neutrophils are recruited to the site and serve as the initial line of defense. The second wave of defense is dominated by monocytes that extravasate into the implantation site and differentiate into macrophages. In contrast to normal wound healing, where these responses are transient, the FBR is associated with chronic accumulation of macrophages, which adhere to the surface of the implant and frequently undergo fusion to generate multinucleated giant cells. The indigestible or poorly digestible material is surrounded or encapsulated within a thin layer of macrophage and foreign body giant cells (FBGC), a hallmark of the FBR. In contrast to Langhan's-type MGC, FBGCs are characterized morphologically by a large number of nuclei arranged in an irregular fashion throughout the cytoplasm.

A role for P2 receptor signaling has been recently implied during the FBR to implanted biomaterials (Figure 6.4B). In a model of subcutaneous implantation of dermal sheep collagen disks in rats, expression of purinergic $P2X_7$ receptors and, to a lesser extent, $P2Y_1$ and $P2Y_2$ receptors, was observed on macrophages and FBGC located in close proximity to the explants.[198] At later stages, expression of $P2X_7$, $P2Y_2$, and to a lesser extent, $P2Y_1$ receptors, was mostly located on giant cells. The chemotactic properties of macrophages have been related to $P2Y_2$ and $P2Y_1$ receptor function,[199] and it is quite possible that engagement of these receptors by either ATP or UTP produced at sites of lesions is critical for the recruitment of macrophages into the collagen disks. Purinergic signaling may be particularly relevant in the case of subcutaneous implantation because the chemokine monocyte chemoattractant protein-1 (CCL2/MCP-1), known to be crucial for the recruitment of monocytes in several models of inflammation,[200] is not required for monocyte recruitment when biomaterials are implanted in the subcutaneous space.[201] However, CCL2/MCP-1 appears to be important for the formation of FBGC since CCL2-null mice with subcutaneous collagen implant displayed a reduction in both FBGC formation and implant damage.[201] Reportedly, $P2X_7$ receptor activation upregulates MCP-1 expression in primary cultures of astrocytes.[202] Therefore, as shown in Figure 6.4B, it is tempting to speculate that while activation of $P2Y_2$ and $P2Y_1$ receptors by macrophages might contribute to macrophages recruitment in subcutaneous implant, stimulation of $P2X_7$ receptors could trigger FBGC formation either directly or through upregulation of other molecules, such as CCL2.

Depending on the type of FBR reaction, other factors may also participate at one stage or another during the process leading to FBGC formation. Notably, the mannose receptor[168] β_2 integrin,[168] the fusion receptor SHPS-1,[201,203] and CD44[203,204] have been linked to FBGC formation. Therefore, physical interaction between $P2X_7$ receptors and β2 integrin[174] or any of these molecules might be necessary for the generation of FBCG associated with different FBR.

Despite numerous observations that FBGC and Langhan's giant cells (LGC) display distinct cell morphology and that their respective formation requires different set of cytokines and receptors,[161] the $P2X_7$ receptor has been implicated in the formation of both types of MGC.[162,198,205] All together, the evidence available thus far supports a primary role for $P2X_7$ receptors at a common step crucial for MGC formation during various granulomatous reactions aimed at eliminating the offending agent(s). At later stages of the FBR-to-subcutaneous collagen implant, expression of $P2X_7$ and $P2Y_2$ receptors was mostly located on FBGC,[198] supporting the notion that MGC are involved in further recruitment and fusion of additional macrophages leading to large FBGC with numerous nuclei. This is consistent with the observations of ongoing fusion between giant cells themselves and giant cells with mononuclear cells during *in vitro* cultures of macrophages.[163]

In contrast to LGC, which are thought to play a protective role during infection, the presence of FBGC at the tissue-implant interface has been associated with degeneration of the biomaterial surface that leads to tissue fibrosis and chronic inflammatory response.[161,206] Consistent with this, a morphological feature of FBGC is the presence of podosomal structures, which are the major adhesive structures present at their ventral and peripheral surfaces.[207] Such attributes imply functional polarization and frustrated phagocytosis via the formation of a closed compartment between the FBGCs and the underlying substrate where degradative enzymes, reactive oxygen intermediates, or other products are secreted. Furthermore, due to their phagocytic activity, FBGC can generate particulate debris that can contribute to the persistence of inflammation. In this context, activation of both $P2Y_1$ and $P2Y_2$ receptors present on FBGC may stimulate phagocytosis[199,208] and reactive oxygen intermediates.[116–118] Furthermore, engagement of $P2X_7$ receptors on FBGC may contribute to sustained MGC formation and initiation of cytokine production, thus perpetuating inflammation and biomaterial damage. In support of this, FBGCs derived from rat alveolar macrophages *in vitro* express high levels of TNF-α and TGF-β (I. Lemaire, unpublished observations) while FBGCs induced *in vivo* by subcutaneous injection of nitrocellulose reportedly produce IL-1, TNF-α, and profibrotic TGF-β.[209]

The foreign body reaction also occurs in response to implantation of prostheses used in total hip and knee replacements. This type of FBR induced by wear particles generated at the articulating surface leads to periprosthetic bone destruction and loosening of the prosthesis.[210] In addition to macrophages and FBGC, osteoclasts are key effector cells in periprosthetic granulomatous osteolysis, particularly in the final osteolytic step (Figure 6.4C). Osteoclasts, the major bone-resorbing cells, are MGCs that form from the fusion of osteoclast progenitors derived from hematopoietic cells of the mononuclear phagocyte lineage.[211,212] Multinucleated osteoclasts also arise from differentiation of mononuclear cells, tissue-derived macrophages,[213] and some macrophages derived from periprosthetic tissues.[214–217] Hence, the osteoclast can be described as a specialized monocyte/macrophage polykaryon.

The mechanisms involved in the formation of osteoclasts are not completely understood. The primary role of osteoblasts in regulating osteoclast formation, differentiation, and activity is well established. MCSF and RANKL secreted by osteoblasts, macrophages, and other stromal cells have been identified as essential molecules for osteoclastogenesis.[218,219] Recently, it has become apparent that purinergic signaling through P2 receptors plays an important role in the regulation of the osteoclast population[220,221] (Figure 6.4C). Relatively low concentrations of ATP that could potentially be generated at sites of prosthetic osteolysis have been shown to increase osteoclast formation and activation.[222] It is also known that osteoblasts release nucleotides in the surrounding environment via a nonlytic release mechanism,[223] providing the source of ATP to trigger purinergic receptors present on osteoclasts. Both P2X and P2Y receptors are expressed on normal rat,[222,224] rabbit,[225,226] and human osteoclasts.[227] In particular, $P2Y_1$, as well as $P2X_4$ and $P2X_7$, receptors are present as functional receptors on these cells. Hoebertz et al.[228] demonstrated that extracellular ADP, probably acting at the $P2Y_1$ receptor, stimulates osteoclast formation from hematopoietic precursors and activates mature osteoclasts. However, it is not clear whether these effects result from a direct action of the nucleotide on osteoclasts since osteoblasts also express $P2Y_1$ receptors.[229] It is possible that ADP stimulates resorption by an indirect mechanism via osteoblasts, in view of the findings that ATP upregulates the expression of RANKL by osteoblasts and that its stimulating effect on osteoclast resorption requires the presence of osteoblasts.[227]

There is convincing evidence that purinergic signaling through $P2X_7$ receptors present on osteoclasts regulates directly critical aspects of osteoclast response (Figure 6.4C). The $P2X_7$ receptor is expressed on osteoclasts and their precursors *in vitro*[229–231] and on human osteoclasts lining the surface of bone.[221] It has been demonstrated that activation of $P2X_7$ receptors on osteoclasts is required for intercellular communication through calcium signaling between osteoblasts and osteoclasts, and among osteoclasts.[232] Moreover, oATP and a $P2X_7$ receptor monoclonal antibody known to specifically block $P2X_7$ receptor function[233] inhibited the fusion of osteoclast precursors to form multinucleated osteoclasts from cell cultures treated with MCSF and RANKL.[221] These observations suggest that MCSF and RANKL require the presence of functional $P2X_7$ receptors for the effective generation of osteoclasts. In the latter study, $P2X_7$ receptor antagonism also decreased osteoclastic resorption. In accordance with this, multinucleated osteoclasts displayed sustained expression of functional $P2X_7$ receptors with time in culture, suggesting their participation in other processes beyond the initial fusion event. Consistent with such assumptions, the potent $P2X_7$ agonist BzATP was shown to activate NF-kB in osteoclasts from normal mice but not $P2X_7$ receptor–deficient mice.[234] The transcription factor NF-kB regulates gene expression of many inflammatory cytokines, and $P2X_7$ receptor expression and activation have been linked to cytokine release.[110,153,193,235] Therefore, it is conceivable that coupling of $P2X_7$ receptors to activation of NF-kB in osteoclasts might contribute to upregulation of the osteolytic cytokines IL-1 and TNF-α.[236,237]

The complex nature of $P2X_7$ receptor–mediated signaling in osteoclasts is underlined by numerous reports of paradoxical effects attributed to $P2X_7$ receptors. Thus, mice lacking the $P2X_7$ receptor exhibit excessive resorption with increased osteoclast numbers, and these observations led to the suggestion that $P2X_7$ receptors promote osteoclast apoptosis.[238] On the other hand, specific activation of $P2X_7$ receptors on osteoclasts from wild-type mice induced anti-apoptotic signaling.[239] Furthermore, antagonism of the $P2X_7$ receptor in cultures of osteoclasts resulted in an inhibition of osteoclastic resorption but an increase of apoptosis.[240] As suggested previously for lymphoid cells,[241] these conflicting observations may be related to the level of expression/activation of the $P2X_7$ receptor. Clearly, more work is needed to delineate the mechanisms of action of $P2X_7$ receptors on osteoclasts and bone metabolism. Nevertheless, a wealth of evidence suggests that extracellular nucleotides acting via P2 receptor subtypes exert an overall potent osteolytic activity[220] that could be relevant to sterile inflammation associated with periprosthetic osteolysis.

6.6 CONCLUSIONS

A large body of *in vitro* data has accumulated in support of an immunomodulatory function of extracellular nucleotides and P2 receptors. Although *in vivo* evidence for such a role is at the moment rather preliminary, an in-depth elucidation of the several pathways modulated by extracellular nucleotides will certainly expand our knowledge of immunity. Given the close interrelationship between bone and the immune system, we believe that this will also provide a better comprehension of bone pathophysiology.

ACKNOWLEGEMENTS

I.L. was a visiting professor at the Department of Experimental and Diagnostic Medicine at the University of Ferrara supported by the Canadian Institutes of Health Research (CIHR)–National Research Council (CNR) of Italy Visiting Scientist Exchange Program. This author is indebted to the colleagues in the Section of General Pathology, University of Ferrara, for their support and stimulating discussion.

REFERENCES

1. Rho, J., Takami, M., and Choi, Y., Osteoimmunology: interactions of the immune and skeletal systems, *Mol. Cell*, 17, 1, 2004.
2. Roodman, G.D., Mechanisms of bone metastasis, *N. Engl. J. Med.*, 350, 1655, 2004.
3. Rodan, G.A. and Martin, T.J., Therapeutic approaches to bone diseases, *Science*, 289, 1508, 2000.
4. Wong, B.R.et al., TRANCE, a TNF family member, activates Akt/PKB through a signaling complex involving TRAF6 and c-Src, *Mol. Cell*, 4, 1041, 1999.

5. Wong, B.R., Josien, R., and Choi, Y., TRANCE is a TNF family member that regulates dendritic cell and osteoclast function, *J. Leukoc. Biol.*, 65, 715, 1999.
6. Takayanagi, H. et al., T-cell–mediated regulation of osteoclastogenesis by signalling cross-talk between RANKL and IFN-γ, *Nature*, 408, 600, 2000.
7. Drury, A.N. and Szent-Gyorgy, A., The physiological activity of adenine compounds with special reference to their action upon the mammalian heart, *J. Physiol.*, 68, 213, 1929.
8. Donaldson, S.H. et al., Basal nucleotide levels, release, and metabolism in normal and cystic fibrosis airways, *Mol. Med.*, 6, 969, 2000.
9. Beigi, R. et al., Detection of local ATP release from activated platelets using cell surface–attached firefly luciferase, *Am. J. Physiol.*, 276, C267, 1999.
10. Coade, S.B. and Pearson, J.D., Metabolism of adenine nucleotides in human blood, *Circ. Res.*, 65, 531, 1989.
11. Forrester, T., An estimate of adenosine triphosphate release into the venous effluent from exercising human forearm muscle, *J. Physiol.*, 224, 611, 1972.
12. Joseph, S.M., Buchakjian, M.R., and Dubyak, G.R., Colocalization of ATP release sites and ecto-ATPase activity at the extracellular surface of human astrocytes, *J. Biol. Chem.*, 278, 23331, 2003.
13. Burnstock, G., The past, present and future of purine nucleotides as signalling molecules, *Neuropharmacology*, 36, 1127, 1997.
14. Ralevic, V. and Burnstock, G., Receptors for purines and pyrimidines, *Pharmacol. Rev.*, 50, 413, 1998.
15. Zimmermann, H., Extracellular metabolism of ATP and other nucleotides, *N. Schmied. Arch. Pharmacol.*, 362, 299, 2000.
16. Burnstock, G. and Knight, G.E., Cellular distribution and functions of P2 receptor subtypes in different systems, *Int. Rev. Cytol.*, 240, 31, 2004.
17. Matzinger, P., The danger model: a renewed sense of self, *Science*, 296, 301, 2002.
18. Skoberne, M., Beignon, A.S., and Bhardwaj, N., Danger signals: a time and space continuum, *Trends Mol. Med.*, 10, 251, 2004.
19. Gallucci, S. and Matzinger, P., Danger signals: SOS to the immune system, *Curr. Opin. Immunol.*, 13, 114, 2001.
20. Van Dyke, K. et al., An analysis of nucleotides and catecholamines in bovine medullary granules by anion exchange high pressure liquid chromatography and fluorescence. Evidence that most of the catecholamines in chromaffin granules are stored without associated ATP, *Pharmacology*, 15, 377, 1977.
21. Lages, B. and Weiss, H.J., Secreted dense granule adenine nucleotides promote calcium influx and the maintenance of elevated cytosolic calcium levels in stimulated human platelets, *Thromb. Haemost.*, 81, 286, 1999.
22. Meyers, K.M., Holmsen, H., and Seachord, C.L., Comparative study of platelet dense granule constituents, *Am. J. Physiol.*, 243, R454, 1982.
23. Abbracchio, M.P. et al., Characterization of the UDP-glucose receptor (re-named here the $P2Y_{14}$ receptor) adds diversity to the P2Y receptor family, *Trends Pharmacol. Sci.*, 24, 52, 2003.
24. Di Virgilio, F. et al., Nucleotide receptors: an emerging family of regulatory molecules in blood cells, *Blood*, 97, 587, 2001.
25. Di Virgilio, F. et al., Leukocyte P2 receptors: a novel target for anti-inflammatory and anti-tumor therapy, *Curr. Drug Targets. Cardiovasc. Haematol. Disord.*, 5, 85, 2005.
26. la Sala, A. et al., Extracellular ATP induces a distorted maturation of dendritic cells and inhibits their capacity to initiate Th1 responses, *J. Immunol.*, 166, 1611, 2001.

27. Wilkin, F. et al., Extracellular adenine nucleotides modulate cytokine production by human monocyte-derived dendritic cells: dual effect on IL-12 and stimulation of IL-10, *Eur. J. Immunol.*, 32, 2409, 2002.
28. Schnurr, M. et al., Extracellular ATP and TNF-α synergize in the activation and maturation of human dendritic cells, *J. Immunol.*, 165, 4704, 2000.
29. Ni, H. et al., Extracellular mRNA induces dendritic cell activation by stimulating tumor necrosis factor–alpha secretion and signaling through a nucleotide receptor, *J. Biol. Chem.*, 277, 12689, 2002.
30. Mellman, I. and Steinman, R.M., Dendritic cells: specialized and regulated antigen processing machines, *Cell*, 106, 255, 2001.
31. la Sala, A. et al., Dendritic cells exposed to extracellular adenosine triphosphate acquire the migratory properties of mature cells and show a reduced capacity to attract type 1 T lymphocytes, *Blood*, 99, 1715, 2002.
32. la Sala, A. et al., Alerting and tuning the immune response by extracellular nucleotides, *J. Leukoc. Biol.*, 73, 339, 2003.
33. Di Virgilio, F., The P2Z purinoceptor: an intriguing role in immunity, inflammation and cell death, *Immunol. Today*, 16, 524, 1995.
34. Griffiths, R.J. et al., ATP induces the release of IL-1 from LPS-primed cells *in vivo*, *J. Immunol.*, 154, 2821, 1995.
35. Perregaux, D. and Gabel, C.A., Interleukin-1 β maturation and release in response to ATP and nigericin. Evidence that potassium depletion mediated by these agents is a necessary and common feature of their activity, *J. Biol. Chem.*, 269, 15195, 1994.
36. Ferrari, D. et al., Mouse microglial cells express a plasma membrane pore gated by extracellular ATP, *J. Immunol.*, 156, 1531, 1996.
37. Perregaux, D.G. et al., ATP acts as an agonist to promote stimulus-induced secretion of IL-1β and IL-18 in human blood, *J. Immunol.*, 165, 4615, 2000.
38. Sanz, J.M. and Di Virgilio, F., Kinetics and mechanism of ATP-dependent IL-1 β release from microglial cells, *J. Immunol.*, 164, 4893, 2000.
39. MacKenzie, A. et al., Rapid secretion of interleukin-1β by microvesicle shedding, *Immunity*, 15, 825, 2001.
40. Wilson, H.L. et al., Secretion of intracellular IL-1 receptor antagonist (type 1) is dependent on $P2X_7$ receptor activation, *J. Immunol.*, 173, 1202, 2004.
41. Hide, I. et al., Extracellular ATP triggers tumor necrosis factor–alpha release from rat microglia, *J. Neurochem.*, 75, 965, 2000.
42. Suzuki, T. et al., Production and release of neuroprotective tumor necrosis factor by $P2X_7$ receptor–activated microglia, *J. Neurosci.*, 24, 1, 2004.
43. Walsh, M.C. and Choi, Y., Biology of the TRANCE axis, *Cytokine Growth Factor Rev.*, 14, 251, 2003.
44. Kostenuik, P.J. and Shalhoub, V., Osteoprotegerin: a physiological and pharmacological inhibitor of bone resorption, *Curr. Pharm. Des.*, 7, 613, 2001.
45. Kong, Y.Y., Boyle, W.J., and Penninger, J.M., Osteoprotegerin ligand: a regulator of immune responses and bone physiology, *Immunol. Today*, 21, 495, 2000.
46. Teitelbaum, S.L., Bone resorption by osteoclasts, *Science*, 289, 1504, 2000.
47. Ihara, H. et al., ATP-stimulated interleukin-6 synthesis through P2Y receptors on human osteoblasts, *Biochem. Biophys. Res. Commun.*, 326, 329, 2005.
48. Kahlenberg, J.M. and Dubyak, G.R., Mechanisms of caspase-1 activation by $P2X_7$ receptor–mediated K^+ release, *Am. J. Physiol.*, 286, C1100, 2004.
49. Mariathasan, S. et al., Differential activation of the inflammasome by caspase-1 adaptors ASC and Ipaf, *Nature*, 430, 213, 2004.

50. Andrei, C. et al., Phospholipases C and A2 control lysosome-mediated IL-1 β secretion: implications for inflammatory processes, *Proc. Natl. Acad. Sci. USA*, 101, 9745, 2004.
51. Wilson, H.L. et al., Secretion of intracellular IL-1 receptor antagonist (type 1) is dependent on $P2X_7$ receptor activation, *J. Immunol.*, 173, 1202, 2004.
52. Perregaux, D.G. et al., ATP acts as an agonist to promote stimulus-induced secretion of IL-1 β and IL-18 in human blood, *J. Immunol.*, 165, 4615, 2000.
53. Rampe, D., Wang, L., and Ringheim, G.E., $P2X_7$ receptor modulation of beta-amyloid– and LPS-induced cytokine secretion from human macrophages and microglia, *J. Neuroimmunol.*, 147, 56, 2004.
54. Hide, I. et al., Extracellular ATP triggers tumor necrosis factor–α release from rat microglia, *J. Neurochem.*, 75, 965, 2000.
55. Ferrari, D. et al., The P2 purinergic receptors of human dendritic cells: identification and coupling to cytokine release, *FASEB J.*, 14, 2466, 2000.
56. Suzuki, T. et al., Production and release of neuroprotective tumor necrosis factor by $P2X_7$ receptor–activated microglia, *J. Neurosci.*, 24, 1, 2004.
57. Hasko, G. et al., ATP suppression of interleukin-12 and tumour necrosis factor–α release from macrophages, *Br. J. Pharmacol.*, 129, 909, 2000.
58. Seo, D.R., Kim, K.Y., and Lee, Y.B., Interleukin-10 expression in lipopolysaccharide-activated microglia is mediated by extracellular ATP in an autocrine fashion, *Neuroreport*, 15, 1157, 2004.
59. John, G.R. et al., Extracellular nucleotides differentially regulate interleukin-1β signaling in primary human astrocytes: implications for inflammatory gene expression, *J. Neurosci.*, 21, 4134, 2001.
60. Idzko, M. et al., Stimulation of P2 purinergic receptors induces the release of eosinophil cationic protein and interleukin-8 from human eosinophils, *Br. J. Pharmacol.*, 138, 1244, 2003.
61. Warny, M. et al., $P2Y_6$ nucleotide receptor mediates monocyte interleukin-8 production in response to UDP or lipopolysaccharide, *J. Biol. Chem.*, 276, 26051, 2001.
62. Mason, N.J., Artis, D., and Hunter, C.A., New lessons from old pathogens: what parasitic infections have taught us about the role of nuclear factor–kB in the regulation of immunity, *Immunol. Rev.*, 201, 48, 2004.
63. Ferrari, D. et al., Extracellular ATP activates transcription factor NF-kB through the P2Z purinoreceptor by selectively targeting NF-kB p65, *J. Cell. Biol.*, 139, 1635, 1997.
64. Korcok, J. et al., Extracellular nucleotides act through $P2X_7$ receptors to activate NF-kB in osteoclasts, *J. Bone Miner. Res.*, 19, 642, 2004.
65. Budagian, V. et al., Signaling through $P2X_7$ receptor in human T cells involves p56lck, MAP kinases, and transcription factors AP-1 and NF-kB, *J. Biol. Chem.*, 278, 1549, 2003.
66. Aga, M. et al., Modulation of monocyte signaling and pore formation in response to agonists of the nucleotide receptor $P2X_7$, *J. Leukoc. Biol.*, 72, 222, 2002.
67. von Albertini, M. et al., Extracellular ATP and ADP activate transcription factor NF-kB and induce endothelial cell apoptosis, *Biochem. Biophys. Res. Commun.*, 248, 822, 1998.
68. Ferrari, D., Stroh, C., and Schulze-Osthoff, K., $P2X_7$/P2Z purinoreceptor-mediated activation of transcription factor NFAT in microglial cells, *J. Biol. Chem.*, 274, 13205, 1999.
69. Burnstock, G., P2X receptors in sensory neurones, *Br. J. Anaesth.*, 84, 476, 2000.
70. Tsuda, M. et al., $P2X_4$ receptors induced in spinal microglia gate tactile allodynia after nerve injury, *Nature*, 424, 778, 2003.

71. Chessell, I.P. et al., Disruption of the $P2X_7$ purinoceptor gene abolishes chronic inflammatory and neuropathic pain, *Pain*, 114, 386, 2005.
72. Steinberg, T.H. and Silverstein, S.C., Extracellular ATP^{4-} promotes cation fluxes in the J774 mouse macrophage cell line, *J. Biol. Chem.*, 262, 3118, 1987.
73. Sipka, S. et al., Effects of suramin on phagocytes *in vitro*, *Ann. Hematol.*, 63, 45, 1991.
74. Verghese, M.W., Kneisler, T.B., and Boucheron, J.A., P2U agonists induce chemotaxis and actin polymerization in human neutrophils and differentiated HL60 cells, *J. Biol. Chem.*, 271, 15597, 1996.
75. McCloskey, M.A., Fan, Y., and Luther, S., Chemotaxis of rat mast cells toward adenine nucleotides, *J. Immunol.*, 163, 970, 1999.
76. Honda, S. et al., Extracellular ATP or ADP induce chemotaxis of cultured microglia through $G_{i/o}$-coupled P2Y receptors, *J. Neurosci.*, 21, 1975, 2001.
77. Idzko, M. et al., Nucleotides induce chemotaxis and actin polymerization in immature but not mature human dendritic cells via activation of pertussis toxin–sensitive P2Y receptors, *Blood*, 100, 925, 2002.
78. Idzko, M. et al., Characterization of the biological activities of uridine diphosphate in human dendritic cells: Influence on chemotaxis and CXCL8 release, *J. Cell. Physiol.*, 201, 286, 2004.
79. Satterwhite, C.M., Farrelly, A.M., and Bradley, M.E., Chemotactic, mitogenic, and angiogenic actions of UTP on vascular endothelial cells, *Am. J. Physiol.*, 276, H1091, 1999.
80. Chaulet, H. et al., Extracellular nucleotides induce arterial smooth muscle cell migration via osteopontin, *Circ. Res.*, 89, 772, 2001.
81. Goepfert, C. et al., Disordered cellular migration and angiogenesis in cd39-null mice, *Circulation*, 104, 3109, 2001.
82. Saiag, B. et al., Lack of uptake, release and action of UTP at sympathetic perivascular nerve terminals in rabbit ear artery, *Eur. J. Pharmacol.*, 358, 139, 1998.
83. Lazarowski, E.R. and Harden, T.K., Quantitation of extracellular UTP using a sensitive enzymatic assay, *Br. J. Pharmacol.*, 127, 1272, 1999.
84. Harden, T.K. and Lazarowski, E.R., Release of ATP and UTP from astrocytoma cells, *Prog. Brain Res.*, 120, 135, 1999.
85. Lazarowski, E.R. and Boucher, R.C., UTP as an extracellular signaling molecule, *News Physiol. Sci.*, 16, 1, 2001.
86. Zalavary, S. et al., Platelets enhance Fc(γ) receptor–mediated phagocytosis and respiratory burst in neutrophils: the role of purinergic modulation and actin polymerization, *J. Leukoc. Biol.*, 60, 58, 1996.
87. Mutini, C. et al., Mouse dendritic cells express the $P2X_7$ purinergic receptor: characterization and possible participation in antigen presentation, *J. Immunol.*, 163, 1958, 1999.
88. Beutler, B., Innate immunity: an overview, *Mol. Immunol.*, 40, 845, 2004.
89. Li, Q. and Verma, I.M., NF-kB regulation in the immune system, *Nat. Rev. Immunol.*, 2, 725, 2002.
90. Akira, S., Takeda, K., and Kaisho, T., Toll-like receptors: critical proteins linking innate and acquired immunity, *Nat. Immunol.*, 2, 675, 2001.
91. Vasselon, T. and Detmers, P.A., Toll receptors: a central element in innate immune responses, *Infect. Immun.*, 70, 1033, 2002.
92. Zasloff, M., Antimicrobial peptides of multicellular organisms, *Nature*, 415, 389, 2002.
93. Hancock, R.E., Cationic peptides: effectors in innate immunity and novel antimicrobials, *Lancet Infect. Dis.*, 1, 156, 2001.

94. Oppenheim, F.G. et al., Histatins, a novel family of histidine-rich proteins in human parotid secretion. Isolation, characterization, primary structure, and fungistatic effects on *Candida albicans*, *J. Biol. Chem.*, 263, 7472, 1998.
95. Koshlukova, S.E. et al., Salivary histatin 5 induces non-lytic release of ATP from *Candida albicans* leading to cell death, *J. Biol. Chem.*, 274, 18872, 1999.
96. Todorov, L.D. et al., Neuronal release of soluble nucleotidases and their role in neurotransmitter inactivation, *Nature*, 387, 76, 1997.
97. Roman, R.M. et al., Hepatocellular ATP-binding cassette protein expression enhances ATP release and autocrine regulation of cell volume, *J. Biol. Chem.*, 272, 21970, 1997.
98. Mizukawa, N. et al., Defensin-1, an antimicrobial peptide present in the saliva of patients with oral diseases, *Oral Dis.*, 5, 139, 1999.
99. Sawaki, K. et al., Immunohistochemical study on expression of α–defensin and β–defensin-2 in human buccal epithelia with candidiasis, *Oral Dis.*, 8, 37, 2002.
100. Lehrer, R.I. et al., Correlation of binding of rabbit granulocyte peptides to *Candida albicans* with candidacidal activity, *Infect. Immun.*, 49, 207, 1985.
101. Boyum, R. and Guidotti, G., Glucose-dependent, cAMP-mediated ATP efflux from *Saccharomyces cerevisiae*, *Microbiology*, 143, 1901, 1997.
102. Chowdhury, B. et al., Isolation and properties of an ATP transporter from a strain of *Aspergillus niger*, *Eur. J. Biochem.*, 247, 673, 1997.
103. Finlay, B.B. and Hancock, R.E., Can innate immunity be enhanced to treat microbial infections? *Nat. Rev. Microbiol.*, 2, 497, 2004.
104. Lehrer, R.I., Primate defensins, *Nat. Rev. Microbiol.*, 2, 727, 2004.
105. Oppenheim, J.J. et al., Roles of antimicrobial peptides such as defensins in innate and adaptive immunity, *Ann. Rheum. Dis.*, 62, Suppl 2, 17, 2003.
106. Elssner, A. et al., A novel $P2X_7$ receptor activator, the human cathelicidin-derived peptide LL37, induces IL-1 β processing and release, *J. Immunol.*, 172, 4987, 2004.
107. Bals, R. et al., Augmentation of innate host defense by expression of a cathelicidin antimicrobial peptide, *Infect. Immun.*, 67, 6084, 1999.
108. Ferrari, D. et al., The antibiotic polymyxin B modulates $P2X_7$ receptor function, *J. Immunol.*, 173, 4652, 2004.
109. Hogquist, K.A., Unanue, E.R., and Chaplin, D.D., Release of IL-1 from mononuclear phagocytes, *J. Immunol.*, 147, 2181, 1991.
110. Ferrari, D. et al., Extracellular ATP triggers IL-1 β release by activating the purinergic P2Z receptor of human macrophages, *J. Immunol.*, 159, 1451, 1997.
111. Gargett, C.E. and Wiley, J.S., The isoquinoline derivative KN-62 a potent antagonist of the P2Z-receptor of human lymphocytes, *Br. J. Pharmacol.*, 120, 1483, 1997.
112. Perregaux, D.G. et al., Antimicrobial peptides initiate IL-1 β posttranslational processing: a novel role beyond innate immunity, *J. Immunol.*, 168, 3024, 2002.
113. Fredholm, B.B., Purines and neutrophil leukocytes, *Gen. Pharmacol.*, 28, 345, 1997.
114. Nakanishi, M. et al., Extracellular ATP itself elicits superoxide generation in guinea pig peritoneal macrophages, *FEBS Lett.*, 282, 91, 1991.
115. Gozal, E., Forman, H.J., and Torres, M., ADP stimulates the respiratory burst without activation of ERK and AKT in rat alveolar macrophages, *Free Radic. Biol. Med.*, 31, 679, 2001.
116. Schmid-Antomarchi, H. et al., Extracellular ATP and UTP control the generation of reactive oxygen intermediates in human macrophages through the opening of a charybdotoxin-sensitive Ca^{2+}-dependent K^+ channel, *J. Immunol.*, 159, 6209, 1997.
117. Boarder, M.R. and Hourani, S.M., The regulation of vascular function by P2 receptors: multiple sites and multiple receptors, *Trends Pharmacol. Sci.*, 19, 99, 1998.

118. Hoyal, C.R. et al., Modulation of the rat alveolar macrophage respiratory burst by hydroperoxides is calcium dependent, *Arch. Biochem. Biophys.*, 326, 166, 1996.
119. Waldo, G.L. and Harden, T.K., Agonist binding and G_q-stimulating activities of the purified human $P2Y_1$ receptor, *Mol. Pharmacol.*, 65, 426, 2004.
120. Bowler, J.W. et al., $P2X_4$, $P2Y_1$ and $P2Y_2$ receptors on rat alveolar macrophages, *Br. J. Pharmacol.*, 140, 567, 2003.
121. Chen, Y. et al., A putative osmoreceptor system that controls neutrophil function through the release of ATP, its conversion to adenosine, and activation of A2 adenosine and P2 receptors, *J. Leukoc. Biol.*, 76, 245, 2004.
122. Meshki, J. et al., Molecular mechanism of nucleotide-induced primary granule release in human neutrophils: role for the $P2Y_2$ receptor, *Am. J. Physiol.*, 286, C264, 2004.
123. Sikora, A. et al., Cutting edge: purinergic signaling regulates radical-mediated bacterial killing mechanisms in macrophages through a $P2X_7$-independent mechanism, *J. Immunol.*, 163, 558, 1999.
124. Beigi, R.D. et al., Oxidized ATP (oATP) attenuates proinflammatory signaling via P2 receptor–independent mechanisms, *Br. J. Pharmacol.*, 140, 507, 2003.
125. Di Virgilio, F., Novel data point to a broader mechanism of action of oxidized ATP: the $P2X_7$ receptor is not the only target, *Br. J. Pharmacol.*, 140, 441, 2003.
126. Murgia, M. Oxidized ATP. An irreversible inhibitor of the macrophage purinergic P2Z receptor, *J. Biol. Chem.*, 268, 8199, 1993.
127. Lammas, D.A., ATP-induced killing of mycobacteria by human macrophages is mediated by purinergic P2Z($P2X_7$) receptors, *Immunity*, 7, 433, 1997.
128. Molloy, A., Laochumroonvorapong, P., and Kaplan, G., Apoptosis, but not necrosis, of infected monocytes is coupled with killing of intracellular bacillus Calmette-Guerin, *J. Exp. Med.*, 180, 1499, 1994.
129. Kusner, D.J. and Adams, J., ATP-induced killing of virulent *Mycobacterium tuberculosis* within human macrophages requires phospholipase D, *J. Immunol.*, 164, 379, 2000.
130. Stober, C.B. et al., Characterization of the $P2X_7$ receptor–mediated signalling pathway associated with rapid killing of intracellular mycobacteria within human macrophages, *Drug Dev. Res.*, 53, 105, 2001.
131. el Moatassim, C. and Dubyak, G.R., Dissociation of the pore-forming and phospholipase D activities stimulated via P2z purinergic receptors in BAC1.2F5 macrophages. Product inhibition of phospholipase D enzyme activity, *J. Biol. Chem.*, 268, 15571, 1993.
132. Melendez, A. et al., FcγRI coupling to phospholipase D initiates sphingosine kinase-mediated calcium mobilization and vesicular trafficking, *J. Biol. Chem.*, 273, 9393, 1998.
133. Smith, R.A., Alvarez, A.J., and Estes, D.M., The $P2X_7$ purinergic receptor on bovine macrophages mediates mycobacterial death, *Vet. Immunol. Immunopathol.*, 78, 249, 2001.
134. Stober, C.B. et al., ATP-mediated killing of *Mycobacterium bovis* bacille Calmette-Guerin within human macrophages is calcium dependent and associated with the acidification of mycobacteria-containing phagosomes, *J. Immunol.*, 166, 6276, 2001.
135. Fairbairn, I.P. et al., ATP-mediated killing of intracellular mycobacteria by macrophages is a $P2X_7$-dependent process inducing bacterial death by phagosome-lysosome fusion, *J. Immunol.*, 167, 3300, 2001.
136. Gomes, M.S. et al., Survival of *Mycobacterium avium* and *Mycobacterium tuberculosis* in acidified vacuoles of murine macrophages, *Infect. Immun.*, 67, 3199, 1999.
137. Bellamy, R. et al., Variations in the NRAMP1 gene and susceptibility to tuberculosis in West Africans, *N. Engl. J. Med.*, 338, 640, 1998.

138. Saunders, B.M. et al., A loss-of-function polymorphism in the human $P2X_7$ receptor abolishes ATP-mediated killing of mycobacteria, *J. Immunol.*, 171, 5442, 2003.
139. Coutinho-Silva, R. et al., Modulation of P2Z/$P2X_7$ receptor activity in macrophages infected with *Chlamydia psittaci*, *Am. J. Physiol.*, 280, C81, 2001.
140. Coutinho-Silva, R. et al., Inhibition of chlamydial infectious activity due to $P2X_7$R-dependent phospholipase D activation, *Immunity*, 19, 403, 2003.
141. Myers, A.J. et al., The purinergic $P2X_7$ receptor is not required for control of pulmonary *Mycobacterium tuberculosis* infection, *Infect. Immun.*, 73, 3192, 2005.
142. Balcewicz-Sablinska, M.K. et al., Pathogenic *Mycobacterium tuberculosis* evades apoptosis of host macrophages by release of TNF-R2, resulting in inactivation of TNF-α, *J. Immunol.*, 161, 2636, 1998.
143. Hueffer, K. and Galan, J. E., Salmonella-induced macrophage death: multiple mechanisms, different outcomes, *Cell. Microbiol.*, 6, 1019, 2004.
144. Byrne, G.I. and Ojcius, D.M., Chlamydia and apoptosis: life and death decisions of an intracellular pathogen, *Nat. Rev. Microbiol.*, 2, 802, 2004.
145. Sinai, A.P. et al., Mechanisms underlying the manipulation of host apoptotic pathways by *Toxoplasma gondii*, *Int. J. Parasitol.*, 34, 381, 2004.
146. Gasparoto, T.H. et al., Apoptosis of phagocytic cells induced by *Candida albicans* and production of IL-10, *FEMS Immunol. Med. Microbiol.*, 42, 219, 2004.
147. Chiapello, L.S. et al., Immunosuppression, interleukin-10 synthesis and apoptosis are induced in rats inoculated with *Cryptococcus neoformans* glucuronoxylomannan, *Immunology*, 113, 392, 2004.
148. Zaborina, O. et al., Secretion of ATP-utilizing enzymes, nucleoside diphosphate kinase and ATPase, by *Mycobacterium bovis* BCG: sequestration of ATP from macrophage P2Z receptors?, *Mol. Microbiol.*, 31, 1333, 1999.
149. Schlosser, S.F., Burgstahler, A.D., and Nathanson, M.H., Isolated rat hepatocytes can signal to other hepatocytes and bile duct cells by release of nucleotides, *Proc. Natl. Acad. Sci. USA*, 93, 9948, 1996.
150. Arcuino, G. et al., Intercellular calcium signaling mediated by point-source burst release of ATP, *Proc. Natl. Acad. Sci. USA*, 99, 9840, 2002.
151. Stout, C.E. et al., Intercellular calcium signaling in astrocytes via ATP release through connexin hemichannels, *J. Biol. Chem.*, 277, 10482, 2002.
152. Perregaux, D.G. et al., ATP acts as an agonist to promote stimulus-induced secretion of IL-1 β and IL-18 in human blood, *J. Immunol.*, 165, 4615, 2000.
153. Mehta, V.B., Hart, J., and Wewers, M.D., ATP-stimulated release of interleukin (IL)-1β and IL-18 requires priming by lipopolysaccharide and is independent of caspase-1 cleavage, *J. Biol. Chem.*, 276, 3820, 2001.
154. Sluyter, R., Dalitz, J.G., and Wiley, J.S., $P2X_7$ receptor polymorphism impairs extracellular adenosine 5′–triphosphate–induced interleukin-18 release from human monocytes, *Genes Immun.*, 5, 588, 2004.
155. Falzoni, S. et al., The purinergic P2Z receptor of human macrophage cells. Characterization and possible physiological role, *J. Clin. Invest.*, 95, 1207, 1995.
156. Blanchard, D.K., McMillen, S., and Djeu, J.Y., IFN-γ enhances sensitivity of human macrophages to extracellular ATP-mediated lysis, *J. Immunol.*, 147, 2579, 1991.
157. Schnurr, M. et al., Extracellular nucleotide signaling by P2 receptors inhibits IL-12 and enhances IL-23 expression in human dendritic cells: a novel role for the cAMP pathway, *Blood*, 105, 1582, 2005.
158. Rook, G.A. et al., The role of γ–interferon, vitamin D3 metabolites and tumour necrosis factor in the pathogenesis of tuberculosis, *Immunology*, 62, 229, 1987.

159. Reiner, N.E., Altered cell signaling and mononuclear phagocyte deactivation during intracellular infection, *Immunol. Today*, 15, 374, 1994.
160. Murch, A.R. et al., Direct evidence that inflammatory multinucleate giant cells form by fusion, *J. Pathol.*, 137, 177, 1982.
161. Anderson, J.M., Multinucleated giant cells, *Curr. Opin. Hematol.*, 7, 40, 2000.
162. Mizuno, K., Okamoto, H., and Horio, T., Heightened ability of monocytes from sarcoidosis patients to form multi-nucleated giant cells *in vitro* by supernatants of concanavalin A–stimulated mononuclear cells, *Clin. Exp. Immunol.*, 126, 151, 2001.
163. Chiozzi, P. et al., Spontaneous cell fusion in macrophage cultures expressing high levels of the P2Z/P2X$_7$ receptor, *J. Cell Biol.*, 138, 697, 1997.
164. Falzoni, S. et al., P2X$_7$ receptor and polykarion formation, *Mol. Biol. Cell*, 11, 3169, 2000.
165. Most, J., Neumayer, H.P., and Dierich, M.P., Cytokine-induced generation of multinucleated giant cells *in vitro* requires interferon-γ and expression of LFA-1, *Eur. J. Immunol.*, 20, 1661, 1990.
166. Fais, S. et al., Multinucleated giant cells generation induced by interferon-γ. Changes in the expression and distribution of the intercellular adhesion molecule-1 during macrophages fusion and multinucleated giant cell formation, *Lab. Invest.*, 71, 737, 1994.
167. McInnes, A. and Rennick, D.M., Interleukin 4 induces cultured monocytes/macrophages to form giant multinucleated cells, *J. Exp. Med.*, 167, 598, 1988.
168. McNally, A.K. and Anderson, J.M., β1 and β2 integrins mediate adhesion during macrophage fusion and multinucleated foreign body giant cell formation, *Am. J. Pathol.*, 160, 621, 2002.
169. Byrd, T.F., Multinucleated giant cell formation induced by IFN-γ/IL-3 is associated with restriction of virulent *Mycobacterium tuberculosis* cell to cell invasion in human monocyte monolayers, *Cell. Immunol.*, 188, 89, 1998.
170. Lemaire, I. et al., Differential effects of macrophage- and granulocyte-macrophage colony-stimulating factors on cytokine gene expression during rat alveolar macrophage differentiation into multinucleated giant cells (MGC): role for IL-6 in type 2 MGC formation, *J. Immunol.*, 157, 5118, 1996.
171. Lemaire, I. et al., M-CSF and GM-CSF promote alveolar macrophage differentiation into multinucleated giant cells with distinct phenotypes, *J. Leukoc. Biol.*, 60, 509, 1996.
172. Gasser, A. and Most, J., Generation of multinucleated giant cells *in vitro* by culture of human monocytes with *Mycobacterium bovis* BCG in combination with cytokine-containing supernatants, *Infect. Immun.*, 67, 395, 1999.
173. Blanchard, D.K., McMillen, S., and Djeu, J.Y., IFN-γ enhances sensitivity of human macrophages to extracellular ATP-mediated lysis, *J. Immunol.*, 147, 2579, 1991.
174. Kim, M. et al., Proteomic and functional evidence for a P2X$_7$ receptor signalling complex, *EMBO J.*, 20, 6347, 2001.
175. Di Virgilio, F. et al., ATP receptors and giant cell formation, *J. Leukoc. Biol.*, 66, 723, 1999.
176. Monck, J.R. and Fernandez, J.M., The fusion pore and mechanisms of biological membrane fusion, *Curr. Opin. Cell Biol.*, 8, 524, 1996.
177. Stanton, L.A. et al., Immunophenotyping of macrophages in human pulmonary tuberculosis and sarcoidosis, *Int. J. Exp. Pathol.*, 84, 289, 2003.
178. Ernst, J.D., Macrophage receptors for *Mycobacterium tuberculosis*, *Infect. Immun.*, 66, 1277, 1998.
179. Haworth, R. et al., The macrophage scavenger receptor type A is expressed by activated macrophages and protects the host against lethal endotoxic shock, *J. Exp. Med.*, 186, 1431, 1997.

180. Fraser, I., Hughes, D., and Gordon, S., Divalent cation-independent macrophage adhesion inhibited by monoclonal antibody to murine scavenger receptor, *Nature*, 364, 343, 1993.
181. Peiser, L. et al., Macrophage class A scavenger receptor-mediated phagocytosis of *Escherichia coli*: role of cell heterogeneity, microbial strain, and culture conditions *in vitro*, *Infect. Immun.*, 68, 1953, 2000.
182. Platt, N. et al., Role for the class A macrophage scavenger receptor in the phagocytosis of apoptotic thymocytes *in vitro*, *Proc. Natl. Acad. Sci. USA*, 93, 12456, 1996.
183. Mahoney, J.A. et al., Cloning and characterization of CPVL, a novel serine carboxypeptidase, from human macrophages, *Genomics*, 72, 243, 2001.
184. Hill, J.O., CD^{4+} T cells cause multinucleated giant cells to form around *Cryptococcus neoformans* and confine the yeast within the primary site of infection in the respiratory tract, *J. Exp. Med.*, 175, 1685, 1992.
185. Smythies, L.E. et al., T cell–derived cytokines associated with pulmonary immune mechanisms in mice vaccinated with irradiated cercariae of *Schistosoma mansoni*, *J. Immunol.*, 148, 1512, 1992.
186. Suga, S. et al., Identification of fusion regulatory protein (FRP)-1/4F2 related molecules: cytoskeletal proteins are associated with FRP-1 molecules that regulate multinucleated giant cell formation of monocytes and HIV-induced cell fusion, *Cell Struct. Funct.*, 20, 473, 1995.
187. Hunninghake, G.W. et al., ATS/ERS/WASOG statement on sarcoidosis. American Thoracic Society/European Respiratory Society/World Association of Sarcoidosis and other Granulomatous Disorders, *Sarcoidosis Vasc. Diffuse. Lung Dis.*, 16, 149, 1999.
188. Newman, L.S., Rose, C.S., and Maier, L.A., Sarcoidosis, *N. Engl. J. Med.*, 336, 1224, 1997.
189. Wirostko, E., Johnson, L., and Wirostko, B., Sarcoidosis associated uveitis. Parasitization of vitreous leucocytes by mollicute-like organisms, *Acta Ophthalmol.*, 67, 415, 1989.
190. Marshall, T.G. and Marshall, F.E., Sarcoidosis succumbs to antibiotics—implications for autoimmune disease, *Autoimmun. Rev.*, 3, 295, 2004.
191. Humphreys, B.D. and Dubyak, G.R., Modulation of $P2X_7$ nucleotide receptor expression by pro- and anti-inflammatory stimuli in THP-1 monocytes, *J. Leukoc. Biol.*, 64, 265, 1998.
192. Terao, I., Hashimoto, S., and Horie, T., Effect of GM-CSF on TNF-α and IL-1-β production by alveolar macrophages and peripheral blood monocytes from patients with sarcoidosis, *Int. Arch. Allergy Immunol.*, 102, 242, 1993.
193. Solle, M. et al., Altered cytokine production in mice lacking $P2X_7$ receptors, *J. Biol. Chem.*, 276, 125, 2001.
194. El Euch, D. et al., Sarcoidosis in a child treated successfully with allopurinol, *Br. J. Dermatol.*, 140, 1184, 1999.
195. Antony, F. and Layton, A.M., A case of cutaneous acral sarcoidosis with response to allopurinol, *Br. J. Dermatol.*, 142, 1052, 2000.
196. Lieberman, J. and Zakria, F., Effect of captopril and enalapril medication on the serum ACE test for sarcoidosis, *Sarcoidosis*, 6, 118, 1989.
197. Mizuno, K., Okamoto, H., and Horio, T., Inhibitory influences of xanthine oxidase inhibitor and angiotensin I–converting enzyme inhibitor on multinucleated giant cell formation from monocytes by downregulation of adhesion molecules and purinergic receptors, *Br. J. Dermatol.*, 150, 205, 2004.
198. Luttikhuizen, D.T. et al., Expression of P2 receptors at sites of chronic inflammation, *Cell Tissue Res.*, 317, 289, 2004.

199. Oshimi, Y., Miyazaki, S., and Oda, S., ATP-induced Ca^{2+} response mediated by P2U and P2Y purinoceptors in human macrophages: signalling from dying cells to macrophages, *Immunology*, 98, 220, 1999.
200. Kuziel, W.A. et al., Severe reduction in leukocyte adhesion and monocyte extravasation in mice deficient in CC chemokine receptor 2, *Proc. Natl. Acad. Sci. USA*, 94, 12053, 1997.
201. Kyriakides, T.R. et al., The CC chemokine ligand, CCL2/MCP1, participates in macrophage fusion and foreign body giant cell formation, *Am. J. Pathol.*, 165, 2157, 2004.
202. Panenka, W. et al., $P2X_7$-like receptor activation in astrocytes increases chemokine monocyte chemoattractant protein–1 expression via mitogen-activated protein kinase, *J. Neurosci.*, 21, 7135, 2001.
203. Saginario, C. et al., MFR, a putative receptor mediating the fusion of macrophages, *Mol. Cell Biol.*, 18, 6213, 1998.
204. Bonnema, H. et al., Distribution patterns of the membrane glycoprotein CD44 during the foreign-body reaction to a degradable biomaterial in rats and mice, *J. Biomed. Mater. Res A*, 64, 502, 2003.
205. Lemaire, I. and Leduc, N., Purinergic $P2X_7$ receptor function in lung alveolar macrophages: pharmacologic characterization and bidirectional regulation by Th1 and Th2 cytokines, *Drug Dev. Res.*, 59, 118, 2003.
206. Zhao, Q. et al., Foreign-body giant cells and polyurethane biostability: *in vivo* correlation of cell adhesion and surface cracking, *J. Biomed. Mater. Res.*, 25, 177, 1991.
207. Honma, T. and Hamasaki, T., Ultrastructure of multinucleated giant cell apoptosis in foreign-body granuloma, *Virchows Arch.*, 428, 165, 1996.
208. Ichinose, M., Modulation of phagocytosis by P2-purinergic receptors in mouse peritoneal macrophages, *Jpn. J. Physiol.*, 45, 707, 1995.
209. Hernandez-Pando, R. et al., Inflammatory cytokine production by immunological and foreign body multinucleated giant cells, *Immunology*, 100, 352, 2000.
210. Ingham, E. and Fisher, J., The role of macrophages in osteolysis of total joint replacement, *Biomaterials*, 26, 1271, 2005.
211. Burger, E.H. et al., *In vitro* formation of osteoclasts from long-term cultures of bone marrow mononuclear phagocytes, *J. Exp. Med.*, 156, 1604, 1982.
212. MacDonald, B.R. et al., Formation of multinucleated cells that respond to osteotropic hormones in long term human bone marrow cultures, *Endocrinology*, 120, 2326, 1987.
213. Udagawa, N. et al., Origin of osteoclasts: mature monocytes and macrophages are capable of differentiating into osteoclasts under a suitable microenvironment prepared by bone marrow–derived stromal cells, *Proc. Natl. Acad. Sci. USA*, 87, 7260, 1990.
214. Neale, S.D. et al., Human bone-derived cells support formation of human osteoclasts from arthroplasty-derived cells *in vitro*, *J. Bone Joint Surg. Br.*, 82, 892, 2000.
215. Itonaga, I. et al., Effect of osteoprotegerin and osteoprotegerin ligand on osteoclast formation by arthroplasty membrane derived macrophages, *Ann. Rheum. Dis.*, 59, 26, 2000.
216. Haynes, D.R. et al., The osteoclastogenic molecules RANKL and RANK are associated with periprosthetic osteolysis, *J. Bone Joint Surg. Br.*, 83, 902, 2001.
217. Sabokbar, A., Kudo, O., and Athanasou, N.A., Two distinct cellular mechanisms of osteoclast formation and bone resorption in periprosthetic osteolysis, *J. Orthop. Res.*, 21, 73, 2003.
218. Khosla, S., Minireview: the OPG/RANKL/RANK system, *Endocrinology*, 142, 5050, 2001.
219. Boyle, W.J., Simonet, W.S., and Lacey, D.L., Osteoclast differentiation and activation, *Nature*, 423, 337, 2003.

220. Hoebertz, A., Arnett, T.R., and Burnstock, G., Regulation of bone resorption and formation by purines and pyrimidines, *Trends Pharmacol. Sci.*, 24, 290, 2003.
221. Gartland, A. et al., P2 receptors in bone—modulation of osteoclast formation and activity via $P2X_7$ activation, *Crit. Rev. Eukaryot. Gene Expr.*, 13, 237, 2003.
222. Morrison, M.S. et al., ATP is a potent stimulator of the activation and formation of rodent osteoclasts, *J. Physiol.*, 511, 495, 1998.
223. Romanello, M. et al., Mechanically induced ATP release from human osteoblastic cells, *Biochem. Biophys. Res. Commun.*, 289, 1275, 2001.
224. Hoebertz, A. et al., ATP and UTP at low concentrations strongly inhibit bone formation by osteoblasts: a novel role for the $P2Y_2$ receptor in bone remodeling, *J. Cell Biochem.*, 86, 413, 2002.
225. Naemsch, L.N., Dixon, S.J., and Sims, S.M., Activity-dependent development of $P2X_7$ current and Ca^{2+} entry in rabbit osteoclasts, *J. Biol. Chem.*, 276, 39107, 2001.
226. Yu, H. and Ferrier, J., Mechanisms of ATP-induced Ca^{2+} signaling in osteoclasts, *Cell Signal.*, 6, 905, 1994.
227. Buckley, K.A. et al., Adenosine triphosphate stimulates human osteoclast activity via upregulation of osteoblast-expressed receptor activator of nuclear factor–k B ligand, *Bone*, 31, 582, 2002.
228. Hoebertz, A. et al., Extracellular ADP is a powerful osteolytic agent: evidence for signaling through the $P2Y_1$ receptor on bone cells, *FASEB J.*, 15, 1139, 2001.
229. Hoebertz, A. et al., Expression of P2 receptors in bone and cultured bone cells, *Bone*, 27, 503, 2000.
230. Modderman, W.E. et al., Permeabilization of cells of hemopoietic origin by extracellular ATP^{4-}: elimination of osteoclasts, macrophages, and their precursors from isolated bone cell populations and fetal bone rudiments, *Calcif. Tissue Int.*, 55, 141, 1994.
231. Gartland, A. et al., Expression of a $P2X_7$ receptor by a subpopulation of human osteoblasts, *J. Bone Miner. Res.*, 16, 846, 2001.
232. Jorgensen, N.R. et al., Intercellular calcium signaling occurs between human osteoblasts and osteoclasts and requires activation of osteoclast $P2X_7$ receptors, *J. Biol. Chem.*, 277, 7574, 2002.
233. Buell, G. et al., Blockade of human $P2X_7$ receptor function with a monoclonal antibody, *Blood*, 92, 3521, 1998.
234. Korcok, J. et al., Extracellular nucleotides act through $P2X_7$ receptors to activate NF-kB in osteoclasts, *J. Bone Miner. Res.*, 19, 642, 2004.
235. Perregaux, D.G. and Gabel, C.A., Post-translational processing of murine IL-1: evidence that ATP-induced release of IL-1α and IL-1β occurs via a similar mechanism, *J. Immunol.*, 160, 2469, 1998.
236. Wei, S. et al., IL-1 mediates TNF-induced osteoclastogenesis, *J. Clin. Invest.*, 115, 282, 2005.
237. Kwan, T.S. et al., IL-6, RANKL, TNF-α/IL-1: interrelations in bone resorption pathophysiology, *Cytokine Growth Factor Rev.*, 15, 49, 2004.
238. Ke, H.Z. et al., Deletion of the $P2X_7$ nucleotide receptor reveals its regulatory roles in bone formation and resorption, *Mol. Endocrinol.*, 17, 1356, 2003.
239. Korcok, J. et al., Extracellular nucleotides act through $P2X_7$ receptors to activate NF-kB in osteoclasts, *J. Bone Miner. Res.*, 19, 642, 2004.
240. Gartland, A. et al., Blockade of the pore-forming $P2X_7$ receptor inhibits formation of multinucleated human osteoclasts *in vitro*, *Calcif. Tissue Int.*, 73, 361, 2003.
241. Adinolfi, E. et al., $P2X_7$ receptor expression in evolutive and indolent forms of chronic B lymphocytic leukemia, *Blood*, 99, 706, 2002.

Future Directions

Geoffrey Burnstock and Tim Arnett

Now that there is growing knowledge of the roles of purinergic receptors in normal bone and cartilage physiology, there is an increasing interest in exploring the pathophysiology of purinergic signaling in bone and possible purine-related therapeutic interventions in diseases such as rheumatoid arthritis, osteoporosis, tumor-induced osteolysis, and periodontitis.[1–3] Therapeutic strategies include not only the development of selective P2 receptor subtype agonists and antagonists, but also agents that control ATP release and degradation.

Much is now known about the ectonucleotidases that break down ATP released from neurons and nonneuronal cells.[4] Several enzyme families are involved: ecto-nucleoside triphosphate diphosphohydrolases (E-NTPDases), of which NTPDase1, 2, 3, and 8 are extracellular; ectonucleotide pyrophosphatase (E-NPP) of 3 subtypes; alkaline phosphatases; ecto-5′-nucleotidase and ecto-nucleoside diphosphokinase (E-NDPK). NTPDase1 hydrolyzes ATP directly to AMP and UTP to UDP, whereas NTPDase 2 hydrolyzes ATP to ADP and 5′-nucleotidase AMP to adenosine.

In cartilage, the metabolism of nucleotides released by chondrocytes constitutively, or as a result of mechanical stimulation or inflammation, appears to be tightly controlled. Ectonucleotidases in cartilage can regulate the levels of extracellular phosphate and pyrophosphate, which promote and inhibit mineral crystal deposition, respectively. The association between articular cartilage mineralization (chondrocalcinosis) and osteoarthritis is increasingly recognized.[5,6] Paracrine signaling, via nucleotides and other agents, may be especially important for chondrocytes, which lack direct intercellular contacts. Available data on the receptor-mediated actions of extracellular nucleotides on cartilage breakdown are somewhat contradictory and may require clarification; for example, ATP stimulates proteoglycan release from nasal cartilage explants but inhibits the same process in articular cartilage.[7]

A role for purinergic signaling in rheumatic diseases has been considered for some time. Quinacrine (Atabrine), a drug that binds strongly to ATP, has been used for the treatment of rheumatoid arthritis patients for many years.[8] One of its mechanisms of action is to decrease levels of prostaglandin E_2 and COX-2, which are known to be produced following occupation of P2Y receptors by ATP.[9,10] The articular fluid removed from arthritic joints contains high levels of ATP.[11] Purinergic regulation of bradykinin-induced plasma extravazation and adjuvant-induced arthritis has been reported.[12] ATP and UTP activate calcium-mobilizing $P2Y_2$ or $P2Y_4$ receptors and act synergistically with interleukin-1 to stimulate prostaglandin E_2 release from human rheumatoid synovial cells.[13] When monoarthritis was induced by the injection of complete Freund's adjuvant into the unilateral temporomandibular joint of the rat, the pain produced was associated with an increase in $P2X_3$ receptor–positive small neurons in the trigeminal ganglion.[14] Relief of inflammatory pain by the $P2X_7$ receptor antagonist, oxidized ATP, in arthritic rats has been reported.[15,16] Spinal P1 receptor activation has been claimed to inhibit inflammation and joint destruction in rat adjuvant-induced arthritis, supporting the view that therapeutic strategies that target the CNS might be useful in arthritis.[17,18] Suppression of experimental zymosan-induced arthritis by intraperitoneal administration of adenosine has also been described.[19]

The actions of extracellular nucleotides on the basic functions of bone cells appear to be relatively clear-cut. ATP and ADP exert striking *in vitro* actions at concentrations in the low micromolar range to stimulate the formation and resorptive activity of osteoclasts, whereas ATP and UTP inhibit mineralized bone formation by osteoblasts.[20] These results suggest that ATP could act as a significant negative modulator of bone metabolism *in vivo*, with actions at $P2Y_1$ and $P2Y_2$ receptors; antagonists acting at these receptor subtypes could possibly provide the basis for novel bone-sparing drugs. Recent work indicates that the inhibitory action of ATP and UTP on bone formation by mature osteoblasts chiefly involves reduction of mineralization;[21] this effect could involve $P2Y_2$ and $P2Y_4$ receptors, which are strongly expressed in differentiated osteoblasts[22] but could also be due to pyrophosphate, long known as a potent inhibitor of mineralization, generated from extracellular ATP and UTP by the action of ecto-nucleotidases.

Transgenic animals now offer opportunities to test the potential roles of specific P2 receptor subtypes in skeletal homeostasis. The best documented work to date concerns mice deficient in the $P2X_7$ receptor. In the study of Ke et al.,[23] such animals displayed reduced circumferential growth of long bones, with evidence of increased trabecular bone resorption; however, an independent investigation failed to observe an overt skeletal phenotype in $P2X_7$ receptor-null animals.[24] Preliminary studies show that juvenile mice deficient in $P2X_2$ receptors exhibit significantly increased bone mass, possibly secondary to greater body mass. No obvious skeletal differences were noted in $P2X_3$ receptor-null mice, but animals deficient in both $P2X_3$ and $P2X_2$ receptors were runted.[25] At the time of writing, investigation of the skeletons of $P2Y_1$ and $P2Y_2$ receptor-deficient mice[26,27] is underway in the editors' laboratories. An obvious difficulty presented by knockout studies is that any skeletal changes may be secondary to general alterations in, for example, body mass, physical activity, or the vasculature. Moreover, in view of the multiplicity of P2 receptor subtypes expressed by bone (and other) cells, and the potential for functional overlap between certain receptor subtypes, "single knockout" animals may display only relatively mild skeletal phenotypes.

Evidence gathered over the last 15 years indicates that extracellular nucleotides participate in complex local signaling pathways in the skeleton, functioning also as important modulators of the action of other paracrine and systemic agents. Despite the challenges posed by the ubiquity of nucleotides in living cells and tissues, the potential remains for interesting new classes of drugs that intervene usefully in extracellular nucleotide signaling in cartilage and bone.

REFERENCES

1. Dixon, S.J. and Sims, S.M., P2 purinergic receptors on osteoblasts and osteoclasts: potential targets for drug development, *Drug Dev. Res.*, 49, 187, 2000.
2. Komarova, S.V., Dixon, S.J., and Sims, S.M., Osteoclast ion channels: potential targets for antiresorptive drugs, *Curr. Pharm. Des.*, 7, 637, 2001.
3. Naemsch, L.N. et al., P2 nucleotide receptors in osteoclasts, *Drug Dev. Res.*, 53, 130, 2001.
4. Zimmermann, H., Ectonucleotidases: some recent developments and a note on nomenclature, *Drug Dev. Res.*, 52, 44, 2001.
5. Graff, R.D., Picher, M., and Lee, G.M., Extracellular nucleotides, cartilage stress, and calcium crystal formation, *Curr. Opin. Rheumatol.*, 15, 315, 2003.
6. Picher, M., Graff, R.D., and Lee, G.M., Extracellular nucleotide metabolism and signaling in the pathophysiology of articular cartilage, *Arthritis Rheum.*, 48, 2722, 2003.
7. Brown, C.J. et al., Proteoglycan breakdown from bovine nasal cartilage is increased, and from articular cartilage is decreased, by extracellular ATP, *Biochim. Biophys. Acta*, 1362, 208, 1997.
8. Wallace, D.J., The use of quinacrine (Atabrine) in rheumatic diseases: a reexamination, *Semin. Arthritis Rheum.*, 18, 282, 1989.
9. Needleman, P., Minkes, M.S., and Douglas, J.R., Stimulation of prostaglandin biosynthesis by adenine nucleotides. Profile of prostaglandin release by perfused organs, *Circ. Res.*, 34, 455, 1974.
10. Brambilla, R. et al., Cyclooxygenase-2 mediates P2Y receptor-induced reactive astrogliosis. Special report, *Br. J. Pharmacol.*, 126, 563, 1999.
11. Ryan, L.M., Rachow, J.W., and McCarty, D.J., Synovial fluid ATP: a potential substrate for the production of inorganic pyrophosphate, *J. Rheumatol.*, 18, 716, 191.
12. Green, P.G. et al., Purinergic regulation of bradykinin-induced plasma extravasation and adjuvant-induced arthritis in the rat, *Proc. Natl. Acad. Sci. USA*, 88, 4162, 1991.
13. Loredo, G.A. and Benton, H.P., ATP and UTP activate calcium-mobilizing P2U-like receptors and act synergistically with interleukin-1 to stimulate prostaglandin E_2 release from human rheumatoid synovial cells, *Arthritis Rheum.*, 41, 246, 1998.
14. Shinoda, M. et al., Changes in $P2X_3$ receptor expression in the trigeminal ganglion following monoarthritis of the temporomandibular joint in rats, *Pain*, 116, 42, 2005.
15. Dell'Antonio, G. et al., Relief of inflammatory pain in rats by local use of the selective $P2X_7$ ATP receptor inhibitor, oxidized ATP, *Arthritis Rheum.*, 46, 3378, 2002.
16. Dell'Antonio, G. et al., Antinociceptive effect of a new $P_{(2Z)}/P2X_7$ antagonist, oxidized ATP, in arthritic rats, *Neurosci. Lett.*, 327, 87, 2002.
17. Boyle, D.L. et al., Spinal adenosine receptor activation inhibits inflammation and joint destruction in rat adjuvant-induced arthritis, *Arthritis Rheum.*, 46, 3076, 2002.
18. Sorkin, L.S. et al., Spinal adenosine agonist reduces c-fos and astrocyte activation in dorsal horn of rats with adjuvant-induced arthritis, *Neurosci. Lett.*, 340, 119, 2003.
19. Baharav, E. et al., Suppression of experimental zymosan-induced arthritis by intraperitoneal administration of adenosine, *Drug Dev. Res.*, 57, 182, 2002.
20. Hoebertz, A., Arnett, T.R., and Burnstock, G., Regulation of bone resorption and formation by purines and pyrimidines, *Trends Pharmacol. Sci.*, 24, 290, 2003.
21. Orriss, I.R. et al., Osteoblast responses to nucleotides increase during differentiation, *Bone*, 39, 300, 2006.
22. Orriss, I.R. et al., Extracellular nucleotides inhibit mineralisation of bone nodules formed by rat osteoblasts, *Calcif. Tissue Int.*, 78, S61, 2006.

23. Ke, H.Z. et al., Deletion of the P2X$_7$ nucleotide receptor reveals its regulatory roles in bone formation and resorption, *Mol. Endocrinol.*, 17, 1356, 2003.
24. Gartland, A. et al., Multinucleated osteoclast formation *in vivo* and *in vitro* by P2X$_7$ receptor-deficient mice, *Crit. Rev. Eukaryot. Gene Expr.*, 13, 243, 2003.
25. Orriss, I.R. et al., P2X$_2$ but not P2X$_3$ receptor knockout mice demonstrate increased bone mass and weight, *J. Bone Miner. Res.*, 20, 1293, 2005.
26. Cressman, V.L. et al., Effect of loss of P2Y$_2$ receptor gene expression on nucleotide regulation of murine epithelial Cl$^-$ transport, *J. Biol. Chem.*, 274, 26461, 1999.
27. Leon, C. et al., Defective platelet aggregation and increased resistance to thrombosis in purinergic P2Y$_1$ receptor-null mice, *J. Clin. Invest.*, 104, 1731, 1999.

Index

A

B

C

D

E

F

G

H

I

L

M

N

O

P

Q

R

S

T

U

V

W

Z